DATE DUE			

ADVANCES IN GENETICS

VOLUME 22

Molecular Genetics of Plants

Contributors to This Volume

P. J. J. Hooykaas

R. A. Schilperoort

Ronald R. Sederoff

John C. Sorenson

Steven Spiker

ADVANCES IN GENETICS

Edited by

JOHN G. SCANDALIOS

Department of Genetics
North Carolina State University
Raleigh, North Carolina

E. W. CASPARI

Department of Biology
University of Rochester
Rochester, New York

VOLUME 22

Molecular Genetics of Plants

Edited by

JOHN G. SCANDALIOS

Department of Genetics
North Carolina State University
Raleigh, North Carolina

1984

ACADEMIC PRESS, INC.
(Harcourt Brace Jovanovich, Publishers)
Orlando   San Diego   San Francisco   New York   London
Toronto   Montreal   Sydney   Tokyo   São Paulo

ACADEMIC PRESS, INC.
Orlando, Florida 32887

United Kingdom Edition published by
ACADEMIC PRESS, INC. (LONDON) LTD.
24/28 Oval Road, London NW1 7DX

LIBRARY OF CONGRESS CATALOG CARD NUMBER: 47-30313

ISBN 0-12-017622-X

PRINTED IN THE UNITED STATES OF AMERICA

84 85 86 87   9 8 7 6 5 4 3 2 1

CONTENTS

Structural Variation in Mitochondrial DNA

RONALD R. SEDEROFF

The Structure and Expression of Nuclear Genes in Higher Plants

JOHN C. SORENSON

Chromatin Structure and Gene Regulation in Higher Plants

STEVEN SPIKER

The Molecular Genetics of Crown Gall Tumorigenesis

P. J. J. HOOYKAAS AND R. A. SCHILPEROORT

CONTRIBUTORS TO VOLUME 22

Numbers in parentheses indicate the pages on which the authors' contributions begin.

P. J. J. HOOYKAAS (209), *Laboratory of Biochemistry, University of Leiden, 2333 AL Leiden, The Netherlands*

R. A. SCHILPEROORT (209), *Laboratory of Biochemistry, University of Leiden, 2333 AL Leiden, The Netherlands*

RONALD R. SEDEROFF (1), *Department of Genetics, North Carolina State University, Raleigh, North Carolina 27650*

JOHN C. SORENSON (109), *Experimental Agricultural Sciences, The Upjohn Company, 7000 Portage Road, Kalamazoo, Michigan 49001*

STEVEN SPIKER (145), *Department of Genetics, North Carolina State University, Raleigh, North Carolina 27650*

PREFACE

During the past decade, enormous technological advances have occurred in biology that have led to significant and revolutionary developments. New techniques allow genetic transformations through cell fusion and by the insertion or modification of genetic information through the cloning of DNA. Methods are now available for manipulating organs, tissues, cells, or protoplasts in culture, for regenerating plants, and for testing the genetic basis of novel traits. Whether these techniques can be successfully applied to all species, and not merely to a few, is currently the focus of numerous investigations. However, the potential for exploiting modern molecular biological technology for plant breeding and genetics is vast and virtually untapped.

The recent developments in genetic engineering permit the plant breeder to bypass the various natural breeding barriers that have limited control of the transfer of genetic information in higher organisms. Recombinant DNA technology allows for the selection and production of amplified copies of specific DNA segments that can then be transferred by appropriate vectors into specific plant cells. The *Ti* (tumor-inducing) plasmid carried by *Agrobacterium tumefaciens* has, to date, proven to be the most efficient vector in transferring DNA in higher plants (discussed in this volume). The full benefits of genetic engineering of plants, or any eukaryotes, will be realized only if adequate systems of selection for desired traits and transfer of specific desirable sequences of DNA are effected, and only if these sequences can be properly regulated for expression in the desired tissue at the proper time during development.

In his Preface to Volume 1 (1947), the founding Editor, Dr. Milislav Demerec, stated, "*Advances in Genetics* has been started in order that critical summaries of outstanding genetic problems, written by competent geneticists, may appear in a single publication." The goal set for the series was "to have articles written in such form that they will be useful as reference material for geneticists and also a source of information to nongeneticists." Over the years, the goals set forth have been more than satisfied, and a tradition of excellence was established first under Demerec, and subsequently under Dr. Ernst W. Caspari. I am pleased to have been invited to join Dr. Caspari in carrying on this series.

Because of the diversity of genetics as a science, *Advances in Genetics* has adhered to a policy of publishing in each volume a series of outstanding but largely unrelated articles. Although it is felt that this tradition should be maintained, we will on occasion depart from this format and periodically review a central topic in a highly "topical" or "thematic" volume. We feel this is essential in view of the extremely rapid developments in genetics, which have led to an unparalleled information explosion in recent years. The present volume is to be the first aperiodic "thematic" volume in the series, and is devoted to a review of some current developments in plant molecular genetics.

The intent of this volume is not to present a comprehensive review of the area. Rather, each author was asked to critically summarize and highlight some of the more interesting and, hopefully, significant advances in his own area of expertise. The four articles composing this volume include "state-of-the-art" discussions on the structure and function of mitochondrial genomes, the structure and expression of nuclear genes, chromatin structure and gene regulation, and the molecular genetics of crown gall tumorigenesis. Our purpose is not just to inform but also to stimulate the reader, whether a beginning or an advanced researcher, to explore, question, and, whenever possible, test various hypotheses advanced herein. We hope we have covered some material of lasting value in view of the very rapid developments in this field.

I wish to acknowledge with gratitude the generous cooperation of the individual contributors. I thank numerous colleagues who aided me in the review process, and the Academic Press staff for their gracious cooperation.

It is fitting that I dedicate this volume to the memory of a visionary scientist and great geneticist, the founding Editor, Dr. Milislav Demerec, who was also my scientific mentor and friend.

JOHN G. SCANDALIOS

ADVANCES IN GENETICS

VOLUME 22

Molecular Genetics of Plants

STRUCTURAL VARIATION IN MITOCHONDRIAL DNA

Ronald R. Sederoff

Department of Genetics, North Carolina State University, Raleigh, North Carolina

1

I. Introduction: Structural Variation and Functional Conservation of Mitochondrial Genomes

Although the function of mitochondria in oxidative phosphorylation is common to virtually all eukaryotic systems, the great diversity of the mitochondrial genomes of plants, animals, fungi, and protozoa is both fascinating and bewildering. In these systems, the basic mitochondrial functions appear highly conserved, whereas the size, organization, and structure of the genomes vary greatly. The size and internal structure of mitochondrial genes also vary greatly in different mitochondrial genomes.

This article is intended to review the structural diversity of mitochondrial chromosomes, particularly the variations of size and structure resulting from rearrangements in mitochondrial DNA (mtDNA). The observed diversity reflects both functional constraints and mechanisms of genetic variation that underlie the divergent paths taken by these unusual organelle genomes. In addition to unitary mitochondrial chromosomes, many additional or supernumerary DNA elements, such as plasmids and plasmidlike molecules, are found in mitochondrial genomes. These additional DNA elements are extremely diverse, are found in mtDNA of widely different groups, and may be important in the organization and evolution of mitochondrial genomes. The comparison of these diverse systems is directed toward a better understanding of the evolution of mtDNA and toward suggesting directions for future research.

Many reviews and books have recently covered many aspects of the structure and function of mitochondrial genomes. "Mitochondrial Genes" (Slonimski *et al.,* 1982) and the "12th Bari Conference" (Kroon and Saccone, 1980) are among the most recent. Gray (1982) and Wallace (1982) recently reviewed mitochondrial genome diversity and organelle evolution. Several reviews of plant mitochondrial genome organization and expression have recently been published (Leaver and Gray, 1982; Levings and Sederoff, 1981; Levings *et al.,* 1983). The evolution of mitochondrial DNA with respect to the origin of mitochondria has been reviewed by Gray and Doolittle (1982) and was a major subject of a symposium of the New York Academy of Sciences (Fredrick, 1981). The genetics and biochemistry of yeast mtDNA have been reviewed by Borst and Grivell (1978, 1981), Tzagoloff et al. (1979), Gillham (1978), Tzagoloff (1982), Borst (1981), and Butow and Strausberg (1981). Several recent reviews of kinetoplast DNA structure and function have also been published (Borst et al., 1981b; Englund et al., 1982a; Englund, 1981; Borst and Hoeijmakers, 1979a).

Most animal mitochondrial genomes are found as monomeric circular DNA molecules of about 16 kb (Borst and Flavell, 1976; Wallace, 1982) (Table 1). Throughout the animal kingdom, relatively little variation is found in size or organization. In contrast, mitochondrial DNAs of fungi are diverse in size, ranging from about 18 kb to over 100 kb, and vary greatly in organization (Clark-Walker and Sriprakash, 1982). Major differences in the organization of sequences, or in the amount of DNA in the mitochondrial genome, are found in closely related species. Intervening sequences, absent from animal mtDNA, are common in fungal mtDNA and have important roles in the expression of specific genes (Dujardin et al., 1982; Borst, 1981; Borst and Grivell, 1981). In some species, amplification of specific DNA sequences as small circular molecules may occur in the development of senescence in Podospora anserina (Wright et al., 1982; Esser et al., 1980; Belcour et al., 1981) or as mutations in fungi (Locker et al., 1979; Faugeron-Fonty et al., 1979; Lazowska and Slonimski, 1977; Heyting et al., 1979b; Lazarus et al., 1980a,b).

In higher plants, the diversity of mtDNA molecules is even greater (Levings and Pring, 1978; Levings and Sederoff, 1981; Leaver and Gray, 1982). Estimated genome sizes range from about 100 to 2400 kb in dicotyledonous plants alone (Bendich, 1982). mtDNA populations can be heterogeneous mixtures of linear and circular molecules, some of which may be greatly amplified with respect to other mitochondrial genes (Levings and Sederoff, 1981; Dale, 1982).

Complete linear mitochondrial chromosomes with fixed ends have been found in groups as divergent as Paramecium, Chlamydomonas, and yeast (Grant and Chiang, 1980; Weslowski and Fukuhara, 1981; Cummings and Pritchard, 1982). In some cases, the chromosomes are always linear, e.g., in Paramecium; however, in Chlamydomonas, the linear molecules appear to arise from site-specific cleavage of circular molecules. The most unusual mitochondrial genomes are the kinetoplast DNAs, which are composed of large maxicircles with inter-

TABLE 1
Sizes of Mitochondrial Genomes

Organism	Length (μm)	Md	kb	Structure	Method	References[a]
Vertebrates						
Homo sapiens			16.569	Circular	Sequence	1
Pan troglodytes (chimpanzee)			16.4	Circular	Contour length	2
Cercopithecus aethiops (guenon: African green monkey)				Circular	Contour length	2
Lagothrix cana (wooly monkey)			16.4	Circular	Contour length	2
Galago senegalensis (bush baby)			16.3	Circular	Contour length	2
Macaca mulatta (rhesus)	5.5		16.5	Circular	Contour length	3
Mus musculus (mouse)			16.295	Circular	Sequence	4
Rattus norvegicus (Norwegian rat)			16.4	Circular	Contour length	2
Rattus rattus (rat)			16.4	Circular	Contour length	2
Bos taurus (bovine)			16.338	Circular	Sequence	5
Ovis aries	5.4	10.8		Circular	Contour length	3,6
Mesocricetus auratus (hamster)			16.3	Circular	Contour length	2
Oryctolagus cuniculus (rabbit)			17.3	Circular	Contour length	2
Canis familiaris (dog)	5.0				Contour length	3
Cavia porcellus (guinea pig)	5.6	11		Circular	Contour length	3,10
Gallus domesticus (chicken)	5.4	10.8	16.2	Circular	Contour length	3,7,10
Gallus ferrugineus (jungle fowl)			(16.2)		Restriction	7,8
Acryllium vulturinum (guinea fowl)			(16.2)		Restriction	7
Phasianus torquatus (ring-necked pheasant)			(16.2)		Restriction	7
Meleagris gallipavo (turkey)			(16.2)		Restriction	7
Anas platyrhynchos (Pekin duck)	5.1	10.2		Circular	Contour length	3
Sauromalus ater (chuckwalla)	5.3	10.6		Circular	Contour length	3
Terrapene ornata (turtle)	5.3			Circular	Contour length	3
Cnemidophorus tigris mundus			17.6	Circular	Contour length	9
Cnemidophorus inornatus			17.4	Circular	Contour length	9
Cnemidophorus neomexicanus			17.3	Circular	Contour length	9

4

Species							
Cnemidophorus tesselatus				17.5	Circular	Contour length	9
Rana pipiens (leopard frog)	5.8	11.6	Contour length		Circular	Contour length	10
Xenopus laevis (clawed frog)	5.8	11.7	Contour length		Circular	Contour length	10
Xenopus mulleri (clawed frog)	5.8	11.7	Contour length		Circular	Contour length	3
Siredon mexicanum (axolotl)	4.7	9.4	Contour length		Circular	Contour length	10
Necturus maculosus (mud puppy)	4.8		Contour length		Circular	Contour length	10
Ictalurus punctatus (channel catfish)	5.1		Contour length			Contour length	3
Carrassius carrassius (carp)	5.4		Contour length			Contour length	3
Echinoderms							
Lytechinus pictus (white sea urchin)	4.7	9.4	Contour length	15.7		Contour length	2
Strongylocentrotus purpuratus (purple sea urchin)				15.7		Contour length	2
Insects							
Drosophila							
melanogaster (fruit fly)		12.35	Contour length		Circular	Contour length	12
simulans		11.9	Contour length		Circular	Contour length	12
mauritiana		11.7	Contour length		Circular	Contour length	12
yakuba		10.28	Contour length		Circular	Contour length	12
teissieri		10.04	Contour length		Circular	Contour length	12
erecta		9.92	Contour length		Circular	Contour length	12
lucipennis		10.96	Contour length		Circular	Contour length	12
suzukii		10.37	Contour length		Circular	Contour length	12
takahashii		10.75	Contour length		Circular	Contour length	12
birchii		11.01	Contour length		Circular	Contour length	12
kikkawai		10.68	Contour length		Circular	Contour length	12
auraria		10.81	Contour length		Circular	Contour length	12
denticulata		10.44	Contour length		Circular	Contour length	12
ficusphila		10.68	Contour length		Circular	Contour length	12
eugracilis		10.31	Contour length		Circular	Contour length	12
elegans		10.40	Contour length		Circular	Contour length	12
bipectinata		10.05	Contour length		Circular	Contour length	12
ananassae		10.17	Contour length		Circular	Contour length	12

(continued)

TABLE 1 (*Continued*)

Organism	Length (μm)	Md	kb	Structure	Method	References[a]
saltans		10.04		Circular	Contour length	12
lutea		10.67		Circular	Contour length	12
willistoni		10.13		Circular	Contour length	12
pseudo-obscura		10.19		Circular	Contour length	12
tripunctata		10.32		Circular	Contour length	12
funebris		10.34		Circular	Contour length	12
mercatorum		10.09		Circular	Contour length	12
neohydei		9.90		Circular	Contour length	12
hydei		10.34		Circular	Contour length	12
robusta		10.61		Circular	Contour length	12
americana		10.06		Circular	Contour length	12
montana		10.32		Circular	Contour length	12
virilis		10.12		Circular	Contour length	12
balioptera		9.91		Circular	Contour length	12
hawaiensis		10.00		Circular	Contour length	12
silvarentis		10.01		Circular	Contour length	12
grimshawi		10.00		Circular	Contour length	12
gymnobasis		10.01		Circular	Contour length	12
duncani		10.11		Circular	Contour length	12
buskii		9.95		Circular	Contour length	12
lebanonensis		10.20		Circular	Contour length	12
Locusta migratoria	5.2		~16		Restriction	56
Musca domestica (house fly)	9.0			Circular	Contour length	3
Rhynchosciara hollaenderi				Circular	Contour length	13
Other animals						
Artemia salina (brine shrimp)	5.1	10.2		Circular	Contour length	3
Ascaris lumbricoides (roundworm)	4.8	9.6		Circular	Contour length	3
Ascaris suum (roundworm)	5.0			Circular	Contour length	11

6

Hymenolepis diminuta (tapeworm)	4.8	9.6		Circular	Contour length	3
Urechis caupo	5.9	11.8		Circular	Contour length	3
Fungi						
Aspergillus nidulans			32–33	Circular	Restriction	14,15
nidulans var *echinulatus*			38	Circular	Restriction	15
var *quadrilineatus*			31	Circular	Restriction	15
Brettanomyces anomalus CBS 77	18.15		57.7	Circular	Contour length	16
			56.5		Restriction	16
Brettanomyces custersii			108		Restriction	16
Candida parapsilosis	11.1	23	26.7	Circular	Contour length	8,17
Cephalosporium acremonium			75	Circular	Restriction	55
Dekkera bruxellensis CBS 74			63.5		Restriction	16
Dekkera intermedia CBS 4914			26.7		Restriction	16
Hanseniaspora vineae CBS 2171			55		Restriction	16
Hansenula mrakii	17.5	37	25.5	Linear	Restriction	18
Hansenula wingei	8.2	17	26.5	Circular	Contour length	17
Kloeckera africana CBS 277	8.33		27.1	Circular	Contour length	16
					Restriction	16
Kluyveromyces lactis	11.4	24	37	Circular	Contour length	17
Neurospora crassa			62	Circular	Contour length	3,8
Podospora anserina	31		95		Restriction	20,21
Pachytichospora transvaalensis CBS 2186			41.4	Circular	Restriction	16
Saccharomyces cerevisiae KL 14-4A			77.8	Circular	Restriction	23
Saccharomyces cerevisiae NCYC 74	24.7			Circular	Contour length	22
NCYC 74			68.0		Restriction	23
Saccharomyces cerevisiae (Danish maltese cross)	26.6			Circular	Contour length	22
Saccharomyces exiguus CBS 379	7.47		23.4	Circular	Restriction	16
			23.7		Contour length	16

(continued)

7

TABLE 1 (Continued)

Organism	Length (μm)	Md	kb	Structure	Method	References[a]
Saccharomyces lipolytica (*Candida lipolytica*)		30.5	44		Contour length	19
Saccharomyces telluris CBS 2685			34.8		Restriction	16
Saccharomyces unisporus CBS 398		28	27.4	Circular	Restriction	16
Saprolegina (sp.)	8.63		42	Circular	Contour length	3
Schizosaccharomyces pombe EFL	14		19	Circular	Contour length	24
Schizosaccharomyces pombe 50h⁻			17.3	Circular	Restriction	24
Torulopsis glabrata CBS 138	5.95		18.9	Circular	Contour length	16
Torulopsis glabrata Phaff 71-91			20.3	Circular	Restriction	16
Ustilago cynodontis		50	75	Circular	Restriction	25
Slime molds						
Physarum polycephalum	19	41		Circular	Contour length	26,8
Dictyostelium discoideum		35–40		Circular	Contour length	8
Water mold						
Achlya ambisexualis			49.8	Circular	Restriction	51
Higher plants						
Zea mays						
B37N (maize)			484		Restriction (minimum estimate)	27
cms-T			463		Restriction	27
cms-C			494		Restriction	27
cms-S			489		Restriction	27
WF9-N			650		Cosmid mapping	28
Sorghum bicolor (sorghum)		150–200			Restriction	30
Glycine max (soybean)		150			Restriction	52,53
Linum usitatissimum (flax)		106			Restriction	32
Pisum sativum (pea)		240			Reassociation	29
Vicia faba (broadbean)		190			Restriction	33

8

Organism				Structure	Method	Ref
Vicia villosa		250			Restriction	33
Triticum aestivum (wheat)		140	210		Restriction	31
Citrullus vulgaris (watermelon)		230			Restriction	34
Cucurbita pepo (zucchini)		220			Reassociation; restriction	29
Cucumis sativus (cucumber)		560			Reassociation; restriction	29
Cucumis melo (muskmelon)		1000			Reassociation	29
Nicotiana tabacum (tobacco)		1600			Reassociation	29
Brassica napus N		260–290			Restriction	35
Brassica napus (cms)		136.5			Restriction	36
Solanum tuberosum (potato)		140.3			Restriction	36
Oenothera berteriana (evening primrose)	90	180–190			Restriction	37
Lactuca sativa (lettuce)		140			Reassociation	8
Parthenocissus tricuspidata (Virginia creeper)		165			Restriction	31
Algae						
Chlamydomonas reinhardtii		10	15	Linear/circular	Contour length	38
Chlorella pyrenidosa		53			Renaturation	39
Euglena gracilis		40–50		Circular	Contour length	40,41,42,43
Protozoa						
Acanthamoeba castellanii	12.8	27	40	Circular	Contour length	44
Plasmodium lophurae	10.3	21.5	18.5	Circular	Contour length	54
Tetrahymena pyriformis						
ST		30		Linear	Contour length	45
GL		32.6		Linear	Contour length	45
W		25.8		Linear	Contour length	45
Paramecium aurelia	14		40	Linear	Contour length; restriction	46
Crithidia luciliae	11.3		35	Maxicircle	Contour length; restriction	47

(continued)

TABLE 1 (Continued)

Organism	Length (μm)	Md	kb	Structure	Method	References[a]
Crithidia fasciculata	12		35	Maxicircle	Contour length; restriction	47
Herpetomonas muscarum			32	Maxicircle	Restriction	47
Trypanosoma brucei	6.3		20–22	Maxicircle	Contour length; restriction	47
Trypanosoma cruzi	9.0		40	Maxicircle	Contour length	47
Trypanosoma mega		16	24.5	Maxicircle	Restriction	47
Trypanosoma equiperdum	15.4		24	Maxicircle	Restriction	48
Phytomonas davidi			36.5	Maxicircle	Restriction	49
Leishmania tarentolae		18–20	30.5	Maxicircle	Restriction	50,47

[a] 1. Anderson et al. (1981); 2. Brown et al. (1981); 3. Borst and Flavell (1976); 4. Bibb et al. (1981); 5. Anderson et al. (1982); 6. Upholt and Dawid (1977); 7. Glaus et al. (1980); 8. Wallace (1982); 9. Brown and Wright (1979); 10. Wolstenholme and Dawid (1968); 11. Rodrick et al. (1977); 12. Fauron and Wolstenholme (1976); 13. Handel et al. (1973); 14. Kuntzel et al. (1982); 15. Turner et al. (1982); 16. Clark-Walker and Sriprakash (1982); 17. O'Conner et al. (1975); 18. Weslowski and Fukuhara (1981); 19. Weslowski et al. (1981); 20. Kuck and Esser (1982); 21. Cummings et al. (1979); 22. Christiansen and Christiansen (1976); 23. Sanders et al. (1977); 24. Wolf et al. (1982); 25. Mery-Drugeon et al. (1980); 26. Bohnert (1977); 27. Borck and Walbot (1982); 28. Lonsdale (personal communication); 29. Ward et al. (1981); 30. Pring et al. (1982); 31. Quetier and Vedel (1977); 32. Lockhart and Levings (unpublished); 33. Bendich (1982); 34. Bonen and Gray (1980); 35. Belliard et al. (1979); 36. Vedel et al. (1982); 37. Brennicke (1980); 38. Grant and Chiang (1980); 39. Bayen and Rode (1973); 40. Manning et al. (1971); 41. Talen et al. (1974); 42. Crouse et al. (1974); 43. Fonty et al. (1975); 44. Bohnert (1973); 45. Goldbach et al. (1979); 46. Goddard and Cummings (1975, 1977); 47. Borst et al. (1981); 48. Riou and Saucier (1979); 49. Cheng and Simpson (1978); 50. Simpson (1979); 51. Hudspeth et al. (1983); 52. Synenki et al. (1978); 53. Levings et al. (1979); 54. Kilejian (1975); 55. Minuth et al. (1982); 56. Gellissen et al. (1983). Values in parentheses are for turkey, pheasant, and fowls are presumed values because the size is indistinguishable from that of chicken (Glaus et al., 1980) but values were not presented in that reference.

wound minicircles concatenated in a special structure (Borst and Hoeijmakers, 1979a; Englund, 1981).

In spite of enormous structural diversity, mitochondrial genomes are remarkable in their functional conservation. In all systems studied, mtDNAs code for a common set of RNA and protein molecules that include the large and small rRNA subunits, mitochondrial tRNAs, several specific proteins known to be important components of the mitochondrial inner membrane, and a small number of additional proteins usually of unknown function.

In the 3 mammalian mitochondrial genomes that have been completely sequenced, human (Anderson et al., 1981), bovine (Anderson et al., 1982a,b), and mouse (Bibb et al., 1981), 5 of 13 polypeptide-coding genes have been identified. The protein-coding genes of identified common function in both mammals and fungi are cytochrome b, three subunits of cytochrome oxidase, and subunit 6 of the ATPase complex.

Some mitochondrial genomes code for more than one subunit of the ATPase complex of the mitochondrial inner membrane. In yeast, two ATPase subunits have been identified, ATPase 6 and ATPase 9 (Borst and Grivell, 1978; Tzagoloff et al., 1979). Subunit 9 is a very hydrophobic polypeptide known as the ATPase proteolipid, or the dicyclohexylcarbodiimide (DCCD)-binding protein. Subunit 9 is coded by a mitochondrial gene in yeast, but in Neurospora the active protein is coded by a nuclear gene (Sebald et al., 1979). In mammals, none of the unidentified open reading frames shares homology with ATPase 9; therefore, this subunit is most likely to be nuclear coded in mammals as well (Anderson et al., 1981).

Additional mitochondrial-coded proteins are known in yeast and filamentous fungi. A ribosomal protein associated with the small rRNA subunit is coded by mtDNA. The protein in yeast is called var 1 because of the extensive polymorphisms associated with this protein (Terpstra et al., 1979; Groot et al., 1979). The var 1 protein-coding sequence has been identified in yeast mtDNA (Tzagoloff et al., 1980; Butow et al., 1982; Hudspeth et al., 1982). A related ribosomal protein called S5 is synthesized in mitochondria of Neurospora crassa (Lambowitz et al., 1976; LaPolla and Lambowitz, 1982). The gene coding for S5 has been tentatively located within the ribosomal RNA intron (Burke and RajBhandary, 1982).

In yeast, eight additional protein-coding sequences have been discovered as introns of other structural genes (Borst, 1981; Borst and Grivell, 1981; Dieckmann et al., 1982; Jacq et al., 1982; Grivell et al., 1982). More than one polypeptide (in some cases, several polypeptides) is coded within the boundaries of a structural gene. For example, the cytochrome *b*-coding sequence spans 9 kb of DNA and codes for a protein of only 385 amino acids. In some strains, the gene is present in a long form containing 5 introns that are transcribed and processed to produce the mature message of 2 kb. Two of the introns must be translated into proteins that act as RNA-processing enzymes to generate the mature message (Slonimski *et al.*, 1978a,b; Nobrega and Tzagoloff, 1980a,b; Van Ommen *et al.*, 1979; Lazowska *et al.*, 1980; Dujardin *et al.*, 1982; Weiss-Brummer *et al.*, 1982). This process of autoregulation by intron coding of RNA-processing enzymes has not been found in other systems. In *Saccharomyces carlsbergensis,* cytochrome *b* has only two introns (short form).

Therefore, in yeast, cytochrome *b* is a complex structural gene, found in long and short forms, with additional coding functions. Both forms of the gene generate the same final coding product, which becomes part of a protein complex active in oxidative phosphorylation. In animal mitochondria the same gene has no introns and is translated directly after cleavage of a larger transcript (Ojala *et al.*, 1981). Functionally, cytochrome *b* is highly conserved, but the gene structure and RNA processing systems are extremely divergent.

In plant mtDNA, only one protein-coding gene has been identified by sequence analysis, cytochrome oxidase subunit II (COII) (Fox and Leaver, 1981). Interestingly, maize COII contains an intron of 794 bp whereas COII in yeast does not contain any introns. The ribosomal genes of wheat and maize have been mapped and cloned (Stern *et al.*, 1982; Gray *et al.*, 1982; Iams and Sinclair, 1982). Higher plant mitochondrial ribosomes are unique, in that they contain a 5 S RNA in addition to the large (26 S) and small (18 S) subunit rRNAs common to all mitochondrial ribosomes. The 5 S rRNA has been sequenced in wheat (Gray and Spencer, 1981) and the gene coding for the 5 S rRNA has been sequenced in maize (Chao *et al.*, 1983). Tentative identification has been made of the var 1-equivalent ribosomal protein, COI, and the ATPase proteolipid (subunit 9) (Leaver and Gray, 1982). tRNAs that appear to be unique to plant mitochondria have been isolated and characterized (Guillemaut and Weil, 1975; Leaver and Gray, 1982; Cunningham and Gray, 1977). In wheat, these tRNAs appear to be coded by genes scattered widely over the mitochondrial genome (Bonen and Gray, 1980). The tRNA genes are clustered at two sites in *Neurospora crassa* and *Aspergillus nidulans,* generally clustered in yeast, and dispersed in animal mtDNA.

In vitro protein synthesis of isolated plant mitochondria shows 18 to 20 mitochondrial polypeptides (Forde *et al.*, 1978). Similar patterns were observed in all other higher plants examined (Leaver *et al., 1982*). This number is close to that observed in animal (HeLa cells) mitochondrial systems, in which 26 discrete translation products have been identified (Attardi *et al.*, 1980). The actual number of polypeptide-coding mitochondrial genes in HeLa cells is 13.

If the sizes of the polypeptides labeled in plant mtDNA, calculated as base pairs, are added to the known sizes of ribosomal genes and 20 tRNAs are included, the number of base pairs needed to code for all the identified proteins and functions is less than 22 kb (Leaver *et al.*, 1982). Although 22 kb is larger than the average animal cell mitochondrial genome, some of the equivalent genes are larger. The small subunit of rRNA in animal mitochondria is only 954 bp whereas the same RNA in maize or wheat is about 1968 bp. A 5 S rRNA (122 bp) is found in plant mitochondrial systems.

Additional functions in plant mtDNA are suggested by the strain-specific cytoplasms that cause cytoplasmic male sterility (Duvick, 1965; Edwardson, 1970). In maize, three diverse types of male sterility can be distinguished by different mtDNA restriction patterns (Pring and Levings, 1978). Several mitochondrial coded proteins have been identified that characterize specific male sterile genotypes (Forde and Leaver, 1980; Leaver *et al.*, 1982) but the specific genes involved with variant proteins in these genotypes have not been identified.

Within the mitochondria, the mtDNA is organized in condensed structures and associated with protein in a mitochondrial nucleus (Kuroiwa, 1982). Mitochondria of lower eukaryotes have only one rod-shaped or spherical mitochondrial nucleus whereas those of higher animals and plants contain more than one condensed structure. The nuclei are usually located in the middle of the mitochondrial matrix.

In animal mitochondria, nuclei are relatively electron transparent in thin sections, with a few visible DNA fibers. Logarithmically growing L cells have about 5 or 6 mitochondrial genomes per mitochondrion (Nass, 1966). Rat liver mitochondria have 14 (Nass *et al.*, 1965). In plants, the mitochondrial nuclei are more electron dense. *Physarum polycephalum* (slime mold) has particularly dense, readily visible mitochondrial nuclei, each containing an estimated 32 mitochondrial genomes in actively growing cells, but only 4 to 8 in inactive cells (Kuroiwa, 1982).

The mtDNA is associated with nonhistone proteins, and is thought to form chromatinlike fibrils. In yeast, an abundant DNA-binding protein has been purified from mitochondria (Caron *et al.*, 1979). The protein, called HM, is 20,000 d and is present in amounts at least equal to the DNA content. The protein is lysine rich and slightly basic. It is thought to be nuclear coded, because it is found in several mutants containing large deletions of mtDNA. In *Physarum,* mitochondrial nuclei contain at least six different basic proteins that range from 32,000 to 105,000 in molecular weight (Kuroiwa, 1982). Mitochondria from rat liver also contain basic histonelike proteins (Hillar *et al.,* 1979).

The mitochondrial code is known to be different from the universal code. In the universal code, a minimum of 32 tRNAs is required (Crick, 1966). Only 22 tRNAs have been identified, by sequencing, in mammalian mtDNA (Anderson *et al.,* 1981, 1982a,b; Bibb *et al.,* 1981), 24 in yeast (Bonitz *et al.,* 1980c; Tzagoloff, 1982), and 23 in *Neurospora* (Yin *et al.,* 1981, 1982). The mitochondrial code is simpler, in that more four-codon families have a single tRNA. These tRNAs have a U in the first anticodon position, and an unmodified U can pair with all four bases. A modified U will read only GU wobble (Heckman *et al.,* 1980; Roe *et al.,* 1982). Because the position immediately 3' to the anticodon in many cases contains a modified nucleotide, it has been proposed that this base may modulate the wobble in the anticodon (Roe *et al.,* 1982).

In addition to the differences between mitochondrial codes and the universal code, mitochondrial codes differ from each other. Whereas the universal code uses UGA as a termination codon, mitochondria of mammals and fungi use UGA for tryptophan (Anderson *et al.,* 1981; Macino *et al.,* 1979; Browning and RajBhandary, 1982). In the maize mitochondrial gene for COII, however, no TGA codons were found, suggesting that UGA may not be a tryptophan codon in plant mtDNA (Fox and Leaver, 1981). The codons AGG and AGA code for arginine in the universal code and in mitochondria of yeast, *Neurospora,* and maize, whereas in mammalian mitochondria, AGG and AGA are termination codons (Anderson *et al.,* 1981). Mammals use AUA to code for methionine in mitochondria, both as an initiation codon or as an internal methionine (Anderson *et al.,* 1981; Bibb *et al.,* 1981). In yeast, the AUA codon, assigned earlier to isoleucine, is more likely to be a methionine codon (Hudspeth *et al.,* 1982). In mtDNA of *Neurospora,* AUA codes for isoleucine (Browning and RajBhandary, 1982).

It is surprising that the genetic code has diverged between yeast and *Neurospora.* In addition to the different use of AUA, yeast alone uses the CUN codon family for threonine (Bonitz *et al.,* 1980c). It has been argued (Jukes, 1981) that this difference is due to a comparatively recent event having occurred in the evolution of the ascomycetes. However, evolution of the mitochondrial codes must be complex, because both mammals and yeast use AUA for methionine whereas *Neu-*

rospora uses AUA for isoleucine, as does the universal code. In maize mtDNA, it has been suggested that CGG (normally arginine) codes for tryptophan, indicating the possibility of another variation in the evolution of mitochondrial genetic codes.

It is variously argued that the mitochondrial codes are either more primitive codes or that they are simply divergent due to relaxed translational constraints. Divergence may be increased by a higher mutation rate and selective pressure for an exceedingly economical genome (Jukes, 1981; Lagerkvist, 1981; Wallace, 1982; Gray, 1982). These views are not mutually exclusive, because a relaxation of constraints and pressure for economy could have selected for a more primitive type of code.

A genome with a small number of protein-coding genes would be subject to fewer constraints for the evolution of the code. If a coding change occurred in a tRNA, it would have a relatively small chance of affecting a critical site in the mitochondrial proteins. The effects of any coding change on a nonorganellar nuclear genome, with many thousands of essential translation products, would most likely be lethal. Because the universal code is not the only code possible, other codes may have coexisted earlier. Our present universal code can be regarded as a frozen accident (Jukes, 1981).

II. Structural Variation of Animal Mitochondrial Genomes

A. MAMMALIAN MITOCHONDRIAL GENOMES

1. Information Content and Organization

Mammalian mitochondrial genomes have been extensively characterized. The human, bovine, and mouse mitochondrial genomes have been completely sequenced (Anderson *et al.*, 1981, 1982a; Bibb *et al.*, 1981). All three genomes have virtually identical organization and common structural features (Fig. 1). The genome is small and is characterized by extreme economy in the utilization of functional sequences. Coding sequences, tRNA genes, and rRNA genes are virtually contiguous. One end of a functional gene is usually within a few nucleotide pairs of the adjacent gene.

Each genome contains a single gene for a large (16 S) and a small (12 S) rRNA subunit, 22 tRNAs, genes for 3 cytochrome oxidase subunits (COI, COII, and COIII), ATPase subunit 6, cytochrome *b*, and 8 unidentified open reading frames (URFs). The protein-coding genes that

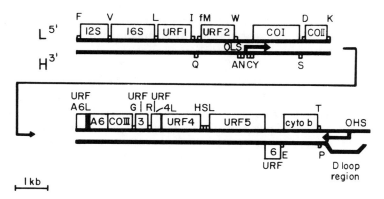

FIG. 1. Physical map of a generalized mammalian mitochondrial genome following the maps of Anderson *et al.* (1981, 1982) and Bibb *et al.* (1981). The map has been divided into two segments. The light strand (L) and heavy strand (H) are shown. Genes are designated on the strand, which is equivalent to the transcript. The 16 S rRNA gene transcribed from the H strand is shown on the L strand, where the sequence of the transcript corresponds to the L strand DNA sequence. tRNA genes are depicted by amino acid abbreviations as small squares on the map. F, Phe; V, Val; L, Leu; I, Ileu; Q, Gln; fM, fMet; A, Ala; N, Asn; C, Cys; Y, Tyr; S, Ser; D, Asp; K, Lys; G, Gly; R, Arg; H, His; S, Ser; L, Leu; E, Glu; T, Thr; P, Pro. The origin of light strand synthesis is designated OLS and OHS is the origin of heavy strand synthesis. Gene designations are standard; 12 S and 16 S for rRNA genes COI, II, and III for subunits of cytochrome oxidase. A6 is the ATPase subunit 6. URFs indicate unidentified open reading frames after Anderson *et al.* (1981, 1982).

have been identified are components of the mitochondrial inner membrane. The URFs are probably functional genes because they are similar in all three genomes, with respect to size, location, and sequence. All but one of these sequences is coded on the same strand. The number of tRNAs and the position of each tRNA are identical in all three species.

In mammalian cells, mitochondrial transcripts are unstable. Most mRNAs have a half-life of about an hour, and the half-lives of the rRNAs are several times longer (Gelfand and Attardi, 1981; Attardi *et al.*, 1982). Blocking mtRNA synthesis greatly extends the lifetime of the existing messages (Lansman and Clayton, 1975). Furthermore, enucleated African green monkey cells continue to synthesize mitochondrial proteins for at least 24 hours with no significant decline (Englund, 1978). Therefore, the amount of mitochondrial protein synthesis is regulated, at least in part, by a mechanism that affects the stability of the mRNA.

Ribosomal rRNAs are present in 30- to 60-fold excess over mRNAs (Attardi *et al.*, 1982). Because all H strand transcripts are controlled by the same promoter, it might be expected that the difference in relative abundance is due to differential stabilities. However, the half-lives differ only by a factor of three; therefore, the abundance of rRNAs must be due to a difference in synthesis. In order to account for a difference in the rate of synthesis of rRNA and mRNA transcripts, it has been proposed that premature termination can occur at the end of the rRNA gene. Transcription of mRNAs occurs at low frequency when termination is bypassed and the rest of the H strand is transcribed (Attardi *et al.*, 1982).

The L strand, like the H strand, is transcribed completely, but the rate of L strand transcription is two to three times more frequent (Cantatore and Attardi, 1980). The L strand contains only 7 tRNAs and URF 6, but RNA molecules of specific sizes containing URF 6 have been found, consistent with a protein-coding function. The H strand is almost completely transcribed to produce rRNAs, most of the poly(A)-containing mRNAs, and tRNAs (Attardi *et al.*, 1980; Ojala *et al.*, 1981).

Detailed mapping of transcripts combined with the complete sequence analysis has provided a precise localization of RNA molecules corresponding to specific genes (Anderson *et al.*, 1981; Montoya *et al.*, 1981). All mRNAs utilized in mammalian mitochondria are endogenous and are translated on mitochondrial polysomes. All mRNAs start directly at or very near to an initiation codon. In most cases the mRNAs are directly flanked on the 5' end by a tRNA gene without any intervening nucleotide (Montoya *et al.*, 1981). Similarly, the 3' ends of all mRNA-coding sequences are immediately contiguous to a tRNA or to another coding sequence. These results have prompted a tRNA punctuation model of RNA processing (Ojala *et al.*, 1981).

According to the model, individual rRNAs, tRNAs, and mRNAs are generated by RNA-processing activity that is a similar to RNase P, which processes 5' ends of tRNAs in *Escherichia coli* (Lund and Dahlberg, 1977; Altman, 1978). This enzyme is known to recognize secondary and tertiary structure rather than specific sequences.

Every transcript, except RNA 7 and RNA 14, is consistent in size with its open reading frame and shows no evidence of discontinuity related to intervening sequences (Anderson *et al.*, 1981; Montoya *et al.*, 1981). In two cases, however, overlapping genes are associated with RNA molecules that may be multicistronic transcripts. In human, bovine, and mouse mitochondrial genomes, the 3' end of URFA6L overlaps the 5' end of ATPase 6 by about 40 bp into the 5' end of

ATPase 6. This region is represented by transcript RNA 14 (Ojala *et al.*, 1980), which contains both coding sequences. A similar situation exists for RNA 7, which contains coding sequences for both URF4L and URF4. It is not known how these transcripts are translated. Other overlapping sequences are sequences coded on opposite strands, or which represent tRNA sequences that might be completed by post-transcriptional addition (Bibb *et al.*, 1981). Alternatively, some transcripts could be defective. Many of the protein-coding genes end with termination codons in their DNA sequence. Several do not, but instead end in a single T or TA. Because the mRNAs are polyadenylated, addition of poly(A) creates termination codons (Ojala *et al.*, 1981; Bibb *et al.*, 1981) that allow normal termination in the mitochondrial translation system. tRNAs are matured after cleavage by addition of CCA at their 3' ends (Anderson *et al.*, 1981).

tRNA sequences in mammalian mitochondria are unusual compared to prokaryotic and eukaryotic tRNAs. The mitochondrial tRNAs are highly variable in size and sequence. All tRNAs, except one leucine tRNA, have variations in some of the standard bases or have a variable length for the TψC loop (Anderson *et al.*, 1981).

Mammalian mitochondrial tRNAs have a higher AU content than their cytoplasmic counterparts. The mitochondrial tRNAs also contain a higher proportion of nonstandard base pairs (AC and GU). These factors are likely to result in a reduced stability in the overall structure of the molecule; consequently, tertiary interactions may be involved in the stability of the overall structure (Roe *et al.*, 1982). The most unusual tRNA structure is that of serine tRNA, which completely lacks the DHU arm of the standard cloverleaf structure. This unusual structure is found in human, bovine, and mouse mitochondrial serine tRNAs. All other tRNAs can form a cloverleaf secondary structure.

The rate of evolution (nucleotide substitution rate) for tRNA of mitochondria is at least 100 times that of nuclear tRNA genes (Brown *et al.*, 1982). Compared to other mitochondrial gene sequences, the rate of change for tRNAs is less than that for structural regions or functional genes (Bibb *et al.*, 1981; Brown *et al.*, 1982).

2. Structural Properties of Mammalian mtDNA

All animal mitochondrial genomes are circular, and intact mtDNAs are usually covalently closed circular molecules (Borst and Flavell, 1976; Wallace, 1982; Clayton, 1982). In mammalian cells, a major fraction of these molecules has a triple-stranded (D loop) region, about 600 bp long, which is located at a unique site in the molecule (Arnberg *et al.*, 1971; Kasamatsu *et al.*, 1971). The short strand can be displaced by introduction of single nicks in the parental strand (Robberson and Clayton, 1973) or by partially denaturing conditions. The displacement loop sequence is not a uniform population of sequences. In the mouse there are four discrete species, ranging from about 520 to 690 bp in length, in which the differences are attributed to differences in the length of the 3' end. In contrast, human D loop strands are composed of three species, with the length variation being at the 5' end (Brown *et al.*, 1978; Gillum and Clayton, 1978; Tapper and Clayton, 1981; Walberg and Clayton, 1981; Clayton, 1982). Heterogeneity in D loop lengths has been observed in rat (Sekiya *et al.*, 1980) and also in *Xenopus* (Gillum and Clayton, 1978).

Replication of the molecule begins when synthesis extends the H strand beyond the region of the D loop. Synthesis continues in one direction until complete. When H strand synthesis is 67% complete, the origin of the L strand is exposed and L strand initiation occurs. When strand synthesis is complete, closure occurs and about 100 negative superhelical turns are introduced, resulting in the supercoiled form of mtDNA (Clayton, 1982). The subsequent formation of the D loop relaxes the superhelical turns. Synthesis of the complete molecule is slow, requiring about 2 hours for the entire process.

Although D loops have been presumed to be primers for H strand replication, this may not be their only function (Clayton, 1982). More than 95% of the D loop strands made in the cell are lost by turnover and do not prime DNA replication (Bogenhagen and Clayton, 1978a). It is possible that D loops are not extended into full-length strands, but that synthesis of H strands begins directly at the origin and that the short D loop strands are not extended to full length. D loops could have other potential roles. One possible role would be to facilitate formation of a structure used to initiate transcription. The major sites for initiation of transcription of both the H strand and the L strand are located in or near the D loop region (Van Etten *et al.*, 1982; Attardi *et al.*, 1982).

The origin of the light-strand replication is two-thirds of the way around the genome, and synthesis of the light strand does not begin until the synthesis of the new H strand has displaced and exposed the parental H strand. This event creates a single-stranded region in the H strand, in a region of dyad symmetry that can form a stem loop structure. The stem loop structure contains a 12-bp stem and a 13-bp loop in mouse (Bibb *et al.*, 1981) and is conserved in structure, but not sequence, in bovine and human DNA (Anderson *et al.*, 1982b; Martens

an insertion or deletion of about 30 bp associated with an *Hpa*II fragment (Hayashi *et al.*, 1981).

Size variation in amphibian mtDNA was noted many years ago by Wolstenholme and Dawid (1968). Comparisons of urodele and anuran amphibians showed significant size differences. The sizes of *Xenopus laevis* (5.79 μm) and *Rana pipiens* (5.82 μm) were considerably larger than those of the urodeles, *Siredon mexicanum* and *Necturus maculosus*, with sizes of 4.77 and 4.96 μm. All estimated circle sizes of vertebrate mtDNA fall within the range of these two groups of amphibians (Wallace, 1982), corresponding to 14.1 kb (*Siredon*) and 17.6 kb (*Xenopus*). Independent size estimates of *Xenopus laevis* mtDNA are 17.4 kb (Bultmann and Borkowski, 1979) and 17.2 kb (Ramirez and Dawid, 1978).

The smaller values obtained for urodele mtDNA would correspond to a 14-kb mitochondrial genome. This value implies that at least 2 kb of sequence is missing compared to mammalian mtDNAs; these mtDNAs would be the smallest known mitochondrial genomes. Large mtDNAs have also been found in four species of parthenogenetic lizards, of the genus *Cnemidophorus*, ranging between 17.3 and 17.6 kb (Table 1) (Brown and Wright, 1979).

mtDNA from two species of *Xenopus* has been further characterized (Dawid and Rastl, 1979). The location of the ribosomal genes, tRNAs, and major transcripts shows strong conservation of location when compared to mammalian mitochondrial genomes (Wallace, 1982). The size of *X. laevis* mtDNA is 17.4 kb (Bultmann and Borkowski, 1979), somewhat larger than that of *X. borealis,* estimated to be from 17.0 to 17.2 kb. *X. laevis* is known to have a large D loop structure of about 1.5 kb, larger than that of *X. borealis* (1.23 kb) (Ramirez and Dawid, 1978); both are considerably larger than the D loop structure of mammalian mtDNA. Sequence divergence between the two *Xenopus* species is extensive (25%), and denaturation maps of the mtDNAs are radically different (Bultmann and Borkowski, 1979). The general organization, however, is highly conserved. Electron microscopy and nuclease digestion experiments provided no evidence of intervening sequences (Dawid and Rastl, 1979).

In mammalian mtDNA, structural variation is greatly restricted, whereas divergence in nucleotide sequence is extensive. In some cases, the nucleotide divergence has been directly determined from sequence analysis. In comparisons of bovine and human genomes, protein-coding genes are from 63 to 79% homologous (Anderson *et al.*, 1982b). Comparison of mouse and human genomes shows similar homology in nucleotide sequence, ranging from 65 to 77% for protein-coding genes (Bibb *et al.*, 1981). Most of the variation is in the silent third position of the codon (Anderson *et al.*, 1982b).

Nucleotide sequence divergence has been studied in rats, also by restriction-site polymorphisms and by nucleotide sequence analysis (Francisco and Simpson, 1977; Brown *et al.*, 1981; Castora *et al.*, 1980; Goddard *et al.*, 1981; Vos *et al.*, 1980). In both domesticated and wild rats a wide range of restriction-fragment polymorphisms was found (Brown *et al.*, 1981; Brown and Simpson, 1981). These results allowed estimation of sequence divergence within and between *Rattus* species. The maximum divergence between populations of *Rattus rattus* and *Rattus norvegicus* is 18.4% whereas the average divergence between these species was 16.4%. This value is high, but not in great contrast with amounts of homology between more divergent primate species, which may have from 21 to 29% sequence divergence between them. Interestingly, within the *R. rattus* species, divergence was as high as 10%. Most of the variation can be attributed to silent changes in the third position of codons in functional genes.

Sequence analysis of two rat COII genes from *R. rattus* and *R. norvegicus* shows that 94.4% of the sequence changes are silent substitutions (Brown and Simpson, 1982). Analysis of mutational changes indicates a strong bias (8:1) for transition mutations over transversions (transitional bias) and a strong bias for C-T transitions in the light strand (Brown and Simpson, 1982).

Restriction endonucleases have been used to measure sequence relatedness among species of *Peromyscus* (Avise *et al.*, 1979a), pocket gopher (Avise *et al.*, 1979b), and subspecies of the mouse (Yonekawa *et al.*, 1981). In *Peromyscus,* individuals from the same population may differ by 1.5% in nucleotide sequence whereas sibling species can differ by as much as 17%. Different species of *Peromyscus* may vary by 20%. Similarly, in the pocket gopher, genetic distances were estimated between gophers throughout the range of the species. Two major populations (eastern and western) differ by at least 3% in mtDNA sequence. In the mouse, nucleotide divergence between local races was less than 0.4%, whereas divergence between Asian and European subspecies was 7.1 and 5.8%. The times of divergence were estimated using about 3% divergence for 10^6 years.

For a region of 896 bp, in which sequence information is known for several species (Brown *et al.*, 1982), the average sequence difference between primate and mouse is 33%, between primate and cow, 30%, and between mouse and cow, 31% (Brown *et al.*, 1982). A small number of single base-pair addition or deletion events were observed that occurred in or between tRNA genes. In a comparison of different primate sequences, transitions greatly outnumbered transversions (from 92 to 8%). As established by earlier results (Brown et al., 1979), the rate of mtDNA evolution is 5- to 10-fold higher than that of nuclear DNA. The sequence data show a high rate of mutation at silent substitution sites that is also 10-fold higher than that established for nuclear genes. For tRNA genes, the mtDNA substitution rate was 100-fold higher than that of nuclear tRNA genes (Brown et al., 1982). Similarly, Miyata et al. (1982) have calculated a rate for silent substitution mutations in mtDNA comparing mouse and rat. The rate $\geq 35 \times 10^{-9}$/site/year is at least sixfold higher than the silent substitution rate for nuclear genes.

The average nucleotide substitution rate in protein-coding genes comprises the replacement rate (which results in an altered amino acid) and the silent rate (which maintains the same amino acid sequence). For known protein-coding genes, most variation between mouse and human is silent. The silent base change frequency varies from 43 to 74% in different known protein-coding genes (average is 56%). The silent nucleotide substitution rate for two protein-coding mitochondrial genes was estimated to be about 10% per 10^6 years (Brown et al., 1982).

The ribosomal genes are 74% (16 S) and 75% (12 S) homologous between human and mouse (Bibb et al., 1981). Similarly, the overall homology of the rRNA sequences of human and bovine is 77% (Anderson et al., 1982b). In comparison with rRNA sequences of Escherichia coli, the rRNAs in mammalian mitochondria are considerably shorter. Several internal regions of the molecule are missing, suggesting that they were lost under selective pressure to reduce mitochondrial genome size. Secondary structure models that conserve common features can be generated between human and bovine rRNA sequences (Anderson et al., 1982b), following the models that have been proposed for E. coli (Noller and Woese, 1981).

The small and large rRNA subunits from rat mtDNA have been sequenced (Kobayashi et al., 1981; Saccone et al., 1982). Homology of the small subunit between rat and mouse was 91.9%, whereas the homology between rat and human was 78.1%. The large subunit showed only 77% homology with mouse and 70% with the human sequence. Many tRNAs have been sequenced in rat mtDNA (Grosskopf and Feldman, 1981a; Kobayashi et al., 1981). The tRNAs are structurally similar to other mammalian tRNAs and have an average homology of 77% between rat and human. The organization of tRNAs and coding sequences in the rat are similar to those in other mammalian systems (Grosskopf and Feldman, 1981b; Wolstenholme *et al.,* 1982), suggesting similar mechanisms of transcription and processing.

It is clear that the animal mitochondrial genome is highly constrained with respect to rearrangements or other structural variations. The extreme economy of nucleotides and the absence of spacers between genes precludes viability for most inversions or rearrangements. What selective pressures in the evolution of metazoa enforce such extreme economy are not known.

B. STRUCTURAL VARIATION IN *Drosophila* SPECIES

Drosophila mtDNA represents an invertebrate mitochondrial genome that shares an ancient common lineage with vertebrate mtDNA but which has been separated for more than 500 million years, since the divergence of protostomes (annelid–arthropod line) and deuterostomes (echinoderm–chordate line). The genus *Drosophila* has over a thousand (1467) distinct species (Wheeler, 1981). The mtDNAs of the genus *Drosophila* are as divergent in size as the mtDNAs of the entire subphylum of vertebrates (Table 1) (Wallace, 1982).

In 39 species of *Drosophila* examined by Fauron and Wolstenholme (1976) mtDNAs vary from 15.7 kb (*D. hydei*) to 19.5 kb (*D. melanogaster*). Most species fall between 15.7 and 16.4 kb. The *melanogaster* species group has a broader range of sizes, with many species having mtDNA larger than 17 kb. Closely related species can vary widely; for example, in the *melanogaster* species subgroup of the *melanogaster* group, the sibling species *melanogaster, simulans,* and *mauritiana* have mitochondrial genome sizes of 19.5, 18.9, and 18.6 kb, respectively, whereas the other members of the subgroup, *yakuba, tessieri,* and *erecta,* have sizes of 16.2, 15.9, and 15.7 kb.

The observed size variation is largely, if not entirely, due to variation in one specific region of the molecule characterized by AT richness (Bultman *et al.,* 1976; Fauron and Wolstenholme, 1976, 1980a,b) that varies from 1 to 5.1 kb in different *Drosophila* species (Goddard *et al.,* 1982; Wolstenholme *et al.,* 1979; Klukas and Dawid, 1976). A part of the AT-rich region has been sequenced in *Drosophila yakuba* and contains only 11.3% GC (Goddard *et al.,* 1982).

The AT-rich region of *Drosophila* mtDNA shares several features with the replication region of mammalian mtDNA. In most *Drosophila* species the AT-rich region is about 1 kb. Both regions contain the origin of replication, lack long open reading frames, are mostly not transcribed, and show extensive divergence between related species.

Heteroduplex analysis of different *Drosophila* species showed that the AT-rich regions usually failed to pair, except for the three sibling species of the *melanogaster* group, even under highly permissive base-pairing conditions (Fauron and Wolstenholme, 1980a,b; Shah and Langley, 1979). Complete homology was observed for all other regions of the molecule in all heteroduplexes examined, with the exception of a small unpaired segment of about 200 bp found between *D. virilis* and *D. yakuba*.

The AT-rich regions of the three sibling species, *D. melanogaster, D. simulans,* and *D. mauritiana,* are large and similar in size (5.1, 4.8, and 4.6 kb, respectively). Most of the AT-rich regions (65%) did not pair, even between sibling species (Fauron and Wolstenholme, 1980a,b). Renaturation analysis (Merten and Pardue, 1981) suggests that the AT-rich region of *D. melanogaster* mtDNA has about 150 bp of unique sequence, flanked by multiple direct repeats on either side. Even within species, major variation in the AT-rich region may be found. In two lines of *D. mauritiana,* a size difference of 700 bp was observed in the AT-rich region (Fauron and Wolstenholme, 1980b). Heteroduplex analysis showed two or three single-stranded regions near the center of the AT-rich region. Size variation in the length of the AT-rich region was also demonstrated within different lines of *D. melanogaster* and within different lines of *D. simulans,* using restriction enzyme analysis (Fauron and Wolstenholme, 1980b; Reilly and Thomas, 1980).

From electron microscopy, the mode of replication in *Drosophila* mtDNA is shown to be different from that of mammalian DNA in some important aspects. Initiation of replication begins in the AT-rich region and proceeds unidirectionally around the molecule. No prevalent class of D loop structures is observed. The orientation of replication of the strand first synthesized is opposite with respect to the orientation of replication and the adjacent ribosomal genes, compared to mammalian cells (Goddard and Wolstenholme, 1980), as though the initiation site had been inverted in the course of evolution.

In contrast with mammalian mtDNA replication, some replicative intermediates appear to complete synthesis of one DNA strand before the synthesis of the complement is initiated. Some totally single-stranded molecules are observed, suggesting that separation of the daughter strand occurred before synthesis of the complementary strand was initiated. Some molecules initiate synthesis of the second strand before synthesis is complete, but the site of synthesis is not well defined. The extent of synthesis that can occur before initiation of synthesis on the second strand varies from 10 to 98%.

In *D. melanogaster* oocytes, two types of covalently closed molecules can be distinguished. One contains few superhelical turns and the other contains about 100 turns. Curiously, cultured cells contain the uncoiled form almost exclusively (Rubenstein *et al.*, 1977). The functional significance of the forms is not known.

Genome organization and transcription mapping of the *D. melanogaster* and *D. virilis* mDNAs have been investigated using R loop hybridization and Northern transfers (Merten and Pardue, 1981). The major transcription products appear similar in size and number to those found in other animal mitochondria. Little, if any, difference was found between *D. melanogaster* and *D. virilis* in the number, size, or map position of transcripts. The abundant transcripts hybridize to most of the genome outside of the AT-rich region, in a closely packed but nonoverlapping fashion. Two sites form R loops with both strands, one region being within the large rRNA transcription unit.

An intriguing similarity between AT-rich regions and nuclear satellite DNA sequences in *Drosophila* has been pointed out (Merten and Pardue, 1981). The sequences are abnormal in base composition, are poorly transcribed, vary greatly between species, and may function in chromosome segregation. The sequences of the *D. yakuba* AT-rich region show some of the same oligonucleotides (Goddard *et al.*, 1982) that characterize the *D. melanogaster* AT satellite, but more sequence information within the same species is needed to test the possibility that the mtDNA sequences are related to the satellite DNA.

An unusual variation of mtDNA size has been found in spermatocytes of the insect *Rhynchosciara hollaenderi,* in which specific mitochondrial differentiation is observed during spermatogenesis (Handel *et al.*, 1973). In contrast with other animal mtDNAs, the spermatocytes have a distinct 9-μm mtDNA circle. It is not known if this result is typical of other tissues or if the molecule is a dimeric form of the usual type of animal mtDNA.

III. Linear Mitochondrial Genomes of Ciliated Protozoa and *Chlamydomonas*

A. Ciliated Protozoa

In several species of *Paramecium,* mtDNA is found as linear molecules of about 40 kb (Goddard and Cummings, 1975, 1977). In the *P.*

aurelia complex, certain species vary considerably in the extent of sequence homology (Cummings, 1980). mtDNA from species 1, for example, is only about 50% homologous with mtDNA of species 4, whereas species 1, 5, and 7 are essentially identical. Much of the homology in the divergent strains occurs in the regions containing the ribosomal genes (Cummings *et al.*, 1979b, 1980a). The rRNA genes are present in single copy and are separated by about 10 kb (Cummings *et al.*, 1980a). The organization of the ribosomal genes is also conserved in these species (Cummings and Laping, 1981). A region of weak homology in divergent species was associated with replication initiation (Cummings and Laping, 1981).

The replication of linear mtDNA is an intriguing problem because of the implications for replication of linear molecules in nuclear chromosomes (Forte and Fangman, 1979). Based on results from electron microscopy, Pritchard and Cummings (1981) and Cummings and Pritchard (1982) proposed that replication is initiated at one specific end at the site of a covalent cross-link (hairpin). Replication is unidirectional and generates a dimer that is processed to generate two daughter molecules. A model has been proposed to explain how replication and processing could reconstitute a cross-linked end (Cummings and Pritchard, 1982). It is proposed that the dimer gets a staggered cut, which is repaired to reconstitute the cross-linked end. The initiation region has been sequenced and is composed of palindromic and nonpalindromic regions. The major part of the nonpalindromic region was composed of AT-rich repeats.

Part of *P. primaurelia* mtDNA coding for the large rRNA subunit and its flanking regions has been sequenced (Seilhamer and Cummings, 1981). The *Paramecium* gene can be aligned with homologous regions of the rRNA genes of yeast mtDNA, mouse mtDNA, and *Escherichia coli*. Somewhat more homology exists between the *Paramecium* gene and the *Escherichia coli* sequence than with the other mitochondrial rRNA genes. No large introns were found in the large rRNA gene of *Paramecium*.

The mtDNA of *Tetrahymena pyriformis* is also linear, but varies considerably in length in different strains (Goldbach *et al.*, 1978a). For example, strain GL contains 49 kb, whereas strain ST has only 42 kb of DNA. In contrast to *Paramecium*, *Tetrahymena* mtDNA duplicates by initiating replication in the middle of the molecule, and replication proceeds bidirectionally to the ends (Goldbach *et al.*, 1979). Cross-hybridization of mtDNA from different strains shows less than 20% homology, and can be largely accounted for by the genes for rRNA (Goldbach *et al.*, 1978b,c).

The organization of the mtDNA is complex, with a duplication of the large ribosomal subunit gene (21 S) at the end, creating a large inverted repeat (Goldbach et al., 1978c). The small rRNA subunit is located about 10 kb from one large subunit gene in the same orientation. The direction of transcription of all the rDNA genes is toward the end of the molecule.

The structure of the ends is unusual. Beyond the inverted repeat that contains the rRNA genes is a terminal region of about 800 bp with no known function. Under conditions of formamide spreading, following partial denaturing conditions, over half the molecules contained duplex loops (lariat structures) at the ends, indicating some tandem duplications near the ends. It has been suggested that the duplication is involved in the replication and completion of the ends (Goldbach et al., 1979).

Another unusual feature of Tetrahymena mtDNA is related to the genes for the tRNAs. Suyama and Hamada (1976) have proposed that tRNA synthetase may act as a transport protein for nuclear tRNAs, because only 10 of the 36 total tRNAs found in Tetrahymena hybridize to mtDNA. It is argued that the protein synthetic machinery of Tetrahymena has taken a different evolutionary path from other mitochondria (Suyama, 1982).

B. Chlamydomonas

mtDNA has been isolated from the unicellular algae Chlamydomonas reinhardtii. The isolation of a cell wall-less mutant permitted the preparation of DNase-impermeable mitochondria and subsequently mtDNA (Ryan et al., 1978). The mtDNA is unexpectedly small (15 kb) compared to other photosynthetic organisms, but similar in size to animal mtDNA.

In the electron microscope, the preponderant molecular form observed is linear, with rare circular molecules observed ($< 1\%$). The physical (restriction) mapping of the mtDNA (Grant and Chiang, 1980) demonstrated a consistent linear map of 15 kb with unique ends. It is not clear if the linear genome is created by site-specific cleavage, from circular precursors, or if the linear molecules are self-replicating. No variations in restriction patterns were observed in different strains or between mating types of C. reinhardtii. Unrestricted mtDNA of a closely related species, C. smithii, was indistinguishable by agarose gel electrophoresis from C. reinhardtii.

IV. Structural Diversity in Yeast mtDNA

A. Variation in Size and Organization of mtDNA in Yeast Species

Mitochondrial genomes of yeasts have a very high level of structural diversity. In different species, mtDNAs vary greatly in size, structure, and gene organization. Sizes of mitochondrial genomes range from 18 kb for *Schizosaccharomyces pombe* (50h$^-$) (Wolf *et al.*, 1982) to 108 kb for *Brettanomyces custersii* (Clark-Walker *et al.*, 1980). The structure of the yeast mitochondrial genome is known in several cases to be circular, based on observation of a major class of circular molecules in the electron microscope or from a circular restriction map. The circular genomes observed by electron microscopy are large (58 kb for *Brettanomyces anomalous*) (Clark-Walker and McArthur, 1978), and rare circles have been observed for *Saccharomyces cerevisiae* equivalent to at least 68 kb (Hollenberg *et al.*, 1970; Christiansen and Christiansen, 1976); 24.7-μm molecules were found for *S. carlsbergensis* and 26.6-μm molecules were found for *S. cerevisiae* (Christiansen and Christiansen, 1976).

The entire range of genome sizes for different yeasts is sixfold, and a threefold range in size is known within the *Saccharomyces* alone. *S. exiguus* has a genome of only 23.4 kb, and *S. cerevisiae* can have up to 78 kb, depending on the strain (Sanders *et al.*, 1977a; Clark-Walker *et al.*, 1981; Borst, 1981).

In several species of yeast, mapping and hybridization analysis indicate great diversity in organization as well as size. Using labeled probes containing specific genes from mtDNA of *S. cerevisiae*, homologous sequences have been located in many other species of yeast (Fig. 2) (Borst and Grivell, 1981; Borst, 1981; Weslowski and Fukuhara, 1981; Weslowski *et al.*, 1981; Clark-Walker and Sriprakash, 1981; Wolf *et al.*, 1982).

If the genetic maps of mtDNA of different yeast species are compared (Fig. 2), the relative position and orientation of specific genes vary so greatly between yeast species that there are few common features to the map. For example, the large and small ribosomal genes are separated by 1 to 2 kb in *Kluyveromyces lactis, Kloeckera africana, Torulopsis glabrata,* and *Hansenula mrakii*. In *Saccharomycopsis lipolytica* they are separated by about 10 kb, and in *S. cerevisiae* they are 27 kb apart. The orientation of the large rRNA genes is reversed for *K. africana* (3' proximal to the small rRNA) and *T. glabrata* (3' end distal to

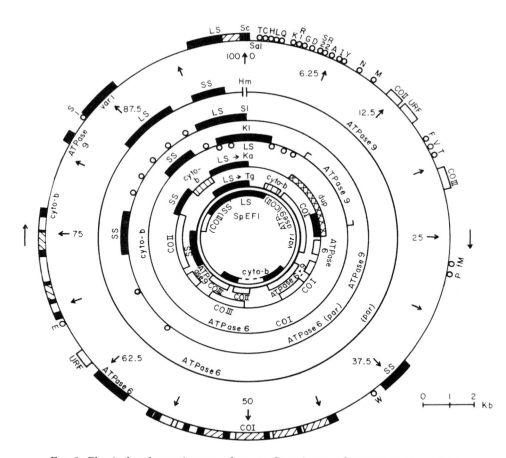

FIG. 2. Physical and genetic maps of yeasts. Genetic maps for seven yeast species are presented as concentric circles drawn approximately to scale where the circumference of each circle is proportional to the length of the genome. The maps are scaled in 100 total units so that all map distances are percentages of the total map, indicated by arrows on the *Saccharomyces cerevisiae* map. The same scale may be used for all the maps, because the full map is always considered to be 100 units. For *Schizosaccharomyces pombe* the scale is reduced by 0.9 compared to other genomes because *Torulopsis glabrata* and *S. pombe* are equivalent in size. Each species is oriented with a *Sal*I site at the top of the map, with the exception of *Hansenula mrakii,* which has a linear map. The ends of the *H. mrakii* map are at the top. Arrows indicate the direction of transcription. Species, in decreasing order with size, are (1) *Saccharomyces cerevisiae* KL14-4A (Sc) ~78 kb, Borst and Grivell (1981); (2) *Hansenula mrakii* (Hm), 55 kb, Weslowski and Fukuhara (1981); (3) *Saccharomycopsis lipolytica* (Sl), 44 kb, Weslowski *et al.* (1981); (4) *Kluyveromyces lactis* (Kl), 37 kb, Weslowski *et al.* (1981); (5) *Kloeckera africana* (Ka), 27.1 kb, Clark-Walker and Sriprakash (1981); (6) *Torulopsis glabrata* CBS 138 (Tg), 18.9 kb, Clark-Walker and Sriprakash (1982); and (7) *Schizosaccharomyces pombe* EFI (Sp), 18.9 kb, Wolf *et al.* (1982). Sc, Tg, and Ka are *petite*-positive yeasts; Kl, Sl, and Sp are *petite*-negative yeasts. tRNAs are indicated by circles on the map. The amino acid code follows that of Fig. 1. Open regions within a block indicate intervening sequences, e.g., LS gene

small rRNA) (Clark-Walker *et al.*, 1981b). In *S. pombe* the rRNA genes are adjacent. In *K. africana* the ribosomal genes are separated by the cytochrome *b* gene, whereas the cytochrome *b* gene is variously at 50 map units (as a percentage of the total map) in *S. pombe* and at 75 in *S. cerevisiae*. Similarly, the gene for COI (*oxi3*) is at 50 for *S. cerevisiae*, at 35 for *K. africana*, and at 18 for *T. glabrata*. Other genes, ATPase 6, ATPase 9, COII, and COIII, show similar variation with respect to other markers. The evidence argues for extensive rearrangement of sequences and for many insertions or deletions to account for the great variation in size. Major changes have occurred in both coding and noncoding regions, because homologous genes can vary greatly in size. For example, the long form of the gene for apocytochrome *b* extends over 7 kb in *S. cerevisiae* whereas the same gene is only 1.5 kb in *T. glabrata*. Similarly, the gene for COI extends over 10 kb in *S. cerevisiae* and covers less than 2 kb in *T. glabrata* or *S. pombe*. Major changes in the structure and expression of introns and intron-coded functions are likely in the evolution of yeasts.

A sequence likely to be involved in promoter functions has been detected in yeast (Osinga *et al.*, 1982). This sequence, ATATAAGTA, is found just preceding the genes for rRNAs in *K. lactis* and *S. cerevisiae*. In both of these distantly related yeasts, the large and small ribosomal genes are transcribed from independent promoters.

Petites are yeast mutations that form small colonies on solid medium, due to a respiratory deficiency. *Petite* mutations may be nuclear (*segregational petites*) or cytoplasmic (*vegetative petites*). Cytoplasmic *petites* have grossly altered mtDNA, or no mtDNA at all. *Vegetative petites* cannot carry out mitochondrial protein synthesis and consequently have pleiotropic defects in the respiratory electron transport chain, including cytochromes $a + a_3$, *b*, and the attendant ATP-generating system (Gillham, 1978).

Yeasts vary in their ability to generate *petite* mutations and in their ability to grow under fermentative conditions without a functional mitochondrial genome. These properties could be related to the size of the genome, particularly if duplicated regions occurred or if additional functions were present (Clark-Walker *et al.*, 1981a). Both *petite*-positive (strains that generate *petites*) and *petite*-negative yeasts vary greatly in size, and there is no indication that the properties of *petite*-

---

or COI gene in *S. cerevisiae*. Abbreviations: COII = *oxi 1* = cytochrome oxidase subunit II; COIII = *oxi 2* = cytochrome oxidase subunit III; COI = *oxi 3* = cytochrome oxidase subunit I; LS = large rRNA subunit; SS = small rRNA subunit; ATPase 6 = *oli 2*; ATPase 9 = *oli 1*; par = paromomycin resistance; and cyto-b = cytochrome *b*.

positive strains are related to extensive duplication of genes (Clark-Walker *et al.*, 1981a). Both *S. cerevisiae* (78 kb) and *T. glabrata* (19 kb) are *petite*-positive yeasts, whereas *S. pombe* (19 kb) and *S. lipolytica* (44 kb) are *petite*-negative yeasts. *Brettanomyces custersii*, with the largest mtDNA, is *petite* positive. The frequency of formation of *petites*, within *petite*-positive yeasts, does generally increase with mtDNA length, but frequencies may vary 10-fold between strains within a species, or between species of a common genus (Clark-Walker *et al.*, 1981a). Therefore, size does not seem to be the major factor related to the formation of *petites*.

In *K. africana*, a *petite*-positive yeast, a long inverted duplication has been found that contains a segment of the large rRNA gene and some tRNA genes (Clark-Walker *et al.*, 1981b). At least 1 kb of the rRNA gene is not repeated; therefore, the duplication of ribosomal sequence is not functional. It has been suggested that inverted repeats may generate high frequencies of *petites* in *S. cerevisiae* (Oakley and Clark-Walker, 1978). Interestingly, the *K. africana* mtDNA is stable and has one of the lowest frequencies of spontaneous formation of *petites* among the *petite*-positive yeasts.

The most intriguing structural variation in these systems is the linear mtDNA of *H. mrakii* (Weslowski and Fukuhara, 1981). Restriction mapping and end-labeling studies indicate that the mtDNA has defined ends that can be labeled at the 5' and 3' terminal nucleotides. One other species of *Hansenula* has been examined, *H. wingei,* which has a circular mtDNA about half the size of *H. mrakii*. The structure and mode of replication of the linear genome are of considerable interest, because of the possibility that *H. mrakii* contains a recently evolved linear mitochondrial genome.

B. VARIATION OF mtDNA IN *Saccharomyces cerevisiae*

1. Size Variation in Strains and Crosses

Restriction analysis of different wild-type strains of *Saccharomyces cerevisiae* shows an unusually high degree of sequence variation. In spite of virtually complete homology between strains measured by hybridization analysis (Groot *et al.*, 1975), and identity of general physical properties of the mtDNA (e.g., buoyant density, pyrimidine isostichs, and GC content), the restriction patterns can have very few bands in common (Prunell *et al.*, 1977; Bernardi *et al.*, 1976; Sanders *et al.*, 1977). If the variation in restriction patterns were due to a high level of base substitution in yeast mtDNA, very little sequence homology would be expected. The high level of observed homology indicates that structural variation is the basis for nonidentity of restriction fragment patterns. The sum of restriction fragments from different strains indicated that *S. carlsbergensis* (often considered a subspecies of *S. cerevisiae*) contained a smaller mitochondrial genome by about 10%. In spite of the high level of variation in restriction patterns, the order of genes is conserved in these strains.

When strains with differently sized genomes and different restriction patterns were crossed and diploid progeny were analyzed, the progeny showed restriction patterns that were similar to but different from parental types—indicating that recombination was frequent in such crosses and that recombination produced the altered structural forms of mtDNA. Two proposals have been offered to explain the extent of variation while preserving sequence homology and gene order. Prunell *et al.* (1977) and Fonty *et al.* (1978) proposed that unequal recombination within internally repetitive AT-rich intergenic spacer regions produced the observed results. In this view, unequal exchange in repetitive regions would alter the length of fragments, when assayed with enzymes that cut at GC-rich restriction sites; such events would not disrupt gene order or homology. Alternatively, Borst and Grivell (1978) proposed that the mitochondrial genomes varied in length due to specific preexisting large insertions or deletions. The deletions, distributed at a few locations in the genome, would result in varying restriction patterns. In crosses, new restriction patterns could result from recombination of homologous sequences that contain different insertions and deletions. In subsequent mapping experiments, the major differences in genome size and restriction patterns between different strains have been explained by large deletions or insertions present in wild-type strains (Sanders *et al.*, 1977; Morimoto and Rabinowitz, 1979).

In a comparison of the physical maps of three *Saccharomyces* strains, Sanders *et al.* (1977) found that four major insertions, ranging in length from 900 to 6000 bp, could account for the size variation between long and short mitochondrial genomes. *S. carlsbergensis* (68 kb) lacked all four insertions that were present in *S. cerevisiae* (KL 14-4A) mtDNA. Similarly, *S. cerevisiae* (JS1-3D) is a relatively short genome, but differs from *S. carlsbergensis* by major insertions of 1500 and 900 bp. Additional smaller deletions of 25 to 50 bp were found in four restriction fragments. Comparison of KL 14-4A to *S. carlsbergensis* showed the same insertions as in JS1-3D, but several additional large deletions were mapped of 2600, 3000, and 1050 bp.

In another study (Morimoto and Rabinowitz, 1979), physical maps of

S. cerevisiae D273-10B (70 kb) and MH41-7B (76 kb) were compared for structural variation. Differences in the number of restriction sites between the strains were located to two regions associated with the *oli* 2 gene (ATPase). Two insertions of 2.7 and 3 kb largely account for the difference in size between the two strains.

Most of these insertion sequences have been found to be intervening sequences in essential mitochondrial genes. Transcription mapping (Van Ommen *et al.*, 1979) indicated that large insertions or deletions that occurred between *S. carlsbergensis* NCYC-74 (a short genome) and *S. cerevisiae* KL 14-4A (a long genome) were present in genetically active regions. Because these sequences represent introns that may or may not be present in normal yeast strains, they may be considered "optional" introns. The best studied cases of such optional introns include those of the genes for cytochrome *b*, COI, and the large rRNA subunit (Borst, 1981; Grivell *et al.*, 1980, 1982; Dujon, 1980).

2. Optional Introns

Both genetic and biochemical evidence have demonstrated that the gene which codes for cytochrome *b* in *S. cerevisiae* has a complex mosaic structure (Slonimski *et al.*, 1978b; Hanson *et al.*, 1979; Nobrega and Tzagoloff, 1980a,b; Grivell *et al.*, 1979; Haid *et al.*, 1979; Alexander *et al.*, 1980). The structure of this gene has been described in detail because of interest in structure of mosaic genes and the role of its introns in RNA processing and gene regulation. The gene is found in two forms due to the presence or absence of introns in the long and short forms of the gene.

In the short form, found in strain D273-10B, two introns are found that vary in base composition and in the presence of open reading frames (Nobrega and Tzagoloff, 1980a; Diekmann *et al.*, 1982). One intron is 1414 bp and has an open reading frame (intron 4), whereas a second intron is 733 bp, is generally rich in AT, and has a GC cluster. In the long form of the gene, there are 5 introns and 6 exons. The first exon of the short form is fragmented into four exons with the three additional introns (Grivell *et al.*, 1980; Lazowska *et al.*, 1980; Van Ommen *et al.*, 1980). Processing of the 7.3-kb primary transcript of the long form begins with the splicing of the first intron, followed by translation of exon 1, exon 2, and the in-phase intron 2 to produce a maturase protein. The maturase is essential for the second splice, which removes intron 2. Subsequent processing steps, including translation of a second maturaselike intron (intron 4), are needed to produce the cytochrome *b* mRNA that is 2 kb (Borst and Grivell, 1981; Jacq *et al.*, 1980; Dujardin *et al.*, 1982).

In addition, intron 4 of the cytochrome *b* gene has trans-acting coding functions that are essential for the synthesis of active cytochrome oxidase (COI) (Jacq *et al.*, 1982; Mahler *et al.*, 1982; Weiss-Brummer *et al.*, 1982). In the short form of the cytochrome *b* gene, the maturase-coding intron 2, intron 1, and intron 3 are deleted, leaving only introns 4 and 5.

By selection for rare pseudo-wild-type revertants of cis-dominant mutations in intron 4, genes have been recovered that lack both introns 4 and 5, but which can make active cytochrome *b*. By recombination of the long form, the short form, and the deleted intron 4 and intron 5 strains, it is possible to add or subtract specific introns (Jacq *et al.*, 1982). In this way an active gene for cytochrome *b* has been constructed that lacks all of the introns. Because a nuclear gene (*NAMX*) can compensate for the intron 4 function needed to make COI, a cell carrying the intronless cytochrome *b* gene and *NAMX* can be respiratory competent (Jacq *et al.*, 1982). The deletion of all introns from the cytochrome *b* gene clearly shows that none of the introns is essential for the synthesis of active cytochrome *b* and, in that sense, all five introns are optional, even though only the first three introns are optional in wild-type strains.

The gene for COI (*oxi 3*) consists of a single large complementation group that extends over about 10 kb of the *S. cerevisiae* mitochondrial genome (Foury and Tzagoloff, 1978; Morimoto *et al.*, 1979; Bonitz *et al.*, 1980a). Transcription mapping together with genetic analysis indicates that the gene has a highly mosaic structure (Van Ommen *et al.*, 1979, 1980). Insertions or deletions occur in the COI gene in long and short mitochondrial genomes (Grivell and Moorman, 1977; Sanders *et al.*, 1977; Netter *et al.*, 1982). The COI gene has seven exons, with one of these provisionally divided into four additional exons in long variants of the genome (Bonitz *et al.*, 1980b; Grivell *et al.*, 1982). Most of the gene is composed of intervening sequences; the exons constitute no more than 16% of the length of the entire gene. The long and short forms of the gene result from the presence or absence of five optional introns (Grivell *et al.*, 1982).

The first four introns of the gene contain long open reading frames that are in phase with their preceding exons. The first two introns are homologous; they show an average of 50% identity of amino acid-coding sequences over the entire length of the two introns. More surprisingly, the fourth intron shows an even higher degree of amino acid homology (70%) with the fourth intron of cytochrome *b* (Bonitz *et al.*, 1980b). These homologies suggest a common origin for certain introns and may indicate related functions.

In all yeast mitochondrial genomes studied, there is a single functional gene for the large rRNA subunit (Borst and Grivell, 1978; Clark-Walker et al., 1982; Wolf et al., 1982; Wallace, 1982). In S. cerevisiae, hybridization analysis and restriction mapping indicate that rRNA hybridizes to two regions separated by about 1100 bp (Bos et al., 1978) of intervening sequence. This intervening sequence corresponds to the insertion VI of Sanders et al. (1977) and to omega plus (ω^+) (Jacq et al., 1977). The insertion is absent in ω^- strains and in S. carlsbergensis (Heyting et al., 1979a,b; Heyting and Menke, 1979; Faye et al., 1979; Bos et al., 1980). The insertion is located about 400 to 500 bp from the 3' of the 21 S rRNA gene (Locker and Rabinowitz, 1981; Bos et al., 1980).

Both ω^+ and ω^- forms of S. cerevisiae are respiratory competent. Crosses between them show a strong deviation from the expected 1:1 segregation ratio in favor of the ω^+ genotype (Netter et al., 1974). The mechanism of ω-directed polarity is not known, but mutants of ω^- have been isolated that lack the polar effect in crosses with ω^+. These mutants, called ω^n for omega neutral allele, show no major structural change compared to ω^- strains (Bos et al., 1980; Dujon, 1980). Indications that the phenomenon is complex comes from deletions in ω^- strains that can generate ω^+ behavior (Devinish et al., 1979; Atchison et al., 1979).

Nucleotide sequence analysis of the 21 S rRNA intron in ω^+, ω^-, and ω^n alleles provides insight into the polarity phenomenon (Dujon, 1980). ω^+ contains a 1143-bp intron and an unexpected "miniinsert" of 66 bp located upstream 156 bp within the exon. The miniinsert has a palindromic sequence that can form a hairpin structure with a stem of 16 bp and a 3-bp loop. The ω^n mutations are attributed to base substitution mutation (Dujon, 1980) at the rib-1-B locus where mutations for chloramphenicol resistance occur. A long open reading frame of 235 amino acids is found within the intron. The precise role of the miniinsert and the mechanism of preferential segregation of ω^+ remain to be determined.

The origin and evolution of introns are of great interest. It is argued that introns come about as a consequence of the organization of coding sequences into new genes (exon shuffling) or, alternatively, that introns represent recent insertions into functional genes by transposition (Gilbert, 1978; Crick, 1979; Lewin, 1982a,b). In mtDNA, the issue is highlighted by introns that code for functional proteins and the fact that even these introns are optional in certain strains of S. cerevisiae (Borst and Grivell, 1981).

Two types of introns are found that have very different properties. Open introns with long coding regions are implicated in splicing or processing of transcripts; other introns lack open reading frames (closed introns) and resemble AT-rich spacer regions (Bernardi, 1982). If introns are dispensible, they should be easily lost. Clearly, strains survive with or without both types of introns. The existence of homology between introns of different genes, or between different introns of the same gene, suggests a common origin (Bontiz et al., 1980b; Dujardin et al., 1982) and duplication of related functions.

Many nuclear and mitochondrial large subunit rRNA genes in other organisms have introns in locations similar to the coding intron of the yeast mitochondrial ribosomal gene (Davies et al., 1982a; Burke and RajBhandary, 1982; Nomiyama et al., 1981a). Although most introns in ribosomal large subunit genes are located near the 3' end, the precise locations are not the same. It has been suggested that this region of the molecule can accept introns without damaging its function even though it may be a highly conserved region of the RNA molecule (Crick, 1979; Gilbert, 1978). In E. coli, the ribosomal subunit interface is in the same region of the molecule that contains the rRNA intron (Brosius et al., 1980; Dujon, 1980). Therefore, the region of intron excision may be at the subunit interface in mitochondrial ribosomes and might allow splicing to take place following ribosome assembly (Dujon, 1980). It is interesting to note that defective mitochondrial genomes that contain the intact ω^+ gene region synthesize an RNA molecule that is the size of the mature RNA in the absence of mitochondrial protein synthesis (Faye et al., 1974, 1975; Merten et al., 1980). Such petites have been used to identify processing precursors and to demonstrate the steps leading to the mature 21 S rRNA molecule (Merten et al., 1980).

3. Variable Genes: var 1 Locus

var 1 is a genetic locus in S. cerevisiae mtDNA that codes for a highly polymorphic protein associated with the 38 S small mitochondrial ribosomal subunit (Perlman et al., 1977; Terpstra and Butow, 1979; Terpstra et al., 1979; Groot et al., 1979). Different strains of yeast show one of a number of different electrophoretic forms in SDS–acrylamide gel analysis.

Six different var 1 polypeptide-size classes have been analyzed and the variation has been attributed to combinations of three different allelic elements (Strausberg et al., 1978; Strausberg and Butow, 1981; Butow, 1982). The different molecular weight polypetides are associated with different sizes of restriction fragments prepared from *var 1* alleles (Strausberg *et al.*, 1978). Nonparental forms arise in crosses and are not associated with reciprocal exchange. The different molecular forms can be accounted for by an asymmetric recombination process. Interestingly, both the gene conversion events of the ω⁺ locus and the *var 1* locus may involve the preferential insertion of a specific allelic form into the alternative mitochondrial chromosome.

The *var 1* coding sequence has been determined and is unusual (Hudspeth *et al.*, 1982). It is 89.6% AT and contains a 46-bp GC-rich palindrome. The GC cluster resembles other GC clusters in the genome, and accounts for 38% of the GC residues in the *var 1* coding sequence. It has been proposed that the size polymorphisms result from "in-frame" insertion of a variable number of AAT (Asn) codons. The supposition that AT-rich regions with GC clusters do not code for proteins should be reconsidered.

A fragment of the *var 1* gene has been recently detected in the yeast nuclear genome (Farrelly and Butow, 1983; Lewin, 1983). This nuclear fragment is homologous with about 115 bp from the central part of the *var 1* gene. Adjacent to this segment are three additional mitochondrial sequences, one from the 3' end of the cytochrome *b* gene (123 bp), one from a section of intron 5 of the cytochrome *b* gene, and a 29-bp sequence containing a sequence for a proposed origin of replication for mtDNA.

This result demonstrates that a segment of mtDNA has been incorporated into the nucleus. The mosaic nature of the inserted sequence, with fragments from different genes, indicates extensive deletion or rearrangement of mitochondrial genes in addition to transposition into the nuclear chromosome. It is proposed that the fragment was a continuous segment of mtDNA that has undergone internal recombination to delete segments and that has become integrated by chance into the nucleus (Farrelly and Butow, 1983; Lewin, 1983).

C. Variation in *Schizosaccharomyces pombe* mtDNA

No other yeast species has been studied as extensively as *Saccharomyces cerevisiae*. Therefore, it is important to learn about variation in other species, such as the *petite-negative* fission yeast *Schizosaccharomyces pombe*, which has a very small mitochondrial genome. Two different strains of *S. pombe* have been mapped (Wolf *et al.*, 1982) by restriction digests and hybridization, using probes containing specific genes from *S. cerevisiae*. In *S. pombe* strain EF-1, the genome is considerably shorter (17 kb) than strain 50h⁻, which is close to 19 kb. The gene for cytochrome *b* has two exons in strain 50h⁻. The intervening region is not homologous with the mtDNA of *S. cerevisiae,* suggesting relative divergence of introns and conservation of exons. In strain EF-1, the cytochrome *b* gene is split into three parts.

An interesting feature of this genome is the discovery of a cytoplasmic gene that induces a high rate of mitochondrial mutations, but not of nuclear mutations. The rate of mutation is increased by 10^2- to 10^3-fold; both deletions and point mutations are induced (Seitz-Mayr and Wolf, 1982). This result implies a new function in a very economical mitochondrial genome.

The size of the mitochondrial genome in *S. pombe* is close to that of several animal mitochondrial genomes (Table 1), yet it still possesses split genes and shows a degree of variation in gene structure and organization not observed in animal mtDNA.

V. Structural Variation and Organization in Filamentous Fungi

The combined physical and genetic maps of mtDNA in the filamentous fungi, *Neurospora, Aspergillus,* and *Podospora,* show variation in genome size and in the relative positions of specific genes (Fig. 3) (Küntzel *et al.,* 1982; Kück and Esser, 1982; Macino, 1980; Köchel *et al.,* 1981; Lazarus *et al.,* 1980a). Between *Aspergillus* and *Podospora,* genome size varies threefold. Comparison of maps shows a similar but not identical map order for at least six different genes. A few inversions can generate a common map for all three species. A single inversion involving the genes for COII and cytochrome *b* would generate a common map for *Neurospora* and *Aspergillus.* The relative organization of genes between *Aspergillus* and *Podospora* can be made common by two different inversions, one involving cytochrome *b* and the ATPase subunit 6 and the other reversing the positions of COI and COII. To reconcile the maps of *Podospora* directly with *Neurospora* requires three events, the first being an inversion of COII and cytochrome *b,* followed by the two inversions that create a common map for *Aspergillus* and *Podospora.*

As in other systems, variation in genome size implies insertion–deletion events. Size variation of mtDNA within closely related species has been found for close relatives of *Aspergillus nidulans* and *Neurospora crassa* (Turner *et al.,* 1982; Earl *et al.,* 1981; Collins *et al.,* 1981). Some of the variation may be accounted for by optional introns within structural genes.

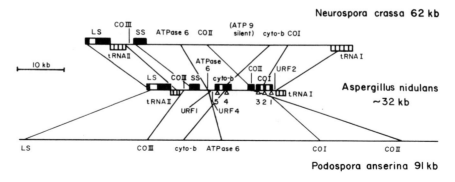

FIG. 3. Physical maps are shown for *Neurospora crassa*, *Aspergillus nidulans*, and *Podospora anserina*, using information and maps from Küch and Esser (1982), Macino (1980), Macino *et al.* (1980), Küntzel *et al.* (1982), Lazarus *et al.* (1980), Earl *et al.* (1981), and Davies *et al.* (1982). Abbreviations follow those in previous figures. tRNA genes are found in two clusters in *N. crassa* and *A. nidulans*. Open blocks in LS, cyto-b, and COI indicate intervening sequences. Small triangles below cyto-b and COI indicate insertion sites for sequences in related species of *Aspergillus*. The *ATP9-silent* region is the location of the "inactive" ATPase proteolipid of *N. crassa*. URFs indicate unidentified open reading frames. The location of URF-1 is given by Davies *et al.* (1982). All maps begin with the ribosomal genes (large subunit) on the left. For *Neurospora* and *Aspergillus*, the orientation of the gene is known because the intron is at the 3′ end of the gene (Bertrand *et al.*, 1982). The small subunit is transcribed in the same direction. tRNA clusters are marked where most, but not all, of the tRNAs are located. In *Neurospora crassa*, most of the tRNAs are transcribed from the same strand as the large and small rRNAs (Heckman and RajBhandary, 1979).

A. *nidulans* mtDNA shows size variation between several related members of the species group. Mapping has been carried out on two varieties of *A. nidulans, A. n. echinulatis* and *A. n. quadrilineatus,* as well as on a second species, *A. rigulosus* (Earl *et al.*, 1981; Turner *et al.*, 1982; Küntzel *et al.*, 1982). *A. n. echinulatis* has a longer mitochondrial genome (40 kb), due to five insertions not present in the *A. nidulans* mitochondrial genome (33 kb). The *A. n. quadrilineatus* genome is smaller than that of *A. nidulans,* with an mtDNA of about 31 kb (Turner *et al.*, 1982) due to deletion of two introns located in the genes for COI and cytochrome *b* (Küntzel *et al.*, 1982). These introns are optional in different strains, similar to the optional introns found in yeast mtDNA (Borst and Grivell, 1978; Borst, 1981).

Pairwise fusion of *A. nidulans* subspecies with long, medium, or short mtDNAs has been carried out (Turner *et al.*, 1982). Fusions of different strains have been examined to determine if variable regions containing insertions or deletions are tolerated in different nuclear backgrounds. It is possible to recover recombinants resulting from fusion of mitochondrial genomes and to determine from the surviving mtDNAs whether a particular insertion or deletion is tolerated in a specific nuclear background. Some, but not all, of the inserts present in the longer mtDNAs appear to be essential for maintaining the mtDNA in its original nuclear background. For example, if the *A. nidulans* mtDNA (medium) is transferred into an *A. n. echinulatus* (long) nuclear background, the most frequent type of mitochondrial genome recovered is largely *A. nidulans,* into which three inserts (of the *A. n. echinulatus* mtDNA) had recombined. A constant feature of the recovered genomes was the inability of the hybrids to lose inserts 1, 2, and 3 in the *A. n. echinulatus* nuclear background. All larger mtDNAs can be transferred to the nuclear background of smaller mitochondrial genomes, indicating that additional introns are tolerated. An alternative interpretation based on unidirectional gene conversion could also explain the results (Turner *et al.,* 1982).

Sequence analysis of several regions of the *A. nidulans* genome has provided information about some specific introns (Küntzel *et al.,* 1982; Davies *et al.,* 1982). The cytochrome *b* gene, which has 61% amino acid homology with *Saccharomyces cerevisiae* and 50% with human, has a large intron of 1050 bp (Waring *et al.,* 1981). The position of the intron (splice points) is precisely the same as that of intron 3 in the yeast gene (Lazowska *et al.,* 1981). In *Aspergillus,* the intron has an open reading frame continuous with the upstream exon for 957 bp, and shows homology with intron 3 of the corresponding yeast gene. Large portions of the open reading frames of both yeast and *Aspergillus* introns show strong amino acid homology (Waring *et al.,* 1982).

In *A. nidulans,* all identified mitochondrial genes are transcribed from the same strand (Küntzel *et al.,* 1982). Inversion of specific genes in *Neurospora* and *Podospora* implies that transcription initiates in different places from opposite strands for at least some of the mitochondrial genes.

In *Neurospora,* wild-type mtDNAs of several strains and species have been compared. Two variant types have been found among laboratory strains (Mannella *et al.,* 1979), and additional structural variants were found in the related species, *N. intermedia* and *N. sitophila* (Collins *et al.,* 1981). The two variant types are characterized by a 2.1-kb fragment that is tandemly repeated and inserted into the mtDNA (Manella *et al.,* 1979). All strains tested showed strong conservation in the rRNA and tRNA region, which spans about one-third of the mitochondrial genome. Variant types usually contain insertions with respect to the wild-type genome, which may be as large as 8 kb for a single insertion (Collins et al., 1981; Collins and Lambowitz, 1983). Many of the strains contain insertions in the regions of the genes coding for cytochrome b and COI. Insertions within these genes may be optional introns. In several wild-type strains, the mtDNAs vary with respect to four different optional insertion/deletion sequences that are thought to be located in the COI-encoding region (Collins and Lambowitz, 1983). Much more structural diversity was observed in comparisons of wild-type strains than had been apparent from previous studies (Manella et al., 1979). Total mitochondrial genome sizes for N. crassa ranged from 60 to 73 kb (Collins and Lambowitz, 1983).

The intron of the large rRNA gene of N. crassa has been studied in some detail (Yin et al., 1982; Green et al., 1981). The intervening sequence is about 2400 bp and contains an open reading frame > 258 amino acids long, reminiscent of the intron within the large rRNA gene of ω^+ strains of S. cerevisiae (Dujon, 1980). The intron is twice the size of that in yeast, but is located within the same part of the gene, in a highly conserved region.

Recent evidence indicates that the intron-coding sequences of the large rRNA subunit genes may code for ribosomal proteins. It has been proposed that the rRNA intron codes for a ribosomal protein (S5) in Neurospora. Sequence analysis indicates that isolated regions of strong homology exist within the coding sequences of the rRNA introns of N. crassa, A. nidulans, and var 1, the ribosomal protein of yeast mtDNA (Hudspeth et al., 1982; Burke and RajBhandary, 1982; Netzler et al., 1982). It is suggested that var 1, S5, and the intron-coded protein of A. nidulans rRNA are related and functionally homologous proteins (Lewin, 1982b).

S5 is one of the major mitochondrial coded proteins in N. crassa and is known to be a component of the small ribosomal subunit (Lambowitz et al., 1976; LaPolla and Lambowitz, 1981). The gene for S5 has not been mapped. The evidence suggesting that it is located in the rRNA intron is based on its size and composition. In contrast to the hydrophobic nature of many mitochondrial membrane proteins, S5 is hydrophilic. In addition, it is highly basic, rich in lysine, asparagine, and leucine, but poor in histidine and tryptophan. The Neurospora rRNA intron-coded sequence resembles S5 in all of these properties. However, proof of the location of S5 will require more direct evidence from protein sequences.

Interestingly, the amino acid composition of S5 and the intron-coded proteins is similar to those of all the open reading frames of the introns of the mitochondrial genes of yeast (Dujon, 1980; Nobrega and Tzagoloff, 1980a; Bonitz *et al.*, 1980b; Lazowska *et al.*, 1980). All of these introns code for polypeptides that are hydrophilic, highly basic, and which resemble S5 in general amino acid composition.

The intron sequences of yeast and *Neurospora* rRNA genes are divergent. However, some regions of the sequence are conserved. A highly conserved segment of 57 bp includes a shorter sequence of 16 bp found in 12 of the 14 sequenced mtDNA introns of fungi and maize, and in all 4 nuclear rRNA genes that have been sequenced. The sequence is ($_G$PyTCA$_{AC}^{GA}$GACTACANG). Within the rRNA genes, the introns of yeast and *N. crassa* mtDNA are at identical sites. In *Physarum,* one of the two nuclear introns is also at exactly the same site as the mitochondrial introns (Nomiyama *et al.*, 1981a,b).

Interestingly, the consensus sequence is far removed from the splicing sites in the primary sequence. However, in at least two of the yeast introns, the consensus sequence is necessary for RNA splicing. In intron 4 of cytochrome *b* (DeLaSalle *et al.*, 1982; Anziano *et al.*, 1982; Weiss-Brummer *et al.*, 1982) and in intron 4 of yeast COI (Netter *et al.*, 1982) mutations within the consensus sequence, in cis, block RNA splicing. It is suggested that the consensus sequence plays an important role as a splicing signal in widely divergent organisms, and for both nuclear and organellar genes.

Splicing of the rRNA intron of *Neurospora* is under the control of at least three different nuclear genes (Bertrand *et al.*, 1982). These genes were detected because mutations in these genes cause defects in RNA splicing. These nuclear mutations cause size alterations in other mitochondrial RNAs also, suggesting that nuclear genes are involved, in a general way, in processing of mitochondrial RNAs in *Neurospora.*

A recent model for RNA splicing in fungal mitochondria has been proposed on the basis of nucleotide sequence analysis and the properties of mutations that affect splicing (Davies *et al.*, 1982a). Four introns from *A. nidulans* and 5 introns from *S. cerevisiae* have been analyzed. All of these intron sequences can form identical RNA secondary structures, which involve four conserved sequences in each intron. It is proposed that this structure brings the ends of each intron together, and allows an internal guide sequence to align the sequences so that splicing can occur. The completed precise alignment is, presumably, what is recognized by the splicing proteins.

Rearrangements have been implicated as a possible explanation of the divergence of intron sequences in both size and information, while the precise location of the intron has been strongly conserved. A cloned fragment of an rRNA intron from yeast, when used as a probe against *Neurospora* mtDNA, hybridizes strongly to a DNA fragment located at a distant region (Burke and RajBhandary, 1982), indicating that sequences have been transposed during the evolution of fungi.

The location of specific tRNAs has been compared in the mtDNA of *Aspergillus* and *Neurospora* (Köchel *et al.*, 1981). In both species, the tRNA genes are tightly clustered within two regions flanking the large rRNA subunit. In *Aspergillus,* the upstream cluster contains 9 genes and the downstream cluster contains 11. All 20 of these tRNAs and the ribosomal genes are transcribed from the same strand. The tRNA genes are separated by short AT-rich spacer sequences. The tRNA genes of *N. crassa* are similarly located on the map, but are not so tightly clustered. In *Aspergillus,* each tRNA cluster extends for about 1 kb, whereas the equivalent cluster in *Neurospora* extends over several times that amount of DNA on the map (Köchel *et al.*, 1981). In yeast (*S. cerevisiae*), 19 tRNAs are located in a region of about 20 kb downstream from the large rRNA subunit gene (Borst and Grivell, 1978).

The identification of specific tRNAs in the downstream clusters of *Neurospora* and *Aspergillus* shows most of the same tRNAs in the same order. In *Aspergillus,* the tRNA cluster is Thr, Glu, Val, fMet (Met 3), Leu 1, Ala, Phe, Leu 2, Gln, and Met, whereas the order in *Neurospora* is Thr, Glu, iLeu, fMet, Leu 1, Ala, Phe, Leu 2, Gln, His, and Met 2. There is conservation in the order and position of most of the tRNA genes, but there is rearrangement of several genes due to transposition, deletion, or duplication of these genes. Although there are two fMet tRNA genes, only one species of fMet tRNA has been found. However, both fMet genes are identical. The elongation Met tRNAs, also present in duplicate, flank the tRNA–rRNA region. The sequences differ by one nucleotide, and both genes are expressed. The two Met tRNA regions show no homology at the flanking 5′ sequences, but are 97% identical for 400 bp at the 3′ end, strongly implicating a recent gene duplication event. No other mitochondrial system contains duplications of both Met tRNA genes for initiation and elongation.

From sequence determination of the tRNA and rRNA regions (Yin *et al.*, 1982) a series of palindromic sequences were found flanking the genes. In the tRNA cluster, downstream from the large subunit ribosomal gene, there are 15 such structures interspersed among 12 tRNA genes. These hairpins consist of a pyrimidine tract, an 18-bp region with 2 *Pst*I sites at the loop, and a purine stretch complementary to the pyrimidines. The entire hairpin varies somewhat in the length of the stem, but a typical hairpin would be from about 70 to 80 nucleotides long with from 20 to 30 paired bases in the stems. The sequence composition of the palindromic sequences is GC rich. Palindromic sequences also flank the large rRNA gene and occur at the boundaries of the intervening sequences, suggesting a role in RNA processing.

The mitochondrial gene for cytochrome oxidase subunit III (COIII) of *N. crassa* has been located and sequenced (Browning and RajBhandary, 1982). The COIII coding sequence is located between the rRNA genes, downstream from the gene for the small subunit rRNA. The gene is immediately flanked by two GC-rich palindromic sequences. Transfer RNA genes are located both upstream and downstream from the COIII gene. The COIII gene, the rRNA genes, and the tRNAs are all transcribed from the same strand. The gene sequence shows strong homology with the corresponding yeast and human amino acid sequences (53 and 47%, respectively). The coding sequence predicts that tryptophan residues are coded by UGA, and that AUA is an isoleucine codon in *N. crassa*. The protein sequence is hydrophobic in character, as expected for a protein thought to be embedded in the mitochondrial inner membrane (Malmstrom, 1979).

It is well established that the gene that codes for the functional ATPase proteolipid in *Neurospora* mitochondria is in the nucleus (Sebald *et al.*, 1977, 1979) in contrast to yeasts, in which the ATPase subunit 9 is coded by the mtDNA (Borst and Grivell, 1978; Clark-Walker and Sriprakash, 1981; Weslowski *et al.*, 1981). Using a yeast gene for the ATPase proteolipid as a probe, van den Boogart *et al.* (1982a,b) have found an homologous gene sequence in the mtDNA of *Neurospora crassa*. DNA sequence analysis was carried out on a region containing COII and a DNA segment that showed homology with the yeast gene for the ATPase proteolipid. The *Neurospora* gene has an open reading frame of 225 nucleotides and shares homology (65%) with the yeast gene. The amino acid sequence coded by this segment is 60% homologous with the coding sequence of the yeast gene and 56% homologous with the protein sequence of the *Neurospora* ATPase proteolipid, itself coded by a nuclear gene (Sebald *et al.*, 1979). The clustering of polar and nonpolar amino acids indicates that the *Neurospora* mitchondrial sequence is typical of an ATPase proteolipid.

A primary transcript is detected as a large RNA for both COII and the adjacent ATPase proteolipid (van den Boogart *et al.*, 1982a). This primary transcript is cleaved to produce a COII mRNA, but no mRNA specific for the ATPase proteolipid is detected. The low degree of homology between the nuclear and mitochondrial genes suggests a separate origin and argues strongly against the idea that the gene has recently been transferred from the mitochondria into the nucleus.

Sequence analysis of some specific regions of *A. nidulans* mtDNA suggests a difference in information content of the mtDNA of this fungus compared to that of *S. cerevisiae*. Two long open reading frames of unidentified function have been found that have not yet been detected in yeast (Davies *et al.*, 1982b) between the genes for cytochrome *b* and ATPase subunit 6. The two reading frames share homology with URF 4 and URF 1 in human DNA (Anderson *et al.*, 1981). The amino acid homology for URF 4 is only 30%, whereas the other sequence, URF 1, has 48% homology (Davies *et al.*, 1982b).

VI. Organization of mtDNA in *Achlya*

The mtDNA of the oomycetous water mold, *Achlya bisexualis*, has recently been examined (Hudspeth *et al.*, 1983). The mtDNA is circular and contains 49.8 kb. Restriction mapping indicates the presence of a large inverted repeat about 12 kb long, containing the rRNA genes. The inverted repeat contains between 19 and 24% of the mitochondrial genome. Recombination between homologous sequences in an inverted repeat can reverse the orientation of the unique sequence segment between the repeats. Hudspeth *et al.* (1983) have found both orientations in *Achlya* mtDNA and in approximately equal proportions. Recombination can also serve as an homogenizing mechanism for the duplicate copies of the ribosomal genes. Duplication of ribosomal genes is viewed as an advantage to the organism; an inverted repeat would be more favorable than a tandem repeat, because recombination between tandem repeats creates deletions. Similar arguments have been made for the presence of inverted repeats in chloroplast DNA (Bedbrook and Bogorad, 1976; Palmer and Thompson, 1981).

VII. Amplified Circular Molecules in Fungal mtDNA

A. *Petites*

In addition to structural diversity of mtDNA caused by rearrangement or insertion–deletion events, a second kind of diversity occurs in mtDNA due to extrachromosomal or supernumerary DNA molecules.

These molecules are often amplified fragments of the mitochondrial genome (see Table 2). The *petite* mutants of yeast serve as the model system for the analysis of supernumerary mitochondrial molecules. Related types of molecules are found in obligate aerobes *Neurospora* and *Aspergillus*. In *Podospora,* extrachromosomal amplification of mtDNA appears to be involved at a specific developmental stage in the process of senescence.

Petites are respiratory-deficient mutants of yeast. *Cytoplasmic petites* are of two types, depending on their transmissibility in crosses with wild type. *Neutral petites* are not transmitted from crosses with wild type, and *suppressive petites* are transmitted to diploid progeny in proportions that vary from 1 to 99%, depending on the specific *petite.*

A large number of studies, which included restriction analysis, demonstrated that *petites* usually result from large deletions of wild-type mtDNA. Each *petite* is characterized by a specific deletion of the genome, which may range from 30% to greater than 99.7% of the mtDNA. *Petite* mutants have been isolated that have no detectable mtDNA. Discussion of the genetics and biochemistry of *petites* may be found in Gillham (1978), Locker *et al.* (1979), Bernardi *et al.* (1980), and Borst and Grivell (1978).

The mtDNAs of many *petite* mutations have been well characterized. Although mtDNA from yeast is known to have a circular genetic map and a circular physical map based on restriction analysis, isolated mtDNA from both wild-type and *petite* yeast is primarily linear. MtDNA from *petites* contains additional populations of circular molecules.

The circular molecules found for specific *petites* can be different in size and sequence for different strains; some are homogeneous whereas others are highly variable. Different strains vary greatly in the sequence complexity of circular mitochondrial molecules (Faye *et al.,* 1973; Locker *et al.,* 1974a,b; Lazowska and Slonimski, 1976). In some strains, the sizes of circular molecules conform to a multimeric series, with a specific monomeric size for each mutant. Many *petites* have minor populations of molecules that differ from the main population (Locker *et al.,* 1979). Other *petites* have complex mixtures of molecules.

Restriction analysis of *petites* usually shows a less heterogeneous pattern because monomeric circles, oligomers, and repetitive linear molecules can generate the same restriction pattern. Most *petites* also show heterogeneity, with additional components present in submolar amounts. Subcloning does not eliminate heterogeneity. *Petites* are known to be unstable in long-term cultures (Faye *et al.,* 1973).

TABLE 2

Diverse DNAs, Extrachromosomal DNAs, and Plasmidlike DNAs in Mitochondria

Organism	Length	Md	kb	Structure	References[a]
Higher plants					
Zea mays					
N, *cms*-C, -T, -S			1.9	Minicircle	1
cms-C			1.5	Minicircle	1
cms-C			1.4	Minicircle	1
Zea mays (cell culture)			1.5	Circle (monomeric series)	2
			1.8	Circle (monomeric series)	2
Zea mays cms-S			6.4	Linear (S-1)	3
			5.4	Linear (S-2)	3
			2.3	Linear (S-3)	4
Zea mays type RU			7.5	Linear (R-1)	5
			5.4	Linear (R-2)	5
Zea mays N	21	45 (48%)		Circular class	6
	15	33 (20%)		Circular class	6
	30	66 (14%)		Circular class	6
	4	8.8 (7.6%)		Circular class	6
	7	15 (5.8%)		Circular class	6
	41	91 (3.8%)		Circular class	6
	9	20 (1%)		Circular class	6
Zea mays (*cms*-T)	25	55 (47%)		Major circular class	6
	12	37 (37%)		Major circular class	6
Zea mays (*cms*-S)	17	36 (40%)		Major circular class	6
	36	78 (32%)		Major circular class	6
Zea diploperennis			7.5	Linear (D-1)	7
			5.4	Linear (D-2)	7
Sorghum bicolor (*cms*)			5.7	Linear (N-1)	8
			5.3	Linear (N-2)	8
Glycine max	5.9			Major circular class	9

48

Beta vulgaris	10	Major circular class	9
	12.9	Major circular class	9
	16.6 (most abundant)	Major circular class	9
	20.4	Major circular class	9
	24.5	Major circular class	9
	29.9	Major circular class	9
Nicotiana (cultured cells)	1.3	Circles	10
	1.4	Circles	10
	1.45	Circles	10
	1.5	Circles	10
	7.3	Circles	10
Phaseolus vulgaris (cultured cells)	10.1	Monomeric series	2
	28.8	Monomeric series	2
	1.9	Monomeric series	2
Oenothera berteriana	6.3	Circular class	18
	7.0	Circular class	18
	8.2	Circular class	18
	9.9	Circular class	18
	13.5	Circular class	18
Fungi			
Neurospora crassa (Mauriceville)	3.6	Circle	11
Neurospora intermedia (Labelle)	4.2	Circle	12
Neurospora intermedia (Fiji)	5.2	Circle	12
Podospora anserina			
α event sen DNA	2.6	Monomeric circle	13
β event sen DNA	9.8	Monomeric circle	13
θ event sen DNA	6.3	Monomeric circle	19
Saccharomyces cerevisiae			
cytoplasmic petite mutations		Amplified circles (diverse)	14,15
Aspergillus amstelodami			
ragged mutations		Amplified circles (diverse)	16

(continued)

49

TABLE 2 (*Continued*)

Organism	Length	Md	kb	Structure	References[a]
Protozoa					
Crithidia luciliae			2.5	Kinetoplast minicircle	17
Crithidia fasciculata			2.5	Kinetoplast minicircle	17
Herpetomonas muscarum			1.1	Kinetoplast minicircle	17
Herpetomonas ingenoplastis			36	Defective maxicircle	17
			17	"Inflated" minicircle	17
			23	"Inflated" minicircle	17
Leishmania tarentolae			0.87	Kinetoplast minicircle	17
Phytomonas davidi			1.1	Kinetoplast minicircle	17
Trypanosoma brucei			1.0	Kinetoplast minicircle	17
Trypanosoma cruzi			1.4	Kinetoplast minicircle	17
Trypanosoma mega			2.3	Kinetoplast minicircle	17

[a]1. Kemble and Bedbrook (1980); 2. Dale (1982); 3. Pring *et al.* (1977); 4. Koncz *et al.* (1981); 5. Weissinger *et al.* (1982); 6. Levings *et al.* (1979); 7. Timothy *et al.* (1982); 8. Pring *et al.* (1982); 9. Synenki *et al.* (1978); 10. Powling (1981); 11. Collins *et al.* (1981); 12. Stohl *et al.* (1982); 13. Wright *et al.* (1981); 14. Locker *et al.* (1979); 15. Borst and Grivell (1978); 16. Küntzel *et al.* (1982); 17. Borst *et al.* (1981); 18. Brennicke and Blanz (1982); 19. Belcour *et al.* (1981).

The original mitochondrial sequences that are amplified in the *petites* as repetitive sequences may be organized in two different ways (Lazowska and Slonimski, 1976; Locker and Rabinowitz, 1976). In one type of organization, amplified repeats are repeated in tandem, with circular molecules containing both odd and even multiples of the basic sequence. In the second type, a sequence is present as an inverted repeat that is then amplified as a tandem repeat; therefore, the unit repeat is present in even numbers of the basic sequence in oligomeric circular molecules.

The mechanisms that generate *petites* are not understood, but plausible models have been proposed based on the structure of the *petites* and on related processes that occur in prokaryotic systems (Lazowska and Slonimski, 1977; Locker *et al.*, 1979; Bernardi *et al.*, 1980). Initially, a monomeric circle may be generated from a wild-type genome by excision, either by homologous or illegitimate recombination. Larger than monomeric sequences may occur by extended replication or by recombination between monomeric circles. Generation of a large monomeric inverted repeat can result if a small inverted repeat in a circular monomer recombines with itself in opposing orientations to generate a palindromic dimer. Amplification of this circle by replication then generates a tandem array of the large inverted repeat. Rolling circles with long replicated tails are found in preparations of *petite* mtDNA (Locker *et al.*, 1974a).

It is reasonable to assume that all *petites* must contain an origin for DNA replication. The spectrum of *petites* suggests many origins, because so many different parts of the mitochondrial genome are represented. However, a small number of origins, or even a single origin, could be used for all *petites* if rearrangements were extensive during *petite* formation. Potential origins of replication and transcription, called *ori* sequences, are found in the majority of *petites* (DeZamaroczy *et al.*, 1981; Bernardi, 1982). Seven *ori* sequences were found that are homologous with each other and that resemble in some structural aspects the origin of replication region of mammalian mtDNA. Blanc and Dujon (1982) have isolated three regions from hypersuppressive *petites* called *rep* that are 300 bp and that are active in replication *in vivo*. These sequences can drive replication of recombinant plasmids after transformation in yeast and function as autonomously replicating sequences (*ars*) outside of the mitochondria. The *rep* sequences are present in normal mtDNA, but the inference that these sequences are used in normal replication has not been established directly. Perhaps many different sites within the yeast mitochondrial genome can serve as replication origins. These sites could have different degrees of efficiency for replication or segregation and therefore account for the differences observed in *neutral* and *suppressive petites*.

B. HETEROGENEITY OF mtDNA IN SENESCENT CULTURES OF *Podospora*

Amplified mtDNA sequences are found in the filamentous fungus *Podospora anserina*. The organism is an obligate aerobe that undergoes cellular senescence under conditions of continuous vegetative growth. Races of *P. anserina* grow from the point of spore germination with a genetically defined life span (Marcou, 1961). As mycelial growth is progressively reduced, visible changes in morphology and color accompany the arrest of growth. The life span varies in different races and is controlled by the genotype of the cytoplasm. Control of senescence can be transferred through mycelial anastomoses. In sexual crosses, the control of senescence is maternally inherited (Marcou, 1961; Smith and Rubenstein, 1973).

During senescence, the cytochrome aa_3 content is gradually reduced, indicating an involvement of mitochondrial function (Belcour and Begel, 1978; Cummings *et al.*, 1979c). Furthermore, the onset of senescence may be postponed by ethidium bromide, which inhibits mtDNA and RNA synthesis, or by chloramphenicol, which inhibits mitochondrial protein synthesis.

mtDNA in senescent cultures is different from mtDNA in juvenile cultures. In juvenile mitochondria, the genome size is about 95 kb (Kück and Esser, 1982; Cummings *et al.*, 1979d). In senescent mtDNA, smaller mitochondrial molecules are observed that are amplified with respect to the normal mitochondrial genome (Stahl *et al.*, 1978; Cummings *et al.*, 1978).

The amplified molecules are closed circular DNAs that vary in size, and which are often found as a multimeric series. Several different monomeric sizes have been observed (Kück *et al.*, 1981; Wright *et al.*, 1982; Belcour *et al.*, 1981). These plasmidlike molecules are found exclusively in senescent cultures. It has been proposed that the closed circular molecules amplified during senescence are the causative agent in aging of *Podospora* (Esser *et al.*, 1980; Kück *et al.*, 1981).

The distribution of sizes for circular molecules of mtDNA from senescent cultures resembles the distribution of circular molecules observed in *petite* yeasts. In these senescent cultures, circles can be classified into one or more sets of monomer–oligomer series. The monomer lengths can vary within a preparation, possibly due to multiple events. The monomer lengths also vary in separate cultures of the same strain and between different strains. Some specific sizes are found more frequently than others (Belcour *et al.*, 1981; Cummings *et al.*, 1979c).

The senescent plasmidlike DNAs are mitochondrial in origin and appear to be excised from the large juvenile mtDNA and amplified at a particular time in the vegetative growth cycle of the organism (Jamet-Vierny *et al.*, 1980; Cummings *et al.*, 1980b; Belcour *et al.*, 1981; Kück *et al.*, 1981; Wright *et al.*, 1982). Degeneration and loss of specific regions may occur while other regions are selectively amplified (Kück *et al.*, 1981). Several different forms of plasmidlike DNAs have been characterized, and they appear to be closely associated with specific mitochondrial genes (Wright *et al.*, 1982). One such molecule, with a monomeric size of 9.8 kb and from what is called β *event sen DNA* (Wright *et al.*, 1982), has been assigned by restriction mapping and hybridization experiments to the *oxi-2* locus (COIII). Another molecule, α *event sen DNA*, a 2.6-kb monomer, involves the *oxi-3* gene (COI) (Wright *et al.*, 1982). Other amplified circular molecules have possible associations with the large ribosomal subunit gene and the *oxi-1* gene (COII). Another plasmidlike DNA hybridizes to a region bounded by cytochrome *b* sequences on both sides, raising the possibility that the plasmidlike DNA may contain intron sequences of a mosaic gene (Kück and Esser, 1982).

Evidence indicates that the excision and amplification of plasmidlike DNA plays a direct role in the process of senescence, rather than being a concurrent event. DNA from senescent cultures can transfer senescence to juvenile protoplasts by DNA transformation (Tudzynski *et al.*, 1980; Tudzynski and Esser, 1979). Cloned fragments of plasmidlike DNA in a hybrid molecule with pBR322 can similarly induce senescence in juvenile cultures (Stahl *et al.*, 1982).

One interpretation of the events involved in senescence would be that plasmidlike molecules are created by mechanisms similar to those in yeast that generate *petites,* by excision and amplification of specific sequences. Perhaps excision and amplification confer a biological advantage on the species through programmed senescence. In this regard, cytoplasmic mutants have been found with extended life-spans. One mutant, originally isolated for chloramphenicol resistance, with a wild-type restriction pattern, had a fivefold increase in life-span (Belcour *et al.*, 1982). Nonetheless, the results indicate a temporal control of chromosome rearrangement of mtDNA that generates molecular heterogeneity.

C. AMPLIFIED DNAs IN *Aspergillus* AND *Neurospora*

Amplified mtDNAs have been found in other fungi as cytoplasmic mutations showing similarities to the *petites* of yeast and to the mtDNA associated with senescence in *Podospora*. In *Aspergillus amstelodami, ragged* is a mitochondrial mutation with irregular morphology when cultured on solid media. The mutations arise spontaneously and are characterized by a number of physiological defects, including cytochrome abnormalities and a high level of cyanide-insensitive respiration (Caten, 1972; Handley, 1975).

In addition to a normal mitochondrial genome, *ragged* mutants contain additional amplified sequences (Lazarus *et al.*, 1980b; Lazarus and Küntzel, 1980). These sequences, analyzed by restriction endonucleases and molecular hybridization, are found to be a part of the normal mtDNA amplified as tandem repeats. DNA from a number of *ragged* mutants has been isolated and contains amplified sequences from one of two defined areas of the wild-type mtDNA (Lazarus and Küntzel, 1981). These specific regions have been partially sequenced. Most *ragged* mutants are homologous to one site of less than 3 kb that contains several highly conserved structural genes. Amplified regions vary in length but all contain a common region thought to be an origin of replication (Küntzel *et al.*, 1982). The second region is represented by a single event in which a 900-bp fragment is amplified in a head-to-tail multimeric series.

Heterogeneity may exist in *ragged* amplified sequences due to multiple events. A second amplification appears to have occurred in *ragged 1* (900-bp repeat) that is called *ragged 7* and has a repeat of 1700 bp. Further subculturing of the *ragged* strain results in the loss of *ragged 1* sequences, suggesting that one amplified sequence may be suppressed by another. This type of suppressiveness is similar to that found in *petites* of yeast and may also be related to the efficiency of different origins of replication (DeZamaroczy *et al.*, 1981). In *ragged 6,* excision has occurred totally within the ATPase subunit 6 coding region. Sequence analysis has not revealed any unusual AT- or GC-rich regions or any long direct or reverse repeats (Küntzel *et al.*, 1982). No homologies were found in the sites of excision or flanking sequences, arguing against an excision event determined by intrastrand homology-dependent recombination.

The *ragged* mutants vary from the *petites* of yeast and from the senescent mtDNA in *Podospora* because the intact wild-type mtDNA is still present. Mutant phenotypes may result from increased dosage of specific amplified regions.

Stopper mutations in *Neurospora crassa* are mitochondrial mutations that resemble the *ragged* mutations of *Aspergillus* and, to some extent, the *petites* of yeast. *Stopper* mutants are characterized by irregular growth patterns on solid media, deficiencies in cytochrome *b* and cytochrome aa_3, and female sterility (Bertrand *et al.*, 1976). Characterization of mtDNA in *stopper* mutations shows major structural changes in the DNA. In a characterization of four *stopper* mutants (Bertrand *et al.*, 1980), the mtDNA consisted primarily of defective molecules that retained a 24-kb segment of the wild-type mtDNA (about 40% of the wild-type chromosome). The amplified segment is thought to be a continuous region of the normal mitochondrial chromosome. Some *stopper* mutations show smaller alterations, such as 5-kb deletions in the normal chromosomes. Because *Neurospora* is an obligate aerobe, it is proposed that some normal mitochondrial sequences are retained, and that the defective phenotype results from competition between defective molecules and low concentrations of less defective molecules (Bertrand *et al.*, 1980).

In a characterization of another *stopper* mutation (E35), two populations of aberrant DNA molecules were observed (DeVries *et al.*, 1981). One form has a 4-kb deletion of the normal chromosome. After prolonged growth, a smaller circular molecule of about 23 kb accumulates that may be a derivative of the larger (DeVries *et al.*, 1981). The retained segment in these mutants carries the ribosomal genes and most of the tRNA genes in the adjacent regions. Some *stopper* mutants show deletions as small as 350 bp.

The mitochondrial molecules that are amplified in yeast and in the filamentous fungi form a continuous series of molecular events, from small deletions in a large mitochondrial genome at one extreme to amplified segments of only a few hundred nucleotides in the most extreme *petites*. Deletion and amplification mechanisms may be able to generate similar types of molecules in all of these mtDNAs; however, because yeast is a facultative aerobe, *petites* can be recovered as conditional mutations.

In contrast with the finding of amplified molecules of mtDNA in various mutants, Collins *et al.* (1981) have isolated and characterized circular molecules in the mtDNA fraction of *Neurospora* that show no detectable sequence homology with normal mtDNA. One such plasmid DNA is found in wild-type strain Mauriceville-1c as an oligomeric series with a monomeric length of 3.6 kb. The 3.6-kb plasmid is transcribed and a major transcript of 3.3 kb has been detected in RNA transfer hybridization. Other plasmid DNAs have been found in strains of *Neurospora intermedia* (Stohl *et al.*, 1982). Strain P405-Labelle contains a plasmid with a monomeric circle size of about 4.2 kb whereas strain Fiji N6-6 contains a plasmid with a monomeric size of 5.2 to 5.3 kb. These plasmids have no substantial sequence homology to each other, or to other plasmid DNAs in *Neurospora crassa*, or to normal mtDNA. In sexual crosses, the plasmids show strict maternal inheritance. They are enriched severalfold over other mitochondrial molecules and therefore are found at 100 or more copies per cell. No phenotypic difference or defect in mitochondrial function has yet been associated with these plasmid molecules (Stohl *et al.*, 1982).

VIII. Kinetoplast DNA

A. General Properties of Kinetoplast DNA

Among mtDNAs, the most unusual is that of the kinetoplast DNA found in the flagellated protozoa of the order Kinetoplastidae. The molecular biology and biochemical taxonomy of these protozoa have been extensively reviewed (Englund *et al.*, 1982a; Newton, 1979; Lumsden and Evans, 1976–1979). Kinetoplasts are Feulgen-positive organelles situated at the base of the flagellum in this order of protozoa. The most widely studied genera are the *Trypanosoma, Crithidia,* and *Leishmania*. These parasitic protozoa cause debilitating disease in humans and livestock, particularly in the developing countries of Africa, South America, and Central America. The genus *Crithidia* is parasitic only in insects. *Phytomonas davidi* is a trypanosomatid parasite of euphorbid plants.

T. brucei, which infects mammals, is transmitted by an insect vector (*Glossina*) and has discrete developmental forms in the mammalian bloodstream (bloodstream form), in the insect midgut (procyclic form), and in the insect salivary gland (metacyclic form). In the procyclic form the mitochondrial respiratory system is fully active, whereas some bloodstream forms have completely suppressed mitochondrial functions.

The kinetoplast, a densely staining structure originally presumed to be involved with cell motility, is now considered to be part of a highly organized mitochondrially derived structure that carries out oxidative phosphylation in these organisms. Electron microscopy of the kinetoplast shows a disklike structure within the membrane of the single mitochondrion in the cell.

The structural features of the kinetoplast DNA are unique. This DNA is organized in a network of about 10,000 interlocking (catenated) networks of circular molecules. The circular molecules are composed of two types, a prevalent (95%) class of small molecules (minicircles, from 1 to 2.4 kb) and a larger molecule present in low abundance (maxicircles, from 20 to 33 kb). Each kinetoplast contains one large network of DNA. The maxicircles are apparently the counterpart of the mtDNA of other eukaryotes; the function of the minicircles is unknown. The highly organized network structure of the kinetoplast DNA implies an important biological function maintained by natural selection. However, the function of the kinetoplast DNA network remains a mystery. The properties and replication of kinetoplast DNA have been the subject of several reviews (Borst and Hoeijmakers, 1979b; Englund, 1981; Englund et al., 1982a; Simpson et al., 1980; Stuart and Gelvin, 1980; Borst et al., 1980b, 1981b).

The structural variation in size and sequence organization that occurs in the evolution of kinetoplast DNA has several interesting features. Maxicircle DNA shows variation in size and sequence that is similar to the variation observed for mtDNA in other systems. Minicircles evolve much faster than maxicircles; moreover, minicircles are often found as a heterogeneous population of different but related circular molecules.

Kinetoplast DNA molecules are interlocked; when spread for electron microscopy, the structure can form a massive sheet. The extent of interlocking is sufficient to include most of the molecules; a given circle is usually interlocked with a small number of neighboring molecules to form a "fishnet" structure. Interlocking of many molecules with distant circles would create a more tightly packed structure (Wolstenholme et al., 1974; Borst and Hoeijmakers, 1979b; Englund, 1978; Barker, 1980).

Maxicircles are visualized in the network as a twisted line of distinct fibers (edge loops) near the periphery of the network (Englund, 1978). The kinetoplast network is probably held together by minicircle interactions, not by the maxicircles. Fragmentation of the maxicircles by digestion with a restriction enzyme does not disrupt the structure of the total kinetoplast DNA (Weislogel et al., 1977; Cheng and Simpson, 1978). Some trypanosome strains lack maxicircles, but maintain networks with identical structure except for the loss of edge loops (Fairlamb et al., 1978; Borst and Hoeijmakers, 1979a,b). Such strains have lost the ability to make functional mitochondria and are therefore restricted to growth in the vertebrate host.

Interlocking DNA rings are found in diverse biological systems, usually at low frequency. They have been observed with bacterial plasmids and in animal mtDNA (Kupersztoch and Helinski, 1973; Novick *et al.*, 1973; Flory and Vinograd, 1973; Castora *et al.*, 1982). Interlocking rings (catenanes) may be a normal product of circular molecules produced by the action of cellular topoisomerases. Topoisomerases, such as DNA gyrase (Gellert, 1981) or the omega protein (Wang, 1971), can form catenanes and separate interlocked molecules under standard reaction conditions. Two types of topoisomerases are distinguished; both types have been found in the mitochondria of the rat (Fairfield *et al.*, 1979; Castora *et al.*, 1982). A type I topoisomerase capable of generating catenanes from nicked circles has been isolated. This enzyme can also be described as a nicking–closing enzyme. The type II enzyme will decatenate circular molecules, including a kinetoplast network, and can generate networks from supercoiled plasmid DNAs.

The T4 phage topoisomerase will quantitatively decatenate kinetoplast networks to form covalently closed minicircles and maxicircles. *E. coli* DNA gyrase will carry out the same reaction but at a lower efficiency (Marini *et al.*, 1980; Englund and Marini, 1980). Similarly, DNA gyrase from *Micrococcus luteus* has been shown to decatenate the kinetoplast DNA of trypanosomes (Riou *et al.*, 1982). It seems reasonable to propose that topoisomerases should be found in trypanosomes and that such enzymes are involved in the release and attachment of minicircles in replication (Englund and Marini, 1980).

Replication of kinetoplast DNA is an interesting topological problem. Nonreplicating minicircles are covalently closed, but nicked minicircles are found in growing cells (Englund, 1981). A model for kinetoplast replication (Englund, 1981) proposes that minicircle molecules are released from the network, replicate, and then reattach to the network. Kinetoplast division appears to involve doubling of the size of the disk-shaped network followed by pinching in the middle of the disk, resulting in the formation of two separate networks of kinetoplast DNA (Englund, 1978; Englund *et al.*, 1982b).

B. Minicircles

The small circular interlocking minicircle molecules compose the bulk of kinetoplast DNA and are responsible for the structure of the network. Minicircles range in size from about 700 to 2500 bp (see Table 2). Within a network, they are homogeneous in size but usually heterogenous in sequence (Riou *et al.*, 1975; Kleisen and Borst, 1975; Fouts *et al.*, 1978; Cheng and Simpson, 1978; Donelson *et al.*, 1979; Morel and

Simpson, 1980; Borst and Hoeijmakers, 1979a,b). Digestion of ki-netoplast DNA with restriction enzymes shows a series of fragments with sequence complexities that greatly exceed the size of the circles. Neither methylation nor heterogeneity of cell populations can explain the results (Borst and Hoeijmakers, 1979b; Kleisen *et al.*, 1976a). Mini-circle DNA appears to evolve very rapidly because there is essentially no homology between minicircles from different species (Steinert *et al.*, 1976a,b). Variation in size of minicircles cut at single sites has been observed within a species (Borst and Hoeijmakers, 1979a,b). Species vary widely in the degree of heterogeneity found in minicircles; some species, *P. davidi* and *T. equiperdum,* have a relatively uniform major class of minicircle (Cheng and Simpson, 1978; Riou and Saucier, 1979; Steinert *et al.*, 1976a,b), whereas *T. brucei* may have 100 different types of minicircles.

There are dramatic differences in the properties of the minicircles between *T. brucei* and its close relative, *T. equiperdum.* In *T. brucei,* restriction analysis indicates a high degree of heterogeneity. Most re-striction enzymes cut a small fraction of the minicircles only once, due to the small minicircle size ($\cong$ 1 kb) and a low GC percentage ($\cong$ 30%) (Englund, 1981; Borst and Hoeijmakers, 1979b; Fairlamb *et al.*, 1978; Donelson *et al.*, 1979; Stuart, 1979; Simpson and Simpson, 1978; Borst *et al.*, 1980a,b).

Reassociation kinetic analysis (Steinert *et al.*, 1976a,b; Steinert and Van Assel, 1980) shows a very complex mixture of molecules. The number of different minicircles was estimated to be about 300, indicat-ing a high degree of divergence within the population. Using cloned minicircles in reassociation tests, Donelson *et al.* (1979) and Stuart (1979) showed that specific minicircles were present in low and differ-ing abundance. For example, a given *Bam*HI-sensitive minicircle was present in 60 copies per network, whereas a *Hind*III-sensitive minicir-cle was present in about 500 copies (Stuart and Gelvin, 1980).

Two cloned minicircles of *T. brucei* have been sequenced (Chen and Donelson, 1980), one 983 bp and the other 1004 bp in length. Both minicircles are AT rich (72%) and share about 27% of their sequences. One region of nearly perfect homology is found for 122 bp. Other small regions of homology of 10 to 15 bp are distributed throughout the molecule. Some of these sequences are repeated several times, but the number and location of the repeats are different in the two minicircles. These results support renaturation analysis, which indicates a general pattern of homology between different minicircles for about one-quar-ter of the sequence, whereas the remaining three-quarters of each minicircle shares sequence homology with but 1 out of 300 minicircles (Donelson *et al.*, 1979). The frequency and distribution of homologous sequences suggest a high frequency of recombination, transposition, or rearrangement of minicircle sequences. Recombination of minicircles had been proposed to explain density labeling results of *Crithidia* kinetoplast DNA (Manning and Wolstenholme, 1976) and the fused circular structures observed in the electron microscope (Brack and Delain, 1975).

The sequence determination provides no evidence for a protein-coding function in these minicircles, due to the absence of any open reading frames. Efforts to detect transcripts of *T. brucei* minicircles have been unsuccessful (Borst and Hoeijmakers, 1979b; Englund, 1981).

T. equiperdum has a predominant 1-kb homogeneous minicircle (Riou and Saucier, 1979) present in about 3000 copies per kinetoplast. In contrast to other minicircle populations of *Trypanosoma,* the minicircle DNA of *T. equiperdum* can generate a unique restriction map (Riou and Barrois, 1979). The minicircle DNA, purified after cleavage with *Bam*HI and sequenced directly, generates a unique sequence (Barrois *et al.*, 1981). The typical minicircle therefore contains 1012 bp, with an AT content of 73%. Open reading frames are restricted to about 20 amino acids. A region of about 130 bp is almost identical to the conserved region of the *T. brucei* minicircle (Chen and Donelson, 1980). The remainder of the molecule shows no relation to *T. brucei* minicircles. In addition, short repeated sequences are distributed throughout the molecule.

The life cycle of *T. equiperdum* is different from that of *T. brucei. T. equiperdum* is a venereally transmitted disease of equines; it is not transmitted through an insect vector and does not require mitochondrial function.

T. cruzi has limited heterogeneity in minicircle DNA, with less than 10 different classes of minicircles (Borst and Hoeijmakers, 1979b). Restriction digests give discrete patterns that can be used to characterize strains (Morel *et al.*, 1980). In addition, *T. cruzi* minicircles have an unusual repetitive sequence organization (Brack and Delain, 1975; Englund, 1981; Riou and Yot, 1977; Leon *et al.*, 1980; Frasch *et al.*, 1981). Restriction enzyme digests of minicircles of *T. cruzi* and *T. rangeli* show a repeating unit that is one-fourth the length of the minicircle. The size of the minicircle is about 1400 bp (Borst and Hoeijmakers, 1979b). This result suggests a series of duplication events in the recent evolution of the minicircle and requires that the duplicated sequence be only about 400 bp. All known minicircles are considerably longer than 400 bp. Therefore, both deletion and duplication may occur in the evolution of these sequences, resulting in repetitive sequences and variation in size.

Kinetoplast DNA from the insect trypanosome, *Crithidia fasciculata*, has at least 10 different types of minicircles, about 2.4 kb in size (Simpson, 1972), that share extensive sequence homology (Kleisen *et al.*, 1976a). *C. fasciculata* minicircles evolve very rapidly. Between species of *Crithidia* minicircles show no conservation of restriction sites, and cross-hybridization is relatively weak. During 2 years of laboratory cultivation, alterations in restriction patterns were observed (Hoeijmakers and Borst, 1982). Minicircle populations were studied by restriction digestion and by heteroduplex analysis in *Crithidia* subspecies or varieties. Although maxicircle fragments are highly conserved, restriction fragments of minicircles show no common fragments beyond chance expectation. However, cross-hybridization shows high levels of homology. The results are due to a high level of segmental rearrangement that results in diverse restriction patterns but which preserves sequence homology (Hoeijmakers and Borst, 1982). Rapidly evolving minicircle populations have been reported for *T. cruzi* and *Leishmania tarentolae* (Simpson *et al.*, 1980; Riou, 1976).

Direct evidence for rearrangements in *C. luciliae* minicircles was obtained by heteroduplex analysis of molecules linearized with *Hind*III (Hoeijmakers *et al.*, 1982b). About 40% of the molecules showed evidence of structural rearrangements. Rearrangements were not randomly distributed, but were preferentially located in specific segments of DNA that appear to rearrange more frequently.

In an earlier study of *Crithidia acanthocephali*, which has minicircles of about the same size (2.3 kb), restriction digests showed that a small fraction of the molecules (from 9 to 12%) were cut by *Hind*III or *Eco*RI (Fouts *et al.*, 1978), suggesting a degree of heterogeneity similar to that observed in other *Crithidia* species. Reassociation kinetics indicated a high degree of homology (Fouts *et al.*, 1975), consistent with the current interpretation of structural rearrangements.

In *C. acanthocephali*, at least 10% of the minicircle component is transcribed (Fouts and Wolstenholme, 1979). Visualization of DNA–RNA hybrids showed duplex formation to single molecules over 10% of their length. Because no evidence for transcription has been obtained with other minicircles of kinetoplast DNA, the significance of this observation is not known.

In *L. tarentolae*, minicircles (750 bp) can be classified into three major sequence classes based on restriction enzyme analysis (Challberg and Englund, 1980). Class I constitutes 70% of the molecules, class II is 15%, and class III is about 7%. Minor classes make up the remaining 8%. Each class is homogeneous in sequence. Heteroduplex analysis between classes I and II shows divergence for 10–15% of base pairs (Englund, 1981). Cloned fragments of minicircles from full-length fragments have been mapped by Simpson et al. (1980), but were different from class I, II, or III restriction maps. These clones may represent minor components or indicate unexpected variation between these strains.

Phytomonas davidi is a trypanosomatid that infects euphorbid plants and is transmitted by an insect vector. Minicircles of this kinetoplast DNA have a mean length of 1064 bp (Cheng and Simpson, 1978). In restriction digests, complexity of fragments is greater than expected molecular weight, indicating heterogeneity. However, reassociation and thermal stability analysis indicated a high level of sequence homogeneity, at least 96%. Based on the comparisons now available from other species, rearrangements of minicircle DNA may be the basis for the high degree of diversity observed in restriction patterns.

C. MAXICIRCLES

Maxicircles were first detected as a complex minor component in kinetoplast DNA. It was subsequently shown that these large circles were intrinsic components of the kinetoplast (Kleisen and Borst, 1975; Kleisen et al., 1976b). The maxicircle DNA can be released from the network, using a restriction enzyme that cuts the maxicircle only once. Maxicircles range in size from about 20 kb (*T. brucei*) to about 38 kb (*C. fasciculata*) (Table 1). Within a given species, the maxicircles are uniform in size and sequence organization (Borst and Hoeijmakers, 1979a,b). Maxicircles are usually present in fewer than 50 copies within a network. No sequence homology with minicircles has been observed.

It is generally agreed that the maxicircles correspond to the functional mtDNA of eukaryotic species. Maxicircles appear to be transcribed. RNA present in a purified kinetoplast fraction hybridizes to maxicircle DNA in *L. tarentolae, P. davidi, T. brucei*, and *C. fasciculata* (Simpson and Simpson, 1978; Simpson et al., 1980, 1982; Cheng and Simpson, 1978; Hoeijmakers and Borst, 1978). The major RNA species found in the kinetoplast are small, 9 and 12 S, corresponding to about 630 and 1230 bp. These small RNAs appear to be the rRNAs because they (1) are most abundant, (2) are present in equimolar amounts, (3) lack poly(A), (4) are the only high-molecular-weight RNAs observed, and (5) are conserved in size among the Kinetoplastidae (Borst *et al.,* 1980b).

Hybridization shows that the rRNAs are transcribed from one site each on the maxicircle. Direct sequence analysis of the DNA indicates homology with the ribosomal genes of human mitochondria and *E. coli* (Simpson *et al.,* 1982). The sequence of the 9 S gene was highly AT rich (80%) (Simpson *et al.,* 1980). These ribosomal transcripts are the smallest ribosomal molecules found in any biological system. The direction of transcription in *T. brucei* has been determined for the 12 and 9 S genes. Both molecules are transcribed from the same strand (Hoeijmakers *et al.,* 1982a,b).

In addition to the ribosomal genes, the maxicircle in *T. brucei* has a number of additional transcripts (Stuart and Gelvin, 1982). The maxicircle is transcribed largely but not completely, in both the bloodstream form, which lacks fully developed mitochondrial respiration, and in the procyclic form (in the insect midgut), which has fully developed respiration (Englund *et al.,* 1982a).

Further evidence for the mitochondrial nature of the maxicircle comes from the localization of a sequence in *T. brucei* with homology to maize COII (Johnson *et al.,* 1982). DNA sequences from yeast that code for COI, COII, COIII, and cytochrome *b* hybridize to defined fragments of maxicircle DNA from *L. tarentolae* (Simpson *et al.,* 1982). In addition, several presumptive poly(A)-containing mRNAs have been detected as maxicircle transcripts in *L. tarentolae* (Simpson *et al.,* 1982).

The maxicircle of *Crithidia* kinetoplast (33 kb) is much larger than that of the *Trypanosoma* (20 kb). In two species of *Crithidia, C. fasciculata* and *C. luciliae,* the maxicircles are similar in size and have identical restriction patterns (Hoeijmakers *et al.,* 1982a). If maxicircle DNA is hybridized with total cellular RNA, transcripts of the maxicircle are detected. The transcribed regions are clustered on one-half of the maxicircle. No transcripts from the minicircle were detected (Hoeijmakers and Borst, 1978).

The strongest areas of hybridization are attributed to the 9 and 12 S rRNA. Other transcripts are found in adjacent regions. Two cloned fragments of *T. brucei* kinetoplast DNA hybridize to the same region. The clones hybridize with continuous regions, that is, they are not interrupted or transposed within the maxicircle. The organization of at least these segments, which cover about 10 kb, would appear to be conserved in these species. The transcription map, compared with that of *T. brucei* (in which most of the molecule is transcribed), implies that the maxicircle of *Crithidia* differs from that of *Trypanosoma* by a large block of continuous nontranscribed DNA.

Comparison of restriction fragment patterns from maxicircle DNA of different strains of *T. brucei* shows several polymorphic sites and variation in size over a 5-kb region (Borst *et al.*, 1981a). The calculated sequence similarity for the different strains is greater than or equal to 97%. Size differences between different maxicircles may be as great as 1.4 kb within a specific region that appears to be a "hot spot" for insertion–deletion events.

Some strains of trypanosomes lack organized kinetoplast networks (Borst *et al.*, 1981b). Some are completely deficient in kinetoplast DNA (Borst and Hoeijmakers, 1979a,b) whereas others contain DNA that is dispersed throughout the mitochondrion (Hajduk, 1978). In some cases, networks of minicircles exist in the absence of maxicircles. Other mutant strains are defective in mitochondrial function, due either to point mutations or to deletion of maxicircle genes (Hajduk and Vickerman, 1981; Borst and Hoeijmakers, 1979b; Borst *et al.*, 1981b). Such strains or mutants with defective respiration exist only as bloodstream forms and cannot infect an insect vector. Therefore, maxicircles are not essential for bloodstream forms of trypanosomes, but are needed for mitochondrial function in the insect. In addition, maxicircles are not necessary for the structure or replication of the kinetoplast network (Borst and Hoeijmakers, 1979a,b; Englund, 1981).

In addition to minicircles and maxicircles, some additional sizes of circular molecules have been observed in kinetoplasts (Borst and Hoeijmakers, 1979a,b). Multicircle oligomers may be either recombination intermediates or oligomeric products of replication. Intermediate circles, heterogeneous in size, have been observed in *C. fasciculata* and *T. mega* (Steinert *et al.*, 1976b; Borst *et al.*, 1977).

Some kinetoplast DNA fragments have anomalous electrophoretic mobility (Borst and Fase-Fowler, 1979; Hoeijmakers *et al.*, 1982a). One fragment has a higher molecular weight in the presence of ethidium (8.6 kb) than in its absence (7.5 kb). The anomalously migrating fragment is located in the nontranscribed portion of the restriction map. Another anomalous fragment, in *L. tarentolae* minicircles, shows unusual behavior on polyacrylamide gels (Challberg and Englund, 1980). This anomaly appears to be a sequence-dependent effect because it is also found in a cloned fragment with the same sequence. The fragment is 490 bp by sequence determination, and migrates as 450 bp in 1% agarose, and as 1380 bp in 12% acrylamide. It has been proposed that the anomaly is due to a systematically bent region of B-DNA (Marini *et al.*, 1982).

Mutants that affect both maxicircles and minicircles suggest important interactions between these components of kinetoplast DNA. Some strains of *Trypanosoma evansi* and *T. equiperdum* completely lack maxicircles (Borst *et al.*, 1980b) but retain networks and minicircles that appear normal (Borst and Hoeijmakers, 1979b). A "remarkable correlation" has been stressed by Hoeijmakers *et al.* (1982b) between the heterogeneity of minicircles and the presence of a functional maxicircle. Minor alterations in the maxicircle are sufficient to abolish minicircle heterogeneity, eventually (Frasch *et al.*, 1980, 1981; Riou and Saucier, 1979; Borst and Hoeijmakers, 1979b). This finding suggests that the absence of a maxicircle affects some process of segregation or replication and consequently results in a homogeneous population of minicircles.

D. Evolution and Function of Minicircle DNA

The results from a variety of trypanosomatid species indicate a rapid rate of divergence for kinetoplast minicircle DNA. A significant amount of the divergence can be accounted for by rearrangement of sequences. Cross-hybridization of minicircles even for closely related species is very weak (Borst and Hoeijmakers, 1979b). Due to the absence of detectable transcription and the great variation within and between species, it is unlikely that minicircles code for a functional protein or RNA. Yet available evidence suggests that the kinetoplast DNA carries out an important sequence-independent function.

First, most of the kinetoplast DNA is minicircle DNA. The replication and segregation of minicircles are complex events, which seem unlikely to have developed without some selective advantage for the kinetoplast structure. It has been variously suggested that the kinetoplast plays a role in cell division related to the close association of the kinetoplast with the kinetosome at the base of the flagellum (see Borst and Hoeijmakers, 1979a,b, for a discussion of these ideas). However, why the kinetoplast DNA includes a highly developed complex network of DNA minicircles is a most important question and suggests a novel role for noncoding DNA.

Interesting systems for further study related to the evolution of the kinetoplast DNA are parasitic protozoa with unconventional networks, such as *Herpetomonas ingenoplastis* (Borst *et al.*, 1981b;

Englund *et al.*, 1982). This insect parasite probably contains 36-kb maxicircles that cross-hybridize with *T. brucei* maxicircles. Instead of minicircles, *H. ingenoplastis* has thousands of large circles that are heterogeneous in size, ranging from 16 to 23 kb. These large heterogeneous circles suggest an evolutionary intermediate in the evolution of the highly developed kinetoplast networks of the flagellated protozoa.

IX. Higher Plant mtDNA Diversity

A. VARIATION IN GENOME SIZE

The most striking features of higher plant mitochondrial genomes are exceptional genome size and molecular diversity (Levings and Pring, 1978; Levings *et al.*, 1979, 1983; Levings and Sederoff, 1981) (see Tables 1 and 2). Mitochondrial genome sizes have been measured by reassocation kinetics, restriction analysis, and electron microscopy in more than a dozen plant species (Levings and Sederoff, 1981; Wallace, 1982). By all estimates, plant mitochondrial genomes are large. In several cases, both restriction analysis and reassociation kinetics have been done and were in reasonable agreement (Ward *et al.*, 1981). These results indicate that these large genomes are composed primarily of unique sequences, rather than repetition of a simple mitochondrial genome. These results are supported by cloning studies in maize, which indicate that most sequences are not repeated in this large mitochondrial genome (Spruill *et al.*, 1980; Lonsdale *et al.*, 1981; Iams and Sinclair, 1982; Stern *et al.*, 1982; Borck and Walbot, 1982).

The size of mitochondrial genomes can vary greatly within a related group of plant species. In four species of cucurbits, mitochondrial genome sizes varied from 330 to 2400 kb (Ward *et al.*, 1981) (Table 1). Variation of mitochondrial genome size in these species is independent of nuclear genome size, which does not vary much among cucurbit species (Ingle *et al.*, 1975). Conversely, in *Vicia faba* and *V. villosa*, which differ in genome size by sevenfold (Bennett and Smith, 1976; Bendich, 1982), the mitochondrial genomes vary less than twofold: *V. faba* (285 kb) and *V. villosa* (375 kb). In contrast with both nuclear and mitochondrial genomes, chloroplast genomes of higher plants are relatively constant in size, ranging from 120 to 180 kb (Wallace, 1982).

Whereas nuclear volume is known to be proportional to genome size, mitochondrial volume per cell does not vary with increasing mitochondrial genome size (Bendich, 1982). The amount of mtDNA per cell in shoot tissue, however, is increased proportionally with the size of the mitochondrial genome over a sevenfold range of mitochondrial genome size (Ward *et al.,* 1981). Therefore, the number of mitochondrial genomes per cell is constant in these species.

Diversity in mtDNA is found in races of maize carrying mutations for cytoplasmic male sterility (*cms*). Cytoplasmic male sterility is frequently found in a wide variety of higher plants (Edwardson, 1970). The character is maternally inherited and affects the normal development of pollen, but has no effect on female fertility. The expression of sterility often depends on the genetic composition of the nucleus. In maize, nuclear genes that act as suppressors of the sterile phenotype, called restorers (*Rf*), are dominant and permit the conditional expression of fertility in the appropriate nuclear background (Duvick, 1965).

In maize, cytoplasmic mutations to male sterility may be classified or divided into three types, based on their response to specific nuclear restorers. Restoration of the Texas male sterile cytoplasm (*cms*-T) requires two genes (*Rf 1* and *Rf 2*). Another sterile, type S (*cms*-S) is restored by *Rf 3,* and *cms*-C is restored by *Rf 4*. Several lines of evidence indicate that the *cms* mutations are coded in the mtDNA (Levings and Pring, 1976, 1979a,b; Pring and Levings, 1978; Levings *et al.,* 1980; Laughnan *et al.,* 1981; Leaver and Gray, 1982). The mtDNAs of the different sterile cytoplasms of maize vary greatly from normal mtDNA and from each other (Levings *et al.,* 1979; Pring and Levings, 1978; Levings and Pring, 1978). Restriction patterns of normal mtDNA and mtDNA from male sterile cytoplasms show many differences (Levings and Pring, 1976, 1978; Spruill *et al.,* 1981; Borck and Walbot, 1982), whereas chloroplast DNAs show almost identical patterns (Pring and Levings, 1978). If visible bands are compared in different maize mtDNAs, 53 to 67% of the restriction fragments comigrate (Borck and Walbot, 1982).

Comparison of restriction digest of mtDNAs between normal maize cytoplasm (N) and *cms*-T using cloned fragments shows polymorphisms for about one-third of the restriction fragments (Spruill *et al.,* 1981). Hybridizations with cloned fragments to mtDNA from N and T show many changes in restriction fragment patterns that might be caused by either rearrangements or by base substitution mutations (Spruill *et al.,* 1981). Stern and Lonsdale (1982) have noted two deletions in some sequences: one in *cms*-T of 100 bp and another in *cms*-C, which is lacking a 1500-bp sequence.

Structural variation occurs in the evolution of mtDNA in maize and in the teosintes. Teosintes, the closest relatives of maize, are wild species from Central America. The mtDNAs of many teosintes have been analyzed by restriction digests, and the patterns are useful for establishing taxonomic relationships (Timothy et al., 1979). Some teosinte mtDNAs are very similar to those of maize and are much less divergent than the mtDNA of male sterile cytoplasms. Other teosintes have mtDNA that is highly divergent.

Hybridization of cloned fragments to mtDNA blots of maize and teosinte showed that about one-third of the restriction fragments were conserved within the genus. The patterns of variation indicated that rearrangements were involved, at least in part, in the divergence of these restriction fragment patterns (Sederoff et al., 1981). Diversity generated by rearrangements could have selective value in organelle DNA, which does not participate in the sexual process. Interestingly, animal mtDNA has a low frequency of rearrangements but a high-base-mutation frequency, perhaps as an alternative mechanism to generate sequence diversity.

B. Circular Molecules in Plant mtDNA

The majority of DNA molecules purified from DNase-treated plant mitochondria are linear and high in molecular weight (Wolstenholme and Gross, 1968), but covalently closed circular molecules can be purified from regions of higher density in CsCl–ethidium bromide density gradients (Quetier and Vedel, 1977; Levings and Pring, 1978; Levings et al., 1979). These circular molecules are heterogeneous and do not correspond to the sizes of the large mitochondrial genomes. Usually, discrete size classes of circles are observed. In normal maize, seven discrete classes of large circles have been observed, ranging from 13 to 136 kb, each with a specific relative abundance (Levings et al., 1979). The sizes and relative abundance of circles did not fit a distribution expected for a simple multimeric series. A heterogeneous distribution of circles was also found in male sterile mitochondria of maize, but the precise size classes and abundance distributions were different for each cytoplasmic genotype (Levings et al., 1979). The predominant circular molecules of normal maize cytoplasm (N mtDNA) are at about 68 kb (48%), whereas cytoplasmic male sterile mtDNA (cms-T) has a predominant circular class of 83 kb (47%). In the S cytoplasm (cms-S), the most frequent class was 54 kb (40%). The second most frequent size class also varies in size and relative abundance: N (50 kb, 20%), S (117 kb, 30%), and T (57 kb, 37%).

In soybeans, similar discrete classes of large circular molecules have been observed, ranging from 19 to 100 kb (Synenki *et al.*, 1978). Evidence for multiple classes of circular molecules has been obtained in flax, also (L. Lockhart and C. S. Levings, personal communication). Examination of circular DNA molecules in cucumber, wheat, potato, and Virginia creeper provided evidence for molecular heterogeneity of mtDNA in several species (Quetier and Vedel, 1977). In order to account for the total sequence complexity of the large mitochondrial genome, it was proposed that mitochondrial genomes contained heterogeneous populations of molecules that often varied in size but which contained different sequences (Levings *et al.*, 1979; Quetier and Vedel, 1977; Levings and Pring, 1978).

Closed circular DNA molecules have been isolated from *Oenothera berteriana* (Brennicke and Blanz, 1982). Five small-size classes were found ranging from 6.3 to 13.5 kb. These molecules are distinct from each other in sequence, with no detectable homology. Restriction analysis indicates that each circle class is not internally repetitive because restriction fragments generally sum to the size of the isolated circles.

In cell cultures of tobacco, bean, and maize (Dale, 1981, 1982; Sparks and Dale, 1980), from 20 to 40% of the total mtDNA can be isolated as supercoiled molecules. The supercoiled molecules are heterogeneous and contain, in part, multimeric series as well as unique fragments. In maize, the predominant circle classes found were 1.5- and 1.8-kb molecules possibly related to minicircles found previously in undigested mtDNA (Levings *et al.*, 1979; Kemble and Bedbrook, 1980). Similarly, in bean, a 1.9-kb circle was the predominant molecule, which was likely to be a monomeric form of an oligomeric series (Dale, 1982). Tobacco mtDNA contained larger circular molecules of 10.1, 20.2, and 28.8 kb. The 28.8-kb class contained unique restriction fragments that differed from those common to the 10.1 and 20.2 circles, which have a monomer–dimer relationship. Therefore, more than one type of circle was contained in the population of molecules of 28.8 kb. Both the small circles and the larger circles hybridized to mtRNA on Northern blots. In maize, the small circles hybridized to RNA molecules far larger than could be coded on the circles alone (Dale, 1982), suggesting that they contain fragments of functional genes or that the larger RNAs are polycistronic.

In other species, such as yeast, the mitochondrial chromosome is circular; however, intact circular molecules of purified mtDNA are observed only rarely (Borst and Grivell, 1978). In higher plants, it is not yet clear what relationship exists between the heterogeneous circular molecules and the functional mtDNA. In maize mtDNA preparations, only about 5% of the molecules are supercoiled. In flax, super-coiled molecules constitute a higher percentage of the mtDNA, about 30% (L. Lockhart and C. S. Levings, personal communication). High percentages (from 20 to 40%) of supercoiled molecules are routinely prepared from cultured cells (Dale, 1981).

Two main hypotheses have been suggested to explain the properties of circular mtDNAs in plants. First, higher plant mitochondrial ge-nomes could be made up of several different large circular DNA mole-cules (Quetier and Vedel, 1977; Levings *et al.*, 1979; Spruill *et al.*, 1980; Dale, 1982). The concept of separate chromosomes in mtDNA is in-teresting because it provides a mechanism for control of the level of gene expression by the control of circle copy number (Levings and Sederoff, 1981). However, attempts to find variation in mtDNA re-striction patterns during development have been negative (Bendich, 1982; Quetier and Vedel, 1980).

Alternatively, the circular mtDNA molecules could represent ampli-fied or autonomous portions of the mitochondrial genome created by recombination or replication (Levings and Pring, 1978; Bendich, 1982). In this case, the circles are analogous to the supernumerary mtDNA molecules found in fungi (Wright *et al.*, 1982; Küntzel *et al.*, 1982). Distributions of circular molecules found in *petite* strains of yeast are striking in their similarity to the distribution of molecules found in higher plants. Although some *petite* mtDNAs consist of a simple oligomeric series, the pattern in plant cells follows more closely that of *petite* mtDNA populations that have arisen from more complex multi-ple events. Mechanisms similar to those that generate *petites* in yeast may operate in plant cells, both in whole plants and in culture, to generate the patterns and complexity of circular molecules that have been observed.

It may be argued that such mechanisms could not generate the speci-ficity observed in plant species or in lines of maize. However, the mech-anisms that generate circles could be sequence dependent or could create different populations of circular molecules by chance. The re-quirement for respiration might preclude the isolation of plant cells with only amplified circular mtDNA molecules.

C. Cosmid Mapping of Maize mtDNA

Application of restriction-mapping techniques to the entire plant mitochondrial genome has proven to be a difficult problem. The large size of the genomes has essentially precluded mapping by standard techniques. Stern *et al.* (1982), Lonsdale *et al.* (1981), and Stern and Lonsdale (1982) have used cosmid cloning with Homer 1 to map the mitochondrial genome. Homer 1 can package fragments as large as 50 kb. Restriction maps have been obtained that extend over large segments of the mitochondrial genome. The estimated size of the maize mitochondrial genome based on the cloned cosmid sequences is over 600 kb, somewhat larger than the 480-kb estimate based on renaturation kinetics and restriction mapping (Ward *et al.*, 1981; Borck and Walbot, 1982). Long stretches, each greater than 60 kb of DNA, have been mapped adjacent to sequences homologous for the S-1 and S-2 DNA elements associated with the S-type of cytoplasmic male sterility in maize. The region of the ribosomal genes has also been mapped over more than 70 kb (Stern *et al.*, 1982; Stern and Lonsdale, 1982).

The restriction maps obtained so far are linear, and suggest that the mitochondrial genome may be organized in larger molecules than predicted from the circular molecules observed in the electron microscope. It is possible, for example, that a large mitochondrial genome may exist as a single chromosome, either linear or circular, with a subpopulation of circular molecules. The resolution to this question requires more information, including a complete physical map.

In the course of cosmid mapping of maize mtDNA (Stern *et al.*, 1982), a sequence of 12 kb has been found that is homologous to a portion of the inverted repeat of the maize chloroplast genome (Stern and Lonsdale, 1982). In the chloroplast DNA, this region contains a 16 S rRNA gene and the coding sequences for isoleucine tRNA and valine tRNA. The evidence indicates that DNA transfer and integration of DNA sequences can take place between organelle genomes. The homology between the mitochondrial contained sequence and the chloroplast sequence is 90%, suggesting that nucleotide sequence divergence has occurred and that the transfer event may have occurred long ago.

More significantly, the result implies that the evolution of the mitochondrial genome is not independent of other DNA sequences in the cell. One obvious implication would be that the large and variable mitochondrial genomes might contain nuclear DNA sequences.

D. PLASMIDLIKE DNAs

The most distinctive feature of the mtDNA of the *cms*-S cytoplasm of maize is the presence of two plasmidlike DNAs (Pring *et al.*, 1977). These molecules, designated S-1 and S-2, are found in a linear conformation and are estimated to be about 6400 and 5400 bp long, respectively. The molecules are inherited maternally with the cms-S cytoplasm and are not found in the chloroplast or the nucleus. These DNAs are characteristic of the S-type cytoplasms of maize (Levings and Pring, 1979a; Laughnan et al., 1981).

The structures of S-1 and S-2 have been studied by hybridization and heteroduplex formation and by restriction analysis. Both molecules have terminal inverted repeats that appear to be homologous in both molecules (Levings and Pring, 1979a). In addition, a region of 1428 bp adjacent to one of the terminal repeats is homologous in both S-1 and S-2 (Levings et al., 1982). Restriction maps of S-2 and S-1 are linear (Koncz et al., 1981; Kim et al., 1982). S-1 and S-2 cannot be derived by random breakage of a circular form. However, it is not known if the molecules are self-duplicating or if they replicate in circular or linear forms. The two plasmidlike molecules are present in roughly equal amounts and amplified about fivefold compared to other mitochondrial sequences, such as ribosomal genes. The complete nucleotide sequence of the S-2 DNA molecule has been determined (Levings and Sederoff, 1983). The molecule is 5452 bp and is terminated by exact 208-bp inverted repeats. Two large open reading frames extend from the ends on opposite strands. These open reading frames are 3294 nucleotides and 1017 nucleotides long and cover 79% of the molecule.

The structure of the ends of the plasmidlike DNAs has been investigated by end labeling and nuclease susceptibility (Kemble and Thompson, 1982). S-1, S-2, and the related short linear DNAs found in other maize cytoplasms have a 5′ terminally associated protein. Extensive deproteinization does not make the 5′ ends accessible to polynucleotide kinase, presumably because the termini are linked covalently to amino acids. The 3′ end, however, is accessible to nuclease digestion and terminal transferase. The DNA–protein complex may be involved in priming replication of these linear DNAs, similar to that found in adenovirus (Challberg and Englund, 1980) or in the bacillus phage φ29 (Yoshikawa and Ito, 1981). Alignment of the terminal nucleotides of S-2 DNA and those of five bacillus phages indicates a high degree of homology (Levings and Sederoff, 1983). For example, φ29 contains a 6-bp terminal inverted repeat sequence (AAAGTA) that is found in the inverted repeat of S-2 beginning at the second nucleotide from each 5′ end (Levings and Sederoff, 1983; Yoshikawa and Ito, 1981).

Plasmidlike DNAs have been found in some other races of maize and in some related species. In a survey of 93 races of Latin American maize (Weissinger et al., 1982, 1983), plasmidlike elements in 18 races were found. Two elements, termed R-1 and R-2, were found in 17 cases. R-1 and R-2 are abundant, linear, double-stranded DNAs with terminal inverted repeats. The larger molecule, R-1, shares considerable homology with S-1 but is about 1000 nucleotides longer and has a sequence of about 2576, which is unique to R-1. S-2 and R-2 are thought to be homologous (Levings et al., 1982). One of the races, 'Conico Norteño,' contained S-1 and S-2 molecules. Eight races produced mtDNA patterns similar to the well-known male sterile cytoplasms C, T, or S.

One species of teosinte, a diploid perennial (Zea diploperennis), also contains plasmidlike DNAs, called D-1 and D-2, which resemble R-1 and R-2 (Timothy et al., 1982). None of the other species or subspecies of teosinte examined thus far (31 accessions) showed plasmidlike DNAs by agarose gel electrophoresis.

Molecules similar to S-1 and S-2 have been found associated with the mtDNA of sorghum (Pring et al., 1982). They are linear duplex molecules, approximately 5700 and 5300 bp, and are associated with a male sterile line (IS1112C cms). All of the maize races containing R-1 and R-2 and the teosintes containing D-1 and D-2 are male fertile. Restriction mapping and heteroduplex analysis of plasmidlike DNAs have suggested a possible relationship of R-1, R-2, S-1, and S-2. A recombination event between R-1 and R-2 can generate a molecule like S-1. Although this proposal is unproven, all of the data are consistent with this explanation (Levings et al., 1982).

In addition to plasmidlike DNAs, mtDNA of maize contains abundant smaller molecules called minicircles and minilinears. In normal and male sterile cytoplasms, a small 1.9-kb circular DNA has been found (minicircle) (Levings et al., 1979; Kemble and Bedbrook, 1980). Small linear molecules (minilinears) of 2.3 kb are found in N, cms-C, and cms-S (Kemble and Bedbrook, 1980; Koncz et al., 1981). In cms-T a 2.1-kb linear molecule is found that is related to the 2.3-kb molecules of N, S, and C cytoplasms, but which presumably has a deletion of 200 bp (Kemble and Thompson, 1982). It is interesting that the predominant circular molecules found in maize tissue culture (Dale, 1982) were 1.5 and 1.8 kb, similar to the sizes of minicircles. Small supercoiled DNA molecules of 1.3, 1.4, 1.45, and 1.5 kb have been found in mtDNA of sugar beet (Powling, 1981). Male sterile lines of sugar beet contained only the 1.5-kb molecule. Additional DNA molecules ranging from 2 to 10 kb were noted as faint bands, but did not correlate with the presence or absence of male sterility.

More direct evidence for rearrangements in the maize mitochondrial genome comes from studies of the S-type of male sterile cytoplasm and the molecular events that occur in reversion to male sterility. Most male sterile cytoplasms are stable, but a specific strain of cms-S, cms-

Vg (M825/Oh07), has a high frequency of reversion (Laughnan and Gabay, 1975a,b). Revertants can be cytoplasmic events that result in male fertile cytoplasms or nuclear mutations that act as new restorer genes. Based on the analysis of the spontaneous revertants to fertility, Laughnan and Gabay (1978) proposed that a male fertility element might move from the cytoplasm into the nucleus. It has recently been shown that plasmidlike DNAs are found in digests of nuclear DNA (Kemble *et al.*, 1983) but no relationship with reversion has been established.

In cytoplasmic revertants, the plasmidlike molecules are usually undetectable as free DNA elements (Levings *et al.*, 1980). Restriction enzyme digests of mtDNA from cytoplasmic revertants showed one or more changes in the visible bands. Southern transfers demonstrated that the altered bands contained sequences that hybridize with S-1 or S-2 DNAs. Independent revertants showed different patterns (Levings *et al.*, 1980). It was proposed that S-1 and S-2 DNAs have inserted into the larger mtDNA as insertion elements, causing both the modification of the restriction pattern and the loss of male sterility (Levings *et al.*, 1980).

If a restriction digest of normal mtDNA is hybridized with sequences from the S-1 and S-2 plasmidlike DNAs, several bands show homology (Thompson *et al.*, 1980; Spruill *et al.*, 1980). Two sites contain most of the hybridizable sequences; one contains S-1-homologous sequences and the other is homologous to S-2 (Lonsdale *et al.*, 1981). Restriction digests and heteroduplex mapping of cloned fragments containing these sites indicate that the sequences are not completely homologous to S-1 and S-2. Both inserted sequences appear to be missing one of the two inverted repeats (Levings *et al.*, 1982). In addition, the region homologous to S-1 shows more homology with the R-1 plasmidlike DNA than with S-1 (Levings *et al.*, 1982). Heterogeneity of inserted sequences has been reported; two different clones containing S-2 sequences were found in restriction-mapping studies by Koncz *et al.* (1981).

Clearly, the sequences found in the plasmidlike DNAs can exist, integrated into normal mtDNA or as free elements. The unintegrated molecules vary in related strains and show structural and functional features resembling transposable elements in prokaryotes and eukaryotes. Studies on the mtDNA sequence relationships of maize and the teosintes suggested that rearrangements were common events in evolution (Sederoff *et al.*, 1981). It is possible that these rearrangements are related to the presence of inserted transposable elements or to that of portions of the plasmidlike DNAs.

E. Variation of Ribosomal Gene Organization

The ribosomal genes of plant mitochondria are different in composition and organization from all other mitochondrial ribosomal genes. A separate 5 S rRNA gene is closely linked to the 18 S small ribosomal subunit gene (Gray *et al.*, 1982; Stern *et al.*, 1982). The 26 S large subunit gene is unlinked in restriction fragment tests (Bonen and Gray, 1980). In maize, the region between the 5 S gene and 18 S gene has been sequenced and the orientation of the genes has been determined (Chao *et al.*, 1983). Restriction mapping of the ribosomal gene region shows that the 26 S gene is separated from the 18 S gene by 16 kb (Stern *et al.*, 1982; Iams and Sinclair, 1982).

The novel organization of these genes is generally conserved, at the level of restriction analysis, in both monocotyledons and dicotyledons. In addition to maize and wheat, broad bean, cucumber, and pea have been examined (Gray *et al.*, 1982). In maize, there is no evidence of heterogeneity of the fragments containing the ribosomal genes (Stern and Lonsdale, 1982). In wheat, three *Sal*I fragments have been identified by hybridization for the 18 and 5 S genes. Two cloned *Sal*I restriction fragments have both the 18 and 5 S genes but differ in the flanking sequences (Gray *et al.*, 1982). Whether the extent of rearrangement and gene duplication may be greater in wheat than in other plant mtDNA remains to be determined.

F. Possible Rearrangements in Cell and Tissue Culture

Protoplast fusion of two varieties of *Nicotiana tabacum* results in hybrid cells and regenerated plant hybrids with altered mtDNA restriction patterns (Belliard *et al.*, 1979). These results also suggest that recombination occurs in the plant mitochondrial genome. More recently, hybrids formed by protoplast fusion between *N. tabacum* and *Nicotiana knightiana* also generated restriction patterns of mtDNA that differed from either parent (Nagy *et al.*, 1981). Using labeled cloned probes to detect specific bands, evidence for rearrangement and recombination was obtained. Novel restriction patterns were explained by structural alteration of parental mtDNA molecules (Nagy *et al.*, 1981).

In maize, plants regenerated from tissue culture in the absence of hybrid cell fusion show alterations in their mtDNA restriction pattern. Cultures of maize tissue derived from plants containing the T-type male sterile cytoplasm (*cms*-T) were exposed to toxin of *Bipolaris maydis* race T, in order to select for lines resistant to the toxin (Gengenbach *et al.,* 1977). Most of the plants recovered were resistant to the toxin. All of the resistant plants were also male fertile. mtDNAs recovered from these resistant plants showed variation in restriction patterns when compared to the *cms*-T line from which they were derived. The restriction patterns suggest the possibility of inversion or rearrangement in the mitochondrial genome (Pring *et al.,* 1981; Gengenbach *et al.,* 1981).

X. Conclusions and Directions of Future Research

A. GENETIC ENGINEERING AND mtDNA

mtDNA is potentially useful for genetic engineering in two ways. First, mtDNA may be used in the construction of vectors for gene transfer (Levings and Pring, 1979b; Atchison *et al.,* 1980; Stahl *et al.,* 1982; Howell, 1982). In addition, it may be possible to insert genes into mitochondria in order to engineer large numbers of specific gene copies within cells. Proteins made in mitochondria would be stored in discrete compartments, which might be of advantage for purposes of purification or for the need to isolate a product from other cellular components. In higher plants, at least certain types of male sterility are caused by the mitochondrial genome. The ability to introduce male sterility in specific lines would be of significant value in crops with small bisexual flowers in a single large inflorescence in or other cases where controlled crosses are difficult (Levings and Pring, 1979b; Leaver and Gray, 1982).

Much information is needed before it will be possible to determine the value or feasibility of mitochondrial systems for genetic engineering. Efforts to introduce foreign genes into mitochondria must overcome several obstacles. A gene must have the appropriate mitochondrial code and carry the correct signals for control and processing. The technology to modify control sequences or to correct for differences in the code is essentially available now. Mitochondria do translate proteins of different hydrophobicity and charge, but we do not know to what extent specific foreign proteins might interfere with normal mitochondrial functions and assembly. However, the mechanisms and constraints of the processes of regulation of gene expression and mitochondrial assembly are not yet well understood in any mitochondrial system, although the recent advances made in animal cells and fungi are impressive in this direction.

The technical capability to manipulate and insert mitochondrial genes would obviously be a major advance for basic research because specific mutations could be created and tested for functional effects. Among the major types of information needed to assess the prospects of engineering the mitochondrial genome, it is necessary to understand the extent and kind of genetic variation that takes place in mitochondrial genomes. The degree of structural variation observed in mtDNA is different in mitochondrial systems of animals, yeast, plants, and protozoa. We do not yet understand why this is so. The answer may tell us whether engineering of animal mitochondrial genomes is possible. In contrast, plant mitochondrial genomes may be more tolerant of additional DNA.

The use of mtDNA sequences in vectors requires that they be non-pathogenic, capable of stable insertion into recipient cells, and capable of carrying genetic determinants expressed in recipient cells (Levings and Pring, 1979b). Although no such system is yet available, several advances have been made in the use of mtDNAs and in the transfer of these DNAs into recipient cells.

One possible strategy for animal mitochondrial genomes would follow that used for the exploitation of small bacterial viruses as cloning vectors. Some small viruses resemble animal mtDNA in the organization and economy of DNA-sequence utilization (Denhardt et al., 1978). Bacterial viruses, such as M13, have been exploited as vectors by engineering cloning sites into the region of the replication origin (Messing, 1982). Similarly, sequences could potentially be cloned into mtDNA, transferred, and expressed in animal cells. Genetic markers for chloramphenicol resistance have been studied in animal mtDNA and could provide appropriate selective markers for the development of transformation systems (Wallace et al., 1982; Kearsey and Craig, 1982).

Mitochondrial sequences as diverse as the *Xenopus* mtDNA replication origin (Zakian, 1981) and yeast mtDNA *petites* (Hyman et al., 1982) have been transformed into yeast. These molecules can be maintained stably in yeast as high-copy-number extrachromosomal elements. These sequences have the properties of autonomously replicating sequences (*ars*) and are able to transform yeast at high efficiency. These properties are important for the construction of stable vector systems. In animal mtDNA, the origins of replication are well defined. In other systems, such as in yeasts, several replication origins have been selected from mtDNA by high-frequency transformation (Stinchcomb et al., 1979; Blanc and Dujon, 1982). Blanc and Dujon

(1982) have used high-frequency transformation and stability of re-combinant plasmids as a means to isolate three different *rep* sequences from yeast mtDNA that are autonomously replicating sequences that may function as replication origins (*rep*) in normal mtDNA. The *rep* sequences are autonomously replicating outside the mitochondria, and have not integrated into chromosomal DNA. The expression of the *URA3* gene, which is part of the plasmid containing one of the *rep* regions, indicates that sequences in these plasmids are transcribed and translated.

Some workers have successfully introduced mtDNA sequences into cells and obtained expression or effects of mitochondrial sequences. Atchison *et al.* (1980) ligated mtDNA from an oligomycin-resistant *petite* together with the yeast 2-μm plasmid DNA. The recombinant molecule was able to transform an oligomycin-sensitive strain of *Saccharomyces cerevisae* at a frequency 50-fold above the spontaneous mutation rate. Two classes of resistant transformants were recovered. The major class did not segregate oligomycin sensitivity, suggesting stable integration of the resistant marker into the mitochondrial genome. In crosses with oligomycin-sensitive strains, the transformants behaved as expected if they carried a mitochondrial gene for oligomycin resistance. A second class, about 20%, segregated oligomycin sensitivity at a low frequency, suggesting unintegrated copies.

A similar mtDNA-hybrid plasmid has been constructed from *Podospora anserina* mtDNA and pBR322 (Stahl *et al.*, 1982). The cloned fragment of 2.4 kb was isolated from senescent cultures in which the sequence is amplified. Protoplasts were infected with the hybrid plasmid and regenerated clones were screened for expression of the senescent phenotype in the wild strain and in a nonsenescing double mutant. The hybrid plasmid is able to transfer senescence to juvenile cultures, providing evidence for the direct role of these plasmidlike DNAs in the expression of senescence. It is not known if the expressed DNA functions in the cytoplasm or in the mitochondria. It has been established that these plasmidlike DNAs are derived from the mito-chondrial chromosome (Kück *et al.*, 1981; Esser *et al.*, 1980; Wright *et al.*, 1982; Belcour *et al.*, 1981).

The linear plasmidlike DNAs of maize mitochondria have been considered as potential vectors to engineer plant genes because they have terminal inverted repeats resembling transposable elements (Levings and Pring, 1979b; Kim *et al.*, 1982) and because they can integrate at new sites in maize mtDNA (Levings *et al.*, 1980). The possibility that these molecules may enter into the nucleus and integrate into chromosomal DNA (Laughnan *et al.*, 1981) has prompted further interest in these molecules as potential vectors.

Another approach to the transfer of mitochondrial genes has been through direct transfer of isolated mitochondria or by protoplast fusion. Microinjection of mitochondria has been carried out in *Paramecium aurelia* using erythromycin-resistant mutants to transfer mitochondria to new nuclear backgrounds. *P. aurelia* is a group of 14 sibling species. Some combinations of sibling species show incompatibility between nuclear and mitochondrial genomes (Beale and Knowles, 1976).

In higher plants, protoplast fusion has been used to create somatic hybrids of *Nicotiana* (Belliard *et al.*, 1979; Nagy *et al.*, 1981; Aviv *et al.*, 1980; Aviv and Galun, 1980). Such fusions have been used both to transfer male sterility and to restore fertility to male sterile cytoplasm (Aviv and Galun, 1980). Similarly, yeast mitochondria have been isolated and fused with protoplasts prepared from a cytoplasmic *neutral petite* (Gunge and Sakaguchi, 1979). The resulting fusion products regenerated at low frequencies into respiration-proficient cells. Use of nuclear and mitochondrial genetic markers confirmed the nuclear and mitochondrial genotype.

B. Conclusions

Analysis and comparison of diverse mitochondrial genomes provide insights into common mechanisms and raise many questions. For example, why is the organization of the animal mitochondrial genome relatively stable in evolution? The economy of animal mtDNA structure would appear to be the result of continuous intense selection in animal cells. The requirement for extreme economy in this system could exclude most of the opportunities for structural variation that depend on nonessential or repeated DNA. Structural changes, such as inversion or transposition, would almost always be lethal in an animal mitochondrial genome. The structure appears to be "frozen" due to strong selection and functional constraints.

In a wide variety of genomes, replication mechanisms and regions that can serve as replication origins appear to be highly variable. In animal mtDNA, the region of the replication origin for the H strand is the most variable part of the genome. In yeast and fungi, several replication origins must exist to explain the diversity of self-replicating molecules and to account for the number of autonomously replicating sequences (*ars*) assayed by stable transformation of yeast plasmids.

The apparent transition between linear and circular mitochondrial genomes in *Hansenula* suggests an unexpected flexibility in replication mechanisms between linear and circular molecules.

The variation in gene size due to optional introns within a species or between closely related species raises the question of how the polymorphism of optional introns is established. The functions of maturase genes are essential for gene expression only in certain strains of yeast. Has maturase function recently become dispensable, or have forms of the specific genes lacking introns been maintained for a long time? In the case of the small rRNA subunit, no introns have been identified, yet gene sizes vary over more than a threefold range in different mitochondrial systems. The size of a kinetoplast small subunit rRNA is only about 630 bp, compared to that of maize mtDNA, which is about 1960 bp. How these greatly different molecules are able to carry out the complex process of translation is most intriguing. It would be interesting to learn about the number and properties of ribosomal proteins in these greatly different systems.

For the higher plants, the basic organization of the mitochondrial genome is not yet clear; cloning studies may soon resolve whether the genome is circular or linear, and the number of functional chromosomes. The extraordinary variation in genome size cannot be explained by simple redundancy, gene amplification, or activity of a transposable element, because the mitochondrial genomes are low in repeated DNA. The recent evidence of DNA transfer into mitochondria may be the explanation of the large and highly variable genomes.

Several independent laboratories have provided evidence for transfer of DNA between organelles or between the mitochondria and the nucleus. In addition to the maize transfer of chloroplast DNA to the mitochondria (Stern and Lonsdale, 1982) and the isolation of mitochondrial sequences in the yeast nucleus (Farrelly and Butow, 1983), evidence for DNA transfer exists for mtDNA of the purple sea urchin (*Strongylocentrotus purpuratus*). Sections of mtDNA have been found in the nuclear genome of the sea urchin, due to a proposed transposition event about 25×10^6 years ago (Jacobs *et al.*, 1982; Lewin, 1983). Sequences complementary to plasmidlike DNA of maize mitochondria have been found in the nucleus (Kemble *et al.*, 1983). In the locust, mtDNA sequences have been found in two fragments of nuclear DNA that contain sequences homologous to the mitochondrial rRNA genes (Gellissen *et al.*, 1983). Similarly, sequences for a large part of the chloroplast genome have been found in the spinach genome (Timmis and Scott, 1983).

The evolutionary status of the higher plant mitochondrial genome remains to be established with respect to prokaryotes and other mitochondria. The close sequence similarity of bacterial rRNA and the mtDNA rRNA sequences of maize and wheat may be due to a slower rate of evolution of mtDNA of plants, compared to mtDNA of animals and fungi. Sufficient data are available to suggest that DNA sequences within a mitochondrial genome or in mtDNA of widely different species can evolve at different rates. Because the rate of evolution and time since divergence are confounding variables, the mitochondrial genomes of more species of plants that are intermediate or primitive species in the evolutionary sense need to be examined.

A major result of these structural comparisons of diverse types of mtDNAs is the parallels between the supernumerary circular molecules of mitochondria of fungi and the populations of circular molecules found in strains and species of higher plant mtDNAs. Small circular molecules recovered in high abundance from cultured plant cells could be amplified due to the stress of culture conditions on plant cells. These molecules may be analogous to those found in *petites* or *ragged* mutants in that they can be derived from a larger mitochondrial molecule and become amplified. It is reasonable to suppose that these molecules could be heterogeneous, strain specific, and sometimes present in an oligomeric series.

According to this view, the heterogeneous classes of circular molecules found in maize and soybean would represent mixed populations. Each class could be part of an oligomeric series derived from different events. The distributions found for heterogeneous populations of mtDNA molecules may then be analogous to those observed when more than one type of mtDNA *petite* molecule occurs in a *petite* yeast cell, resulting in mixtures of different oligomeric series. Alternatively, specific autonomous circular molecules that are stable components of the genome could be the basis for molecular heterogeneity.

At present, we lack answers to many fundamental questions about the mitochondrial genome. Mitochondrial genomes have taken extraordinarily divergent paths, such as those in trypanosomes, fungi, and higher plants and animals. The reasons for this divergence are not understood. Furthermore, as Borst (1972, 1981) has argued, it is not clear why we need separate mitochondrial genomes at all. Most proteins of present-day mitochondria are nuclear coded, and both hydrophilic and hydrophobic protein subunits can be imported into mitochondria (Tzagoloff *et al.*, 1979). However, if all genes could be readily transferred to the nucleus, there would seem to have been enough time for this transfer to have taken place. Perhaps mitochondrial evolution is constrained by mechanisms that maintain compartmentation, such as the mitochondrial genetic code, regulatory requirements for gene expression, or transport control. The maintenance of a separate mitochondrial genome might be another frozen accident.

ACKNOWLEDGMENTS

I wish to thank many colleagues who provided reprints, preprints, and advice for this article. Special thanks are due to Vivian Smith for typing the manuscript.

REFERENCES

Alexander, N. J., Perlman, P. S., Hanson, D. K., and Mahler, H. R. (1980). Mosaic organization of a mitochondrial gene: Evidence from double mutants in the cytochrome b region of *Saccharomyces cerevisiae. Cell* **20,** 199–206.

Altman, S. (1978). Transfer RNA biosynthesis. *In* "Biochemistry of Nucleic Acids II 17" (B. F. Clark, ed.), pp. 19–44. Univ. Park Press, Baltimore, Maryland.

Anderson, S., Bankier, A. T., Barrell, B. G., deBruijn, M. H. L., Coulson, A. R., Drouin, J., Eperon, I. C., Nierlich, D. P., Roe, B. A., Sanger, F., Schreier, P. H., Smith, A. J. H., Staden, R., and Young, I. G. (1981). Sequence and organization of the human mitochondrial genome. *Nature (London)* **290,** 457–464.

Anderson, S., deBruijn, M. H. L., Coulson, A. R., Eperon, I. C., Sanger, F., and Young, I. G. (1982a). Complete sequence of bovine mitochondrial DNA: Conserved features of the mammalian mitochondrial genome. *J. Mol. Biol.* **157,** 683–717.

Anderson, S., Bankier, A. T., Barrell, B. G., deBruijn, M. H. L., Coulson, A. R., Drouin, J., Eperon, I. C., Nierlich, D. P., Roe, B. A., Sanger, F., Schreier, P. H., Smith, A. J. H., Staden, R., and Young, I. G. (1982b). Comparison of the human and bovine mitochondrial genomes. *In* "Mitochondrial Genes" (P. Slonimski, P. Borst, and G. Attardi, eds.), pp. 5–43. Cold Spring Harbor Laboratory, Cold Spring Harbor, New York.

Anziano, P. Q., Hanson, D. K., Mahler, H. R., and Perlman, P. S. (1982). Functional domains in introns: Trans-acting and cis-acting regions of intron 4 of the *cob* gene. *Cell* **30,** 925–932.

Arnberg, A., Van Bruggen, E. F. J., ter Schegget, J., and Borst, P. (1971). The presence of DNA molecules with a displacement loop in standard mitochondrial DNA preparations. *Biochim. Biophys. Acta* **246,** 353–357.

Atchison, B. A., Choo, K. B., Devinish, R. J., Linnane, A. W., and Nagley, P. (1979). Biogenesis of mitochondria 53: Physical map of genetic loci in the 21S ribosomal RNA region of mitochondrial DNA in *Saccharomyces cerevisiae. Mol. Gen. Genet.* **174,** 307–316.

Atchison, B. A., Devenish, R. J., Linnane, A. W., and Nagley, P. (1980). Transformation of *Saccharomyces cerevisiae* with yeast mitochondrial DNA linked to two micron circular yeast plasmid. *Biochem. Biophys. Res. Commun.* **96,** 580–586.

Attardi, G., Cantatore, P., Ching, E., Crews, S., Gelfund, R., Merkel, C., Montoya, J., and Ojala, D. (1980). The remarkable features of gene organization and expression of human mitochondrial DNA. *In* "The Organization and Expression of the Mitochondrial Genome" (A. M. Kroon and C. Saccone, eds.), pp. 103–119. Elsevier, Amsterdam.

Attardi, G., Cantatore, P., Chomyn, A., Crews, S., Gelfand, R., Merkel, C., Montoya, J., and Ojala, D. (1982). A comprehensive view of mitochondrial gene expression in human cells. *In* "Mitochondrial Genes" (P. Slonimski, P. Borst, and G. Attardi, eds.), pp. 51–71. Cold Spring Harbor Laboratory, Cold Spring Harbor, New York.

Avise, J. C., Lansman, R. A., and Shade, R. O. (1979a). The use of restriction endonucleases to measure mitochondrial DNA sequence relatedness in natural populations. I. Population structure and evolution in the genus *Peromyscus. Genetics* **92,** 279–295.

Avise, J. C., Giblin-Davidson, C., Laerm, J., Patton, J. C., and Lansman, R. A. (1979b). Mitochondrial DNA clones and matriarchal phylogeny within and among geographic populations of the pocket gopher, *Geomys pinetis. Proc. Natl. Acad. Sci. U.S.A.* **76,** 6694–6698.

Aviv, D., and Galun, E. (1980). Restoration of fertility in cytoplasmic male sterile (CMS) *Nicotiana sylvestris* by fusion with X-irradiated *N. tabacum* protoplasts. *Theor. Appl. Genet.* **58,** 121–127.

Aviv, D., Fluhr, R., Edelman, M., and Galun, E. (1980). Progeny analysis of the interspecific somatic hybrids: *Nicotiana tabacum* (CMS) + *Nicotiana sylvestris* with respect to nuclear and chloroplast markers. *Theor. Appl. Genet.* **56,** 145–150.

Barker, D. C. (1980). The ultrastructure of kinetoplast DNA with particular reference to the interpretation of dark field electron microscopy images of isolated, purified networks. *Micron* **11,** 21–62.

Barrell, B. G., Anderson, S., Bankier, A. T., deBruijn, M. H. L., Chen, E., Coulson, A. R., Drouin, J., Eperon, I. C., Nierlich, D. P., Roe, B. A., Sanger, F., Schreier, P. H., Smith, A. J. H., Staden, R., and Young, I. G. (1980). Different pattern of codon recognition by mammalian mitochondrial tRNAs. *Proc. Natl. Acad. Sci. U.S.A.* **77,** 3164–3166.

Barrois, M., Riou, G., and Galibert, F. (1981). Complete nucleotide sequence of minicircle kinetoplast DNA from *Trypanosoma equiperdum. Proc. Natl. Acad. Sci. U.S.A.* **78,** 3323–3327.

Bayen, M., and Rode, A. (1973). The 1.700 DNA of *Chlorella pyrenoidosa:* Heterogeneity and complexity. *Plant Sci. Lett.* **1,** 385.

Beale, G. H., and Knowles, J. K. C. (1976). Interspecies transfer of mitochondria in *Paramecium aurelia. Mol. Gen. Genet.* **143,** 197–201.

Bedbrook, J. R., and Bogorad, L. (1976). Endonuclease recognition sites mapped on *Zea mays* chloroplast DNA. *Proc. Natl. Acad. Sci. U.S.A.* **73,** 4309–4313.

Belcour, L., and Begel, O. (1978). Lethal mitochondrial genotypes in *Podospora anserina:* A model for senescence. *Mol. Gen. Genet.* **163,** 113–123.

Belcour, L., Begel, O., Mossé, M.-O., and Vierny, C. (1981). Mitochondrial DNA amplification in senescent cultures of *Podospora anserina*: Variability between the retained, amplified sequences. *Curr. Genet.* **3,** 13–21.

Belcour, L., Begel, O., Keller, A.-M., and Vierny, C. (1982). Does senescence in *Podospora anserina* result from instability of the mitochondrial genome? *In* "Mitochondrial Genes" (P. Slonimski, P. Borst, and G. Attardi, eds.), pp. 415–421. Cold Spring Harbor Laboratory, Cold Spring Harbor, New York.

Belliard, G., Vedel, F., and Pelletier, G. (1979). Mitochondrial recombination in cytoplasmic hybrids of *Nicotiana tabacum* by protoplast fusion. *Nature (London)* **281,** 401–403.

Bendich, A. J. (1982). Plant mitochondrial DNA: The last frontier. *In* "Mitochondrial Genes" (P. Slonimski, P. Borst, and G. Attardi, eds.), pp. 477–481. Cold Spring Harbor Laboratory, Cold Spring Harbor, New York.

Bennett, M. D., and Smith, J. B. (1976). Nuclear DNA amounts in angiosperms. *Philos. Trans. R. Soc. London* **274**, 227–274.

Bernardi, G. (1982). Evolutionary origin and the biological function of noncoding sequences in the mitochondrial genome of yeast. *In* "Mitochondrial Genes" (P. Slonimski, P. Borst, and G. Attardi, eds.), pp. 269–278. Cold Spring Harbor Laboratory, Cold Spring Harbor, New York.

Bernardi, G., Prunell, A., Fonty, G., Kopeck, H., and Strauss, F. (1976). The mitochondrial genome of yeast: Organization, evolution and the petite mutation. *In* "The Genetic Function of Mitochondrial DNA" (C. Saccone and A. M. Kroon, eds.), pp. 185–198. Elsevier, Amsterdam.

Bernardi, G., Baldacci, G., Bernardi, G., Faugeron-Fonty, G., Gaillard, C., Goursot, R., Huyard, A., Mangin, M., Marotta, R., and deZamaroczy, M. (1980). The petite mutation: Excision sequences, replication origins and suppressivity. *In* "The Organization and Expression of the Mitochondrial Genome" (A. M. Kroon and C. Saccone, eds.), pp. 21–36. Elsevier, Amsterdam.

Bertrand, H., Szakacs, N. A., Nargang, F. E., Zagozeski, C. A., Collins, R. A., and Harrigan, J. C. (1976). The function of mitochondrial genes in *Neurospora crassa*. *Can. J. Genet. Cytol.* **18**, 397–409.

Bertrand, H., Collins, R. A., Stohl, L. L., Goewert, R. R., and Lambowitz, A. M. (1980). Deletion mutants of *Neurospora crassa* mitochondrial DNA and their relationship to the "stop-start" growth phenotype. *Proc. Natl. Acad. Sci. U.S.A.* **77**, 6032–6036.

Bertrand, H., Bridge, P., Collins, R. A., Garriga, G., and Lambowitz, A. M. (1982). RNA splicing in *Neurospora* mitochondria. Characterization of new nuclear mutants with defects in splicing the mitochondrial large mRNA. *Cell* **29**, 517–526.

Bibb, M. J., Van Etten, R. A., Wright, C. T., Walberg, M. W., and Clayton, D. A. (1981). Sequence and gene organization of mouse mitochondrial DNA. *Cell* **26**, 167–180.

Blanc, H., and Dujon, B. (1982). Replicator regions of the yeast mitochondrial DNA active *in vivo* and in yeast transformants. *In* "Mitochondrial Genes" (P. Slonimski, P. Borst, and G. Attardi, eds.), pp. 279–294. Cold Spring Harbor Laboratory, Cold Spring Harbor, New York.

Bogenhagen, D., and Clayton, D. A. (1978a). Mechanisms of mitochondrial DNA replication in mouse L-cells; Kinetics of synthesis and turnover of the initiation sequence. *J. Mol. Biol.* **11**, 49–68.

Bogenhagen, D., and Clayton, D. A. (1978b). Mechanism of mitochondrial DNA replication in mouse L-cells: Introduction of superhelical turns into newly replicated molecules. *J. Mol. Biol.* **119**, 69–81.

Bohnert, H. J. (1973). Circular mitochondrial DNA from *Acanthamoeba castellanii* (Neff-strain). *Biochim. Biophys. Acta* **324**, 199–205.

Bohnert, H. J. (1977). Size and structure of mitochondrial DNA from *Physarum polycephalum*. *Exp. Cell Res.* **106**, 426–430.

Bonen, L., and Gray, M. W. (1980). Organization and expression of the mitochondrial genome of plants. I. The genes for wheat mitochondrial ribosomal and transfer RNA: Evidence for an unusual arrangement. *Nucleic Acids Res.* **8**, 319–335.

Bonitz, S. G., Coruzzi, G., Thalenfeld, B. E., and Tzagoloff, A. (1980a). Assembly of the mitochondrial membrane system: Physical map of the Oxi3 locus of yeast mitochondrial DNA. *J. Biol. Chem.* **255**, 11922–11926.

Bonitz, S. G., Coruzzi, G., Thalenfeld, B. E., and Tzagoloff, A. (1980b). Assembly of the mitochondrial membrane system: Structure and nucleotide sequence of the gene coding for subunit 1 of yeast cytochrome oxidase. *J. Biol. Chem.* **255,** 11927–11941.

Bonitz, S. G., Berlani, R., Coruzzi, G., Li, M., Macino, G., Nobrega, F. G., Nobrega, M. P., Thalenfeld, B. E., and Tzagoloff, A. (1980c). Codon recognition rules in yeast mitochondria. *Proc. Natl. Acad. Sci. U.S.A.* **77,** 3167–3170.

Borck, K. S., and Walbot, V. (1982). Comparison of the restriction endonuclease digestion patterns of mitochondrial DNA from normal and male sterile cytoplasms of *Zea Mays* L. *Genetics* **102,** 109–128.

Borst, P. (1972). Mitochondrial nucleic acids. *Annu. Rev. Biochem.* **41,** 333–376.

Borst, P. (1977). Structure and function of mitochondrial DNA. *In* "International Cell Biology. 1976–1977" (B. R. Brinkley and K. R. Porter, eds.), pp. 237–244. Rockefeller Univ. Press, New York.

Borst, P. (1981). Control of mitochondrial biosynthesis. *In* "Cellular Controls in Differentiation" (C. W. Lloyd and D. A. Rees, eds.), pp. 231–254. Academic Press, New York.

Borst, P., and Fase-Fowler, F. (1979). The Maxi-circle of *Trypanosoma brucei* kinetoplast DNA. *Biochim. Biophys. Acta* **565,** 1–12.

Borst, P., and Flavell, R. A. (1976). Properties of mitochondrial DNAs. *Handb. Biochem. Mol. Biol. 3rd Ed. Nucleic Acid* **2,** 363–374.

Borst, P., and Grivell, L. A. (1978). The mitochondrial genome of yeast. *Cell* **15,** 705–723.

Borst, P., and Grivell, L. A. (1981). One gene's intron is another gene's exon. *Nature (London)* **289,** 439–440.

Borst, P., and Hoeijmakers, J. H. J. (1979a). Kinetoplast DNA. *Plasmid* **2,** 20–40.

Borst, P., and Hoeijmakers, J. H. J. (1979b). Structure and function of kinetoplast DNA of the African trypanosomes. *In* "Extrachromosomal DNA: ICN-UCLA Symposia on Molecular and Cellular Biology" (D. J. Cummings, P. Borst, I. B. Dawid, S. M. Weisman, and C. F. Fox, eds.), pp. 515–531. Academic Press, New York.

Borst, P., Fase-Fowler, F., Steinert, M., and Van Assel, S. (1977). Maxicircles in the kinetoplast DNA of *Trypanosoma mega. Exp. Cell Res.* **110,** 167–173.

Borst, P., Fase-Fowler, F., Hoeijmakers, J. H. J., and Frasch, A. C. C. (1980a). Variations in maxi-circle and mini-circle sequences in kinetoplast DNAs from different *Trypanosoma brucei* strains. *Biochim. Biophys. Acta* **610,** 197–210.

Borst, P., Hoeijmakers, J. H. J., Frasch, A. C. C., Snijders, A., Janssen, J. W. G., and Fase-Fowler, F. (1980b). The kinetoplast DNA of *Trypanosoma brucei*: Structure, evolution, transcription, mutants. *In* "The Organization and Expression of the Mitochondrial Genome" (A. M. Kroon and C. Saccone, eds.), pp. 7–19. Elsevier, Amsterdam.

Borst, P., Fase-Fowler, F., and Gibson, W. C. (1981a). Quantitation of genetic differences between *Trypanosoma brucei gambiense, rhodesience* and *brucei* by restriction enzyme analysis of kinetoplast DNA. *Mol. Biochem. Parasitol.* **3,** 117–131.

Borst, P., Hoeijmakers, J. H. J., and Hajduk, S. L. (1981b). Structure, function and evolution of kinetoplast DNA. *Parasitology* **82,** 81–93.

Bos, J. L., Heyting, C., Borst, P., Arnberg, A. C., and Van Bruggen, E. F. J. (1978). An insert in the single gene for the large ribosomal RNA in yeast mitochondrial DNA. *Nature (London)* **275,** 336–338.

Bos, J. L., Osinga, K. A., Van derHorst, G., Hecht, N. B., Tabak, H. F., Van Ommen, G.-J. B., and Borst, P. (1980). Splice point sequence and transcripts of the intervening sequence in the mitochondrial 21S ribosomal RNA gene of yeast. *Cell* **20,** 207–214.

Brack, C., and Delain, E. (1975). Electron microscopic mapping of AT rich regions and of *Escherichia coli* RNA polymerase binding sites of the circular kinetoplast DNA of *Trypanosoma cruzi*. *J. Cell Sci.* **17**, 287–306.

Brennicke, A., and Blanz, P. (1982). Circular mitochondrial DNA species from *Oenothera* with unique sequences. *Mol. Gen. Genet.* **187**, 461–466.

Brennicke, A., and Clayton, D. A. (1981). Nucleotide assignment of alkali-sensitive sites in mouse mitochondrial DNA. *J. Biol. Chem.* **256**, 10613–10617.

Brosius, J., Dull, T. J., and Noller, H. F. (1980). Complete nucleotide sequence of a 23S ribosomal RNA gene from *Escherichia coli*. *Proc. Natl. Acad. Sci. U.S.A.* **77**, 201–204.

Brown, G. G., and Simpson, M. V. (1981). Intra- and interspecific variation of the mitochondrial genome in *Rattus norvegicus* and *Rattus rattus*: Restriction enzyme analysis of variant mitochondrial DNA molecules and their evolutionary relationships. *Genetics* **97**, 125–143.

Brown, G. G., and Simpson, M. V. (1982). Novel features of animal mtDNA evolution as shown by sequences of two rat cytochrome oxidase subunit II genes. *Proc. Natl. Acad. Sci. U.S.A.* **79**, 3246–3250.

Brown, G. G., Castora, F. J., Frantz, S. C., and Simpson, M. V. (1981). Mitochondrial DNA polymorphism: Evolutionary studies on the genus *Rattus*. *In* "Origins and Evolution of Eukaryotic Intracellular Organelles." *Ann. N.Y. Acad. Sci.* **361**, 135–153.

Brown, W. M. (1981). Mechanisms of evolution in animal mitochondrial DNA. *In* "Origins and Evolution of Eukaryotic Intracellular Organelles." *Ann. N.Y. Acad. Sci.* **361**, 119–134.

Brown, W. M., and Wright, J. W. (1979). Mitochondrial DNA analyses and the origin and relative age of parthenogenetic lizards (genus *Cnemidophorus*). *Science* **203**, 1247–1249.

Brown, W. M., Shine, J., and Goodman, H. M. (1978). Human mitochondrial DNA: Analysis of 7S DNA from the origin of replication. *Proc. Natl. Acad. Sci. U.S.A.* **75**, 735–739.

Brown, W. M., George, M., Jr., and Wilson, A. C. (1979). Rapid evolution of animal mitochondrial DNA. *Proc. Natl. Acad. Sci. U.S.A.* **76**, 1967–1971.

Brown, W. M., Prager, E. M., Wang, A., and Wilson, A. C. (1982). Mitochondrial DNA sequences of primates: Tempo and mode of evolution. *J. Mol. Evol.* **18**, 225–239.

Browning, K. S., and RajBhandary, U. L. (1982). Cytochrome oxidase subunit III gene in *Neurospora crassa* mitochondria. *J. Biol. Chem.* **257**, 5253–5256.

Bultmann, H., and Borkowski, J. L. (1979). Contrasting denaturation maps of *Xenopus laevis* and *Xenopus borealis* mitochondrial DNAs. *Biochim. Biophys. Acta* **564**, 352–354.

Bultmann, H., Zakour, R. A., and Sosland, M. A. (1976). Evolution of *Drosophila* mitochondrial DNAs. Comparison of denaturation maps. *Biochim. Biophys. Acta* **454**, 21–44.

Burke, J. M., and RajBhandary, U. L. (1982). Intron within the large rRNA gene of *N. crassa* mitochondria: A long open reading frame and a consensus sequence possibly important in splicing. *Cell* **31**, 509–520.

Butow, R. A., and Strausberg, R. L. (1981). The mitochondrial genome. *In* "Mitochondria and Microsomes" (C. P. Lee, G. Schatz, and G. Dallner, eds.), pp. 67–91. Addison-Wesley, Reading, Massachusetts.

Butow, R. A., Farrelly, F., Zassenhaus, H. P., Hudspeth, M. E. S., Grossman, L. I., and Perlman., P. S. (1982). var 1 determinant region of yeast mitochondrial DNA. *In*

"Mitochondrial Genes" (P. Slonimski, P. Borst, and G. Attardi, eds.), pp. 241–253. Cold Spring Harbor Laboratory, Cold Spring Harbor, New York.

Cantatore, P., and Attardi, G. (1980). Mapping of nascent light and heavy strand transcripts on the physical map of HeLa cell mitochondrial DNA. *Nucleic Acids Res.* **8,** 2605–2624.

Caron, F., Jacq, C., and Rouvière-Yaniv, J. (1979). Characterization of a histone-like protein extracted from yeast mitochondria. *Proc. Natl. Acad. Sci. U.S.A.* **76,** 4265–4269.

Castora, F. J., Arnheim, N., and Simpson, M. V. (1980). Mitochondrial DNA polymorphism: Evidence that variants detected by restriction enzymes differ in nucleotide sequence rather than in methylation. *Proc. Natl. Acad. Sci. U.S.A.* **77,** 6415–6419.

Castora, F. J., Sternglanz, R., and Simpson, M. V. (1982). A new mitochondrial topoisomerase from rat liver that catenates DNA. *In* "Mitochondrial Genes" (P. Slonimski, P. Borst, and G. Attardi, eds.), pp. 143–154. Cold Spring Harbor Laboratory, Cold Spring Harbor, New York.

Caten, C. E. (1972). Vegetative incompatibility and cytoplasmic inheritance in fungi. *J. Gen. Microbiol.* **72,** 221–229.

Challberg, S. S., and Englund, P. T. (1980). Heterogeneity of minicircles in kinetoplast DNA of *Leishmania tarentolae*. *J. Mol. Biol.* **138,** 447–472.

Chao, S., Sederoff, R., and Levings, C. S., III. (1983). Partial sequence of the 5S-18S gene region of the maize mitochondrial genome. *Plant Physiol.* **71,** 190–193.

Chen, K. K., and Donelson, J. E. (1980). Sequences of two kinetoplast DNA minicircles of *Trypanosoma brucei*. *Proc. Natl. Acad. Sci. U.S.A.* **77,** 2445–2449.

Cheng, D., and Simpson, L. (1978). Isolation and characterization of kinetoplast DNA and RNA of *Phytomonas davidi*. *Plasmid* **1,** 207–315.

Christiansen, G., and Christiansen, C. (1976). Comparison of the fine structure of mitochondrial DNA from *Saccharomyces cerevisiae* and *S. carlsbergensis*: Electron microscopy of partially denatured molecules. *Nucleic Acids Res.* **3,** 465–476.

Clark-Walker, G. D., and McArthur, C. R. (1978). Structural and functional relationships of mitochondrial DNAs from various yeasts. *In* "Biochemistry and Genetics of Yeasts" (M. Bacila, B. L. Horecker, and A. O. M. Stoppani, eds.), pp. 255–272. Academic Press, New York.

Clark-Walker, G. D., and Sriprakash, K. S. (1981). Sequence rearrangement between mitochondrial DNAs of *Torulopsis glabrata* and *Kloeckera africana* identified by hybridization with six polypeptide encoding regions from *Saccharomyces cerevisiae* mitochondrial DNA. *J. Mol. Biol.* **151,** 367–387.

Clark-Walker, G. D., and Sriprakash, K. S. (1982). Size diversity and sequence rearrangements in mitochondrial DNAs from yeasts. *In* "Mitochondrial Genes" (P. Slonimski, P. Borst, and G. Attardi, eds.), pp. 349–354. Cold Spring Harbor Laboratory, Cold Spring Harbor, New York.

Clark-Walker, G. D., Sriprakash, K. S., McArthur, C. R., and Azad, A. A. (1980). Mapping of mitochondrial DNA from *Torulopsis glabrata* location of ribosomal and transfer RNA genes. *Curr. Genet.* **1,** 209–217.

Clark-Walker, G. D., McArthur, C. R., and Daley, D. J. (1981). Does mitochondrial DNA length influence the frequency of spontaneous petite mutants in yeasts? *Curr. Genet.* **4,** 7–12.

Clayton, D. A. (1982). Replication of animal mitochondrial DNA. *Cell* **28,** 693–705.

Clayton, D. A., and Smith, C. A. (1975). Complex mitochondrial DNA. *Int. Rev. Exp. Pathol.* **14,** 1–67.

Collins, R. A., and Lambowitz, A. M. (1981). Characterization of a variant *Neurospora crassa* mitochondrial DNA which contains tandem reiterations of a 1.9kb sequence. *Curr. Genet.* **4**, 131–133.

Collins, R. A., and Lambowitz, A. M. (1983). Structural variation and optional introns in the mitochondrial DNAs of *Neurospora* strains isolated from nature. *Plasmid* **9**, 53–70.

Collins, R. A., Stohl, L. L., Cole, M. D., and Lambowitz, A. M. (1981). Characterization of a novel plasmid DNA found in mitochondria of *N. crassa*. *Cell* **24**, 443–452.

Crick, F. (1979). Split genes and RNA splicing. *Science* **204**, 264–271.

Crick, F. H. C. (1966). Codon-anticodon pairing: The wobble hypothesis. *J. Mol. Biol.* **19**, 548–555.

Crouse, E., Vandrey, J., and Stutz, E. (1974). Comparative analysis of chloroplast and mitochondrial DNAs from *Eugena gracilis. Proc. Int. Congr. Photosynth. 3rd* **3**, 1775–1786.

Cummings, D. J. (1980). Evolutionary divergence of mitochondrial DNA from *Paramecium aurelia. Mol. Gen. Genet.* **180**, 77–84.

Cummings, D. J., and Laping, J. L. (1981). Organization and cloning of mitochondrial deoxyribonucleic acid from *Paramecium tetraaurelia* and *Paramecium primaurelia. Mol. Cell. Biol.* **1**, 972–982.

Cummings, D. J., and Pritchard, A. E. (1982). Replication mechanism of mitochondrial DNA from *Paramecium aurelia*: Sequence of the cross-linked origin. *In* "Mitochondrial Genes" (P. Slonimski, P. Borst, and G. Attardi, eds.), pp. 441–447. Cold Spring Harbor Laboratory, Cold Spring Harbor, New York.

Cummings, D. J., Belcour, L., and Grandchamp, C. (1978). Étude au microscope électronique du DNA mitochondrial de *Podospora anserina* et présence d'une série multimérique de molécules circulaires de DNA dans des cultures senescentes. *C.R. Acad. Sci. Ser. D* **287**, 157–160.

Cummings, D. J., Borst, P., Dawid, I. B., Weissman, S. M., and Fox, C. F., eds. (1979a). "Extrachromosomal DNA," Vol. XV. ICN-UCLA Symposia on Molecular and Cellular Biology. Academic Press, New York.

Cummings, D. J., Pritchard, A. E., and Maki, R. A. (1979b). Restriction enzyme analysis of mitochondrial DNA from closely related species of *Paramecium. In* "ICN-UCLA Symposium on Extrachromosomal DNA" (D. J. Cummings, P. Borst, E. B. Dawid, S. M. Weissman, and C. F. Fox, eds.), pp. 35–51. Academic Press, New York.

Cummings, D. J., Belcour, L., and Grandchamp, C. (1979c). Mitochondrial DNA from *Podospora anserina. Mol. Gen. Genet.* **171**, 229–238.

Cummings, D. J., Belcour, L., and Grandchamp, C. (1979d). Mitochondrial DNA from *Podospora anserina. Mol. Gen. Genet.* **171**, 239–250.

Cummings, D. J., Maki, R. A., Conlon, P. J., and Laping, J. (1980a). Anatomy of mitochondrial DNA from *Paramecium aurelia. Mol. Gen. Genet.* **178**, 499–510.

Cummings, D. J., Laping, J. L., and Nolan, P. E. (1980b). Cloning of senescent mitochondrial DNA from *Podospora anserina:* A beginning. *In* "The Organization and Expression of the Mitochondrial Genome" (A. M. Kroon and C. Saccone, eds.), pp. 97–102. Elsevier, Amsterdam.

Cunningham, R. S., and Gray, M. W. (1977). Isolation and characterization of ^{32}P-labelled mitochondrial and cytosol ribosomal RNA from germinating wheat embryos. *Biochim. Biophys. Acta* **475**, 476–491.

Dale, R. M. K. (1981). Sequence homology among different size classes of plant mtDNAs. *Proc. Natl. Acad. Sci. U.S.A.* **78**, 4454–4457.

Dale, R. M. K. (1982). Structure of plant mitochondrial DNAs. *In* "Mitochondrial Genes"

(P. Slonimski, P. Borst, and G. Attardi, eds.), pp. 471–476. Cold Spring Harbor Laboratory, Cold Spring Harbor, New York.

Davies, R. W., Waring, R. B., Ray, J. A., Brown, T. A., and Scazzocchio, C. (1982a). Making ends meet: A model for RNA splicing in fungal mitochondria. *Nature (London)* **300**, 719–724.

Davies, R. W., Scazzocchio, C., Waring, R. B., Lee, S., Grisi, E., Berks, M., and Brown, T. A. (1982b). Mosaic genes and unidentified reading frames that have homology with human mitochondrial sequences are found in the mitochondrial genome of *Aspergillus nidulans*. *In* "Mitochondrial Genes" (P. Slonimski, P. Borst, and G. Attardi, eds.), pp. 405–410. Cold Spring Harbor Laboratory, Cold Spring Harbor, New York.

Dawid, I. B., and Rastl, E. (1979). Structure and evolution of animal mitochondrial DNA. *In* "Extrachromosomal DNA: ICN-UCLA Symposia on Molecular and Cellular Biology" (D. J. Cummings, P. Borst, I. G. Dawid, S. M. Weissman, and C. F. Fox, eds.), pp. 395–407. Academic Press, New York.

De La Salle, H., Jacq, C., and Slonimski, P. P. (1982). Critical sequences within mitochondrial introns: Pleiotropic mRNA maturase and cis-dominant signals of the box intron controlling reductase and oxidase. *Cell* **28**, 721–732.

Del Giudice, L., Wolf, K., Sassone-Corsi, P., and Alvino, C. (1978). Circular molecules of heterogeneous size from mitochondrial fractions of the petite negative yeast *Schizosaccharomyces pombe*. *Mol. Gen. Genet.* **164**, 289–293.

Denhardt, D. T., Dressler, D., and Ray, D. S., eds. (1978). "The Single-Stranded DNA Phages." Cold Spring Harbor Laboratory, Cold Spring Harbor, New York.

Devenish, R. J., Hall, R. M., Linnane, A. W., and Lukins, H. B. (1979). Biogenesis of mitochondria. 52: Deletions in petite strains occurring in the mitochondrial gene for the 21S ribosomal RNA, that affect the properties of mitochondrial recombination. *Mol. Gen. Genet.* **174**, 297–305.

De Vos, W. M., Bakker, H., Saccone, C., and Kroon, A. M. (1980). Further analysis of the type differences of rat-liver mitochondrial DNA. *Biochim. Biophys. Acta* **607**, 1–9.

deVries, H., deJonge, J. C., van't Sant, P., Agsteribbe, E., and Arnberg, A. (1981). A "stopper" mutant of *Neurospora crassa* containing two populations of aberrant mitochondrial DNA. *Curr. Genet.* **3**, 205–211.

deZamaroczy, M., Marotta, R., Faugeron-Fonty, G., Goursot, R., Mangin, M., Baldacci, G., and Bernardi, G. (1981). The origins of replication of the yeast mitochondrial genome and the phemonenon of suppressivity. *Nature (London)* **292**, 75–78.

Dieckmann, C. L., Bonitz, S. G., Hill, J., Homison, G., McGraw, P., Pape, L., Thalenfeld, B. E., and Tzagoloff, A. (1982). Structure of the apocytochrome-b transcripts in *Saccharomyces cerevisiae*. *In* "Mitochondrial Genes" (P. Slonimski, P. Borst, and G. Attardi, eds.), pp. 213–223. Cold Spring Harbor Laboratory, Cold Spring Harbor, New York.

Donelson, J. E., Majiwa, P. A. O., and Williams, R. O. (1979). Kinetoplast DNA minicircles of *Trypanosoma brucei* share regions of sequence homology. *Plasmid* **2**, 572–588.

Dujardin, G., Jacq, C., and Slonimski, P. P. (1982). Single base substitution in an intron of oxidase gene compensates splicing defects of the cytochrome b gene. *Nature (London)* **298**, 628–632.

Dujon, B. (1980). Sequence of the intron and flanking exons of the mitochondrial 21S rRNA gene of yeast strains having different alleles at the ω and rib-1 loci. *Cell* **20**, 185–197.

Duvick, D. N. (1965). Cytoplasmic pollen sterility in corn. *Adv. Genet.* **13**, 1–56.

Earl, A. J., Turner, G., Croft, J. H., Dales, R. B. G., Lazarus, C. M., Lunsdorf, H., and

Kuntzel, H. (1981). High frequency transfer of species specific mitochondrial DNA sequences between members of the *Aspergillaceae*. *Curr. Genet.* **3**, 221–228.

Edwardson, J. R. (1970). Cytoplasmic male sterility. *Bot. Rev.* **36**, 341–420.

Englund, P. T. (1978). The replication of kinetoplast DNA networks in *Crithidia fasciculata*. *Cell* **14**, 157–168.

Englund, P. T. (1981). Kinetoplast DNA. *Biochem. Physiol. Protozoa* **4**, 333–383.

Englund, P. T., and Marini, J. C. (1980). The replication of kinetoplast DNA. *Am. J. Trop. Med. Hyg. Suppl.* **29**, 1064–1069.

Englund, P. T., Hajduk, S. L., and Marini, J. C. (1982a). The molecular biology of trypanosomes. *Annu. Rev. Biochem.* **51**, 695–726.

Englund, P. T., Hajduk, S. L., Marini, J. C., and Plunkett, M. L. (1982b). Replication of kinetoplast DNA. *In* "Mitochondrial Genes" (P. Slonimski, P. Borst, and G. Attardi, eds.), pp. 423–433. Cold Spring Harbor Laboratory, Cold Spring Harbor, New York.

Esser, K., Tudzynski, P., Stahl, U., and Kück, U. (1980). A model to explain senescence in the filamentous fungus *Podospora anserina* by the action of plasmid like DNA. *Mol. Gen. Genet.* **178**, 213–216.

Esser, K., Kück, U., Stahl, U., and Tudzynski, P. (1981). Mitochondrial DNA and senescence in *Podospora anserina*. *Curr. Genet.* **4**, 83.

Fairfield, F. R., Bauer, W. R., and Simpson, M. V. (1979). Mitochondria contain a distinct DNA topoisomerase. *J. Biol. Chem.* **254**, 9352–9354.

Fairlamb, A. H., Weislogel, P. O., Hoeijmakers, J. H. J., and Borst, P. (1978). Isolation and characterization of kinetoplast DNA from bloodstream form of *Trypanosoma brucei*. *J. Cell Biol.* **76**, 293–309.

Farrelly, F., and Butow, R. A. (1983). Rearranged mitochondrial genes in the yeast nuclear genome. *Nature (London)* **301**, 296–301.

Faugeron-Fonty, G., Culard, F., Baldacci, G., Goursot, R., Prunell, A., and Bernardi, G. (1979). The mitochondrial genome of wild-type yeast cells VIII. The spontaneous cytoplasmic "petite" mutation. *J. Mol. Biol.* **134**, 493–537.

Fauron, C. M.-R., and Wolstenholme, D. R. (1976). Structural heterogeneity of mitochondrial DNA molecules within the genus *Drosophila*. *Proc. Natl. Acad. Sci. U.S.A.* **73**, 3623–3627.

Fauron, C. M.-R., and Wolstenholme, D. R. (1980a). Extensive diversity among *Drosophila* species with respect to nucleotide sequences within the adenine + thymine-rich region of mitochondrial DNA molecules. *Nucleic Acids Res.* **8**, 2439–2452.

Fauron, C. M.-R., and Wolstenholme, D. R. (1980b). Intraspecific diversity of nucleotide sequences within the adenine + thymine rich region of mitochondrial DNA molecules of *Drosophila mauritiana, Drosophila melanogaster* and *Drosophila simulans*. *Nucleic Acids Res.* **8**, 5391–5411.

Faye, G., Fukuhara, H., Grandchamp, C., Lazowska, J., Michel, F., Casey, J., Getz, G. S., Locker, J., Rabinowitz, M., Bolotin-Fukuhara, M., Coen, D., Deutsch, J., Dujon, B., Netter, B., and Slonimski, P. P. (1973). Mitochondrial nucleic acids in the petite colony mutants: Deletions and repetitions of genes. *Biochimie* **55**, 779–792.

Faye, G., Kujawa, C., and Fukuhara, H. (1974). Physical and genetic organization of petite and grande yeast mitochondrial DNA. IV. *In vivo* transcription products of mitochondrial DNA and localization of 23S ribosomal RNA in petite mutants of *Saccharomyces cerevisiae*. *J. Mol. Biol.* **88**, 185–203.

Faye, G., Kujawa, C., Dujon, B., Bolotin-Fukuhara, M., Wolf, K., Fukuhara, H., and Slonimski, P. P. (1975). Localization of the gene coding for the mitochondrial 16S ribosomal RNA using rho⁻ mutants of *Saccharomyces cerevisiae*. *J. Mol. Biol.* **99**, 203–217.

Faye, G., Dennebouy, N., Kujawa, C., and Jacq, C. (1979). Inserted sequence in the mitochondrial 23S ribosomal RNA gene of the yeast *Saccharomyces cerevisiae*. *Mol. Gen. Genet.* **168**, 101–109.

Ferris, S. D., Wilson, A. C., and Brown, W. M. (1981). Evolutionary tree for apes and humans based on cleavage maps of mitochondrial DNA. *Proc. Natl. Acad. Sci. U.S.A.* **78**, 2432–2436.

Flory, P. J., Jr., and Vinograd, J. (1973). 5-Bromodeoxyuridine labelling of monomeric and catenated circular mitochondrial DNA in HeLa cells. *J. Mol. Biol.* **74**, 81–94.

Fonty, G., Crouse, E. J., Stutz, E., and Bernardi, G. (1975). The mitochondrial genome of *Euglena gracilis*. *Eur. J. Biochem.* **54**, 367–372.

Fonty, G., Goursot, R., Wilkie, D., and Bernardi, G. (1978). The mitochondrial genome of wild-type yeast cells. *J. Mol. Biol.* **119**, 213–235.

Forde, B. G., and Leaver, C. J. (1980). Nuclear and cytoplasmic genes controlling synthesis of variant mitochondrial polypeptides in male-sterile maize. *Proc. Natl. Acad. Sci. U.S.A.* **77**, 418–422.

Forde, B. G., Oliver, R. J. C., and Leaver, C. J. (1978). Variation in mitochondrial translation products associated with male-sterile cytoplasms in maize. *Proc. Natl. Acad. Sci. U.S.A.* **75**, 3941–3945.

Forte, M. A., and Fangman, W. L. (1979). Yeast chromosomal DNA molecules have strands which are cross-linked at their termini. *Chromosoma* **72**, 131–150.

Foury, F., and Tzagaloff, A. (1978). Assembly of the mitochondrial membrane system. Genetic complementation of *mit⁻* mutations in mitochondrial DNA of *Saccharomyces cerevisiae*. *J. Biol. Chem.* **253**, 3792–3797.

Fouts, D. L., and Wolstenholme, D. R. (1979). Evidence for a partial RNA transcript of the small circular component of kinetoplast DNA of *Crithidia acanthocephali*. *Nucleic Acids Res.* **6**, 3785–3804.

Fouts, D. L., Manning, J. E., and Wolstenholme, D. R. (1975). Physicochemical properties of kinetoplast DNA from *Crithidia acanthocephali*, *Crithidia luciliae*, and *Trypanosoma lewisi*. *J. Cell Biol.* **67**, 378–399.

Fouts, D. L., Wolstenholme, D. R., and Boyer, H. W. (1978). Heterogeneity in sensitivity to cleavage by the restriction endonucleases *Eco*RI and *Hind*III of circular kinetoplast DNA molecules of *Crithidia acanthocephali*. *J. Cell Biol.* **79**, 329–341.

Fox, T. D., and Leaver, C. J. (1981). The *Zea mays* mitochondrial gene coding cytochrome oxidase subunit II has an intervening sequence and does not contain TGA codons. *Cell* **26**, 315–323.

Francisco, J. F., and Simpson, M. V. (1977). The occurrence of two types of mitochondrial DNA in rat populations as detected by *Eco*RI endonuclease analysis. *FEBS Lett.* **79**, 291–294.

Frasch, A. C. C., Hajduk, S. L., Hoeijmakers, J. H. J., Borst, P., Brunel, F., and Davison, J. (1980). The kinetoplast DNA of *Trypanosoma equiperdum*. *Biochim. Biophys. Acta* **607**, 397–410.

Frasch, A. C. C., Goijman, S. G., Cazzulo, J. J., and Stoppani, A. O. M. (1981). Constant and variable regions in DNA mini-circles from *Trypanosoma cruzi* and *Trypanosoma rangeli*: Application to species and stock differentiation. *Mol. Biochem. Parasitol.* **4**, 163–170.

Frederick, J. F., ed. (1981). Origins and evolution of eukaryotic intracellular organelles. *Ann. N.Y. Acad. Sci.* **361**.

Garriga, G., Collins, R. A., Grant, D. M., Lambowitz, A. M., and Bertrand, H. (1982). Mitochondrial RNA splicing in *Neurospora crassa*. *In* "Mitochondrial Genes" (P.

Slonimski, P. Borst, and G. Attardi, eds.), pp. 381–390. Cold Spring Harbor Laboratory, Cold Spring Harbor, New York.

Gelfand, R., and Attardi, G. (1981). Synthesis and turnover of mitochondrial RNA in HeLa cells: The mature ribosomal and messenger RNA species are metabolically unstable. *Mol. Cell. Biol.* **1**, 497–511.

Gellert, M. (1981). DNA topoisomerases. *Annu. Rev. Biochem.* **50**, 879–910.

Gellissen, G., Bradfield, J. Y., White, B. N., and Wyatt, G. R. (1983). Mitochondrial DNA sequences in the nuclear genome of a locust. *Nature (London)* **301**, 631–634.

Gengenbach, B. G., Green, C. F., and Donovan, C. M. (1977). Inheritance of selected pathotoxic resistance in maize plants regenerated from cell cultures. *Proc. Natl. Acad. Sci. U.S.A.* **74**, 5113–5117.

Gengenbach, B. G., Connelly, J. A., Pring, D. R., and Conde, M. F. (1981). Mitochondrial DNA variation in maize plants regenerated during tissue culture selection. *Theor. Appl. Genet.* **59**, 161–167.

Gilbert, W. (1978). Why are genes in pieces? *Nature (London)* **271**, 501.

Gillham, N. W. (1978). "Organelle Heredity." Raven, New York.

Gillum, A. M., and Clayton, D. A. (1978). Displacement-loop replication initiation sequence in animal mitochondrial DNA exists as a family of discrete lengths. *Proc. Natl. Acad. Sci. U.S.A.* **75**, 677–681.

Glaus, K. R., Zassenhaus, H. P., Fechheimer, N. S., and Perlman, P. S. (1980). Avian mtDNA: Structure, organization and evolution. *In* "The Organization and Expression of the Mitochondrial Genome" (A. M. Kroon and C. Saccone, eds.), pp. 131–135. Elsevier, Amsterdam.

Goddard, J. M., and Cummings, D. J. (1975). Structure and replication of mitochondrial DNA from *Paramecium aurelia. J. Mol. Biol.* **97**, 593–609.

Goddard, J. M., and Cummings, D. J. (1977). Mitochondrial DNA replication in *Paramecium aurelia.* Cross linking of the initiation end. *J. Mol. Biol.* **109**, 327–344.

Goddard, J. M., and Wolstenholme, D. R. (1980). Origin and direction of replication in mitochondrial DNA molecules from the genus *Drosophila. Nucleic Acids Res.* **8**, 741–757.

Goddard, J. M., Masters, J. N., Jones, S. S., Ashworth, W. D., Jr., and Wolstenholme, D. R. (1981). Nucleotide sequence variants of *Rattus norvegicus* mitochondrial DNA. *Chromosoma* **82**, 595–609.

Goddard, J. M., Fauron, C. M.-R., and Wolstenholme, D. R. (1982). Nucleotide sequences within the A + T-rich region and the large-rRNA gene of mitochondrial DNA molecules of *Drosophila yakuba. In* "Mitochondrial Genes" (P. Slonimski, P. Borst, and G. Attardi, eds.), pp. 99–193. Cold Spring Harbor Laboratory, Cold Spring Harbor, New York.

Goldbach, R. W., Arnberg, A. C., Van Bruggen, E. F. J., Defize, J., and Borst, P. (1978a). The structure of *Tetrahymena pyriformis* mitochondrial DNA: I. Strain differences and occurrence of inverted repetitions. *Biochim. Biophys. Acta* **477**, 37–50.

Goldbach, R. W., Borst, P., Bollen-De Boer, J. E., and Van Bruggen, E. F. J. (1978b). The organization of ribosomal RNA genes in the mitochondrial DNA of *Tetrahymena pyriformis* strain ST. *Biochim. Biophys. Acta* **521**, 169–186.

Goldbach, R. W., Bollen-De Boer, J. E., Van Bruggen, E. F. J., and Borst, P. (1978c). Conservation of the sequence and position of the ribosomal RNA genes in *Tetrahymena pyriformis* mitochondrial DNA. *Biochim. Biophys. Acta* **521**, 187–197.

Goldbach, R. W., Bollen-De Boer, J. E., Van Bruggen, E. F. J., and Borst, P. (1979). Replication of the linear mitochondrial DNA of *Tetrahymena pyriformis. Biochim. Biophys. Acta* **562**, 400–417.

Grant, D., and Chiang, K.-S. (1980). Physical mapping and characterization of *Chlamydomonas* mitochondrial DNA molecules: Their unique ends, sequence homogeneity, and conservation. *Plasmid* **4**, 82–96.

Gray, M. W. (1982). Mitochondrial genome diversity and the evolution of mitochondrial DNA. *Can. J. Biochem.* **60**, 157–171.

Gray, M. W., and Doolittle, W. F. (1982). Has the endosymbiont hypothesis been proven? *Microbiol. Rev.* **46**, 1–42.

Gray, M. W., and Spencer, D. F. (1981). Is wheat mitochondrial 5S ribosomal RNA prokaryotic in nature? *Nucleic Acids Res.* **9**, 3523–3529.

Gray, M. W., Bonen, L., Falconet, D., Huh, T. Y., Schnare, M. N., and Spender, D. F. (1982). Mitochondrial ribosomal RNAs of *Triticum aestivum* (Wheat): Sequence analysis and gene organization. *In* "Mitochondrial Genes" (P. Slonimski, P. Borst, and G. Attardi, eds.), pp. 483–488. Cold Spring Harbor Laboratory, Cold Spring Harbor, New York.

Green, M. R., Grimm, M. F., Goewert, R. R., Collins, R. A., Cole, M. D., Lambowitz, A. M., Heckman, J. E., Yin, S., and RajBhandary, U. L. (1981). Transcripts and processing patterns for the ribosomal RNA and transfer RNA region of *Neurospora crassa* mitochondrial DNA. *J. Biol. Chem.* **256**, 2027–2034.

Grivell, L. A., and Moorman, A. F. M. (1977). A structural analysis of the oxi-3 region on yeast mtDNA. *In* "Mitochondria 1977: Genetics and Biogenesis of Mitochondria" (W. Bandlow, R. J. Schweyen, K. Wolf, and F. Kandewitz, eds.), pp. 371–385. De Gruyter, Berlin.

Grivell, L. A., Arnberg, A. C., De Boer, P. H., Bos, J. L., Groot, G. S. P., Hecht, N. B., Hensgens, L. A., Van Ommen, G., and Tabak, H. F. (1979). Transcripts of yeast mitochondrial DNA and their processing. *In* "Extra-Chromosomal DNA" (D. Cummings, P. Borst, I. Dawid, S. Weissmann, and C. F. Fox, eds.), pp. 305–324. Academic Press, New York.

Grivell, L. A., Arnberg, A. C., Hensgens, L. A. M., Roosendaal, E., Van Ommen, G. J. B., and Van Bruggen, E. F. J. (1980). Split genes on yeast mitochondrial DNA: Organization and expression. *In* "The Organization and Expression of the Mitochondrial Genome" (A. M. Kroon and C. Saccone, eds.), pp. 37–49. Elsevier, Amsterdam.

Grivell, L. A., Hensgens, L. A. M., Osinga, K. A., Tabak, H. F., Boer, P. H., Crusius, J. B. A., van der Laan, J. C., deHaan, M., van der Horst, G., Evers, R. F., and Arnberg, A. C. (1982). RNA processing in yeast mitochondria. *In* "Mitochondrial Genes" (P. Slonimski, P. Borst, and G. Attardi, eds.), pp. 225–239. Cold Spring Harbor Laboratory, Cold Spring Harbor, New York.

Groot, G. S. P., Flavell, R. A., and Sanders, J. P. M. (1975). Sequence homology of nuclear and mitochondrial DNAs of different yeasts. *Biochim. Biophys. Acta* **378**, 186–194.

Groot, G. S. P., Mason, T. L., and Van Harten-Loosbrock, N. (1979). Var$_1$ is associated with the small ribosomal subunit of mitochondrial ribosomes in yeast. *Mol. Gen. Genet.* **174**, 339–342.

Grosskopf, R., and Feldmann, H. (1981a). tRNA genes in rat liver mitochondrial DNA. *Curr. Genet.* **4**, 191–196.

Grosskopf, R., and Feldmann, H. (1981b). Analysis of a DNA segment from rat liver mitochondria containing the genes for the cytochrome oxidase subunits I, II and III, ATPase subunit 6, and several tRNA genes. *Curr. Genet.* **4**, 151–158.

Guillemant, P., and Weil, J. H. (1975). Aminoacylation of *Phaseolus vulgaris* cytoplasmic, chloroplastic and mitochondrial tRNAs met by homologous and heterologous enzymes. *Biochim. Biophys. Acta* **407**, 240–248.

Gunge, N., and Sakaguchi, K. (1979). Fusion of mitochondria with protoplasts in *Saccharomyces cerevisiae*. *Mol. Gen. Genet.* **170**, 243–247.

Haid, A., Schweyen, R. J., Kaudewitz, F., Solioz, M., and Schatz, G. (1979). The mitochondrial cob region in yeast codes for apocytochrome b and is mosaic. *Eur. J. Biochem.* **94**, 451–464.

Hajduk, S. L. (1978). Influence of DNA complexing compounds on the kinetoplast of Trypanosomatids. *Prog. Mol. Subcell. Biol.* **6**, 158–200.

Hajduk, S. L., and Vickerman, K. (1981). Absence of detectable alteration in the kinetoplast DNA of a *Trypanosoma brucei* clone following loss of ability to infect the insect vector (*Glossina morsitans*). *Mol. Biochem. Parasitol.* **4**, 17–28.

Handel, M. A., Papaconstantinou, J., Allison, D. P., Julku, E. M., and Chin, E. T. (1973). Synthesis of mitochondrial DNA in spermatocytes of *Rhynchosciara hollaenderi*. *Dev. Biol.* **35**, 240–249.

Handley, L. (1975). Ph.D. thesis. University of Birmingham.

Hanson, D. K., Miller, D. H., Mahler, H. R., Alexander, N. J., and Perlman, P. S. (1979). Regulatory interactions between mitochondrial genes. II. Detailed characterization of novel mutants mapping within one cluster in the COB2 region. *J. Biol. Chem.* **254**, 2480–2490.

Hayashi, J.-I., Tagashira, Y., Moriwaki, K., and Yosida, T. H. (1981). Polymorphisms of mitochondrial DNAs in Norway rats (*Rattus norvegicus*): Cleavage site variations and length polymorphism of restriction fragments. *Mol. Gen. Genet.* **184**, 337–341.

Heckman, J. E., and RajBhandary, U. L. (1979). Organization of tRNA and rRNA genes in *N. crassa* mitochondria: Intervening sequence in the large rRNA gene and strand distribution of the RNA genes. *Cell* **17**, 583–595.

Heckman, J. E., Sarnoff, J., Alzner-DeWeerd, B., Yin, S., and RajBhandary, U. L. (1980). Novel features in the genetic code and codon reading patterns in *Neurospora crassa* mitochondria based on sequences of six mitochondrial tRNAs. *Proc. Natl. Acad. Sci. U.S.A.* **77**, 3159–3163.

Heyting, C., and Menke, H. H. (1979). Fine structure of the 21S ribosomal RNA region on yeast mitochondrial DNA: III. Physical location of yeast mitochondrial genetic markers and the molecular nature of ω. *Mol. Gen. Genet.* **168**, 279–291.

Heyting, C., Meijlink, F. C. P. W., Verbeet, M. P., Sanders, J. P. M., Bos, J. L., and Borst, P. (1979a). Fine structure of the 21S ribosomal RNA region on yeast mitochondrial DNA: I. Construction of the physical map and localization of the cistron for the 21S mitochondrial ribosomal RNA. *Mol. Gen. Genet.* **168**, 231–250.

Heyting, C., Talen, J.-L., Weijers, P. J., and Borst, P. (1979b). Fine structure of the 21S ribosomal RNA region on yeast mitochondrial DNA: II. The organization of sequences in petite mitochondrial DNAs carrying genetic markers from the 21S region. *Mol. Gen. Genet.* **168**, 251–277.

Hillar, M., Rangayya, V., Jafar, B. B., Chambers, D., Vitzu, M., and Wyborny, L. E. (1979). Membrane-bound mitochondrial DNA: Isolation, transcription and protein composition. *Arch. Int. Physiol. Biochim.* **87**, 29–49.

Hoeijmakers, J. H. J., and Borst, P. (1978). RNA from the insect trypanosome *Crithidia luciliae* contains transcripts of the maxi-circle and not of the mini-circle component of kinetoplast DNA. *Biochim. Biophys. Acta* **521**, 407–411.

Hoeijmakers, J. H. J., and Borst, P. (1982). Kinetoplast DNA in the insect Trypanosomes *Crithidia luciliae* and *Crithidia fasciculata*. *Plasmid* **7**, 210–220.

Hoeijmakers, J. H. J., Schoutsen, B., and Borst, P. (1982a). Kinetoplast DNA in the insect trypanosomes *Crithidia luciliae* and *Crithidia fasciculata*. *Plasmid* **7**, 199–209.

Hoeijmakers, J. H. J., Weljers, P. J., Brakenhoff, G. J., and Borst, P. (1982b). Kinetoplast DNA in the insect Trypanosomes *Crithidia luciliae* and *Crithidia fasciculata*: III. Heteroduplex analysis of the *C. luciliae* minicircles. *Plasmid* **7**, 221–229.

Hollenberg, C. P., Borst, P., and Van Bruggen, E. F. J. (1970). Mitochondrial DNA: V. A 25-μ closed circular duplex DNA molecule in wild-type yeast mitochondria. Structure and genetic complexity. *Biochim. Biophys. Acta* **209**, 1–15.

Howell, S. H. (1982). Plant molecular vehicles: Potential vectors for introducing foreign DNA into plants. *Annu. Rev. Plant Physiol.* **33**, 609–650.

Hudspeth, M. E. S., Ainley, W. M., Shumard, D. S., Butow, R. A., and Grossman, L. I. (1982). Location and structure of the var 1 gene on yeast mitochondrial DNA: Nucleotide sequence of the 40.0 allele. *Cell* **30**, 617–626.

Hudspeth, M. E. S., Shumard, D. S., Bradford, C. J. R., and Grossman, L. I. (1983). Organization of *Achlya* mtDNA: A population with two orientations and a large inverted repeat containing the rRNA genes. *Proc. Natl. Acad. Sci. U.S.A.* **80**, 142–146.

Huh, T. Y., and Gray, M. W. (1982). Conservation of ribosomal RNA gene arrangement in the mitochondrial DNA of angiosperms. *Plant Mol. Biol.* **1**, 245–249.

Hyman, B. C., Cramer, J. H., and Rownd, R. H. (1982). Properties of a *Saccharomyces cerevisiae* mtDNA segment conferring high-frequency yeast transformation. *Proc. Natl. Acad. Sci. U.S.A.* **79**, 1578–1582.

Iams, K. P., and Sinclair, J. H. (1982). Mapping the mitochondrial DNA of *Zea mays*: Ribosomal gene localization. *Proc. Natl. Acad. Sci. U.S.A.* **79**, 5926–5929.

Ingle, J., Timmis, J. N., and Sinclair, J. (1975). The relationship between satellite deoxyribonucleic acid, ribosomal ribonucleic acid gene redundancy and genome size in plants. *Plant Physiol.* **55**, 496–501.

Jacobs, H. T., Posakony, J. W., Grula, J. W., Roberts, J. W., Xin, J., Britten, R. J., and Davidson, E. H. (1983). Mitochondrial DNA sequences in the nuclear genome of *Strongylocentrotus purpuratus*. *J. Mol. Biol.* **165**, 609–632.

Jacq, C., Kujawa, C., Grandchamp, C., and Netter, P. (1977). Physical characterization of the difference between yeast mitochondrial DNA alleles omega⁺ and omega⁻. *In* "Mitochondria 1977: Genetics and Biogenesis of Mitochondria" (W. Bandlow, R. J. Schweyen, K. Wolf, and F. Kandewitz, eds.), pp. 255–270. De Gruyter, Berlin.

Jacq, C., Lazowska, J., and Slonimski, P. P. (1980). Cytochrome b messenger RNA maturase encoded in an intron regulates the expression of the split gene: I. Physical location and base sequence of intron mutations. *In* "The Organization and Expression of the Mitochondrial Genome" (A. M. Kroon and C. Saccone, eds.), pp. 139–152. Elsevier, Amsterdam.

Jacq, C., Pajot, P., Lazowska, J., Dujardin, G., Claisse, M., Groudinsky, O., de la Salle, H., Grandchamp, C., Labouesse, M., Gargouri, A., Guiard, B., Spyridakis, A., Dreyfus, M., and Slonimski, P. O. (1982). Role of introns in the yeast cytochrome-b gene: Cis- and Trans-acting signals, intron manipulation, expression, and intergenic communications. *In* "Mitochondrial Genes" (P. Slonimski, P. Borst, and G. Attardi, eds.), pp. 155–183. Cold Spring Harbor Laboratory, Cold Spring Harbor, New York.

Jamet-Vierney, C., Begel, O., and Belcour, L. (1980). Senescence in *Podospora anserina*: Amplification of a mitochondrial DNA sequence. *Cell* **21**, 189–194.

Johnson, B. J. B., Hill, G. C., Fox, T. D., and Stuart, K. (1982). The maxicircle of *Trypanosoma brucei* kinetoplast DNA hybridizes with a mitochondrial gene encoding cytochrome oxidase subunit II. *Mol. Biochem. Parasitol.* **5**, 381–390.

Jukes, T. H. (1981). Amino acid codes in mitochondria as possible clues to primitive codes. *J. Mol. Evol.* **18**, 15–17.

Kasamatsu, H., Robberson, D. L., and Vinograd, J. (1971). A novel closed-circular mitochondrial DNA with properties of a replicating intermediate. *Proc. Natl. Acad. Sci. U.S.A.* **68**, 2252–2257.

Kearsey, S. E., and Craig, I. W. (1982). Genetic basis of chloramphenicol resistance in mouse and human cell lines. In "Mitochondrial Genes" (P. Slonimski, P. Borst, and G. Attardi, eds.), pp. 117–120. Cold Spring Harbor Laboratory, Cold Spring Harbor, New York.

Kemble, R. J., and Bedbrook, J. R. (1980). Low molecular weight circular and linear DNA in mitochondria from normal and male-sterile *Zea mays* cytoplasm. *Nature (London)* **284**, 565–566.

Kemble, R. J., and Thompson, R. D. (1982). S1 and S2, the linear mitochondrial DNAs present in a male sterile line of maize, possess terminally attached proteins. *Nucleic Acids Res.* **10**, 8181–8190.

Kemble, R. J., Mans, R. J., Gabay-Laughnan, S., and Laughnan, J. R. (1983). Sequences homologous to episomal mitochondrial DNAs in the maize nuclear genome. *Nature (London)* **304**, 744–747.

Kilejian, A. (1975). Circular mitochondrial DNA from the avian malarial parasite *Plasmodium lophurae*. *Biochim. Biophys. Acta* **390**, 276–284.

Kim, B. D., Mans, R. J., Conde, M. E., Pring, D. R., and Levings, C. S., III. (1982). Physical mapping of homologous segments of mitochondrial episomes from S male-sterile maize. *Plasmid* **7**, 1.

Kleisen, C. M., and Borst, P. (1975). Sequence heterogeneity of the mini-circles of kinetoplast DNA of *Crithidia luciliae* and evidence for the presence of a component more complex than mini-circle DNA in the kinetoplast network. *Biochim. Biophys. Acta* **407**, 473–478.

Kleisen, C. M., Borst, P., and Weijers, P. J. (1976a). The structure of kinetoplast DNA. I. The mini-circles of *Crithidia luciliae* are heterogeneous in base sequence. *Eur. J. Biochem.* **64**, 141–151.

Kleisen, C., Weislogel, P., Fonck, K., and Borst, P. (1976b). The structure of kinetoplast DNA. 2. Characterization of a novel component of high complexity present in the kinetoplast DNA network of *Crithidia luciliae*. *Eur. J. Biochem.* **64**, 153–160.

Klukas, C. K., and Dawid, I. B. (1976). Characterization and mapping of mitochondrial ribosomal RNA and mitochondrial DNA in *Drosophila melanogaster*. *Cell* **9**, 615–625.

Kobayashi, M., Seki, T., Yaginuma, K., and Koike, K. (1981). Nucleotide sequences of small ribosomal RNA and adjacent transfer RNA genes in rat mitochondrial DNA. *Gene* **16**, 297–307.

Köchel, H. G., Lazarus, C. M., Basak, N., and Küntzel, H. (1981). Mitochondrial tRNA gene cluster in *Aspergillus nidulans*: Organization and nucleotide sequence. *Cell* **23**, 625–633.

Koncz, C., János, S., Udvardy, A., Racsmány, M., and Dudits, D. (1981). Cloning of mtDNA fragments homologous to mitochondrial S2 plasmid-like DNA in maize. *Mol. Gen. Genet.* **183**, 449–458.

Kroon, A. M., and Saccone, C., eds. (1980). "The Organization and Expression of the Mitochondrial Genome." Elsevier, Amsterdam.

Kück, U., and Esser, K. (1982). Genetic map of mitochondrial DNA in *Podospora anserina*. *Curr. Genet.* **5**, 143–147.

Kück, U., Stahl, U., and Esser, K. (1981). Plasmid-like DNA is part of mitochondrial DNA in *Podospora anserina*. *Curr. Genet.* **3**, 151–156.

Küntzel, H., and Köchel, H. G. (1981). Evolution of rRNA and origin of mitochondria. *Nature (London)* **293,** 751–755.

Küntzel, H., Köchel, H. G., Lazarus, C. M., and Lunsdorf, H. (1982). Mitochondrial genes in *Aspergillus. In* "Mitochondrial Genes" (P. Slonimski, P. Borst, and G. Attardi, eds.), pp. 391–403. Cold Spring Harbor Laboratory, Cold Spring Harbor, New York.

Kupersztoch, Y. M., and Helinski, D. R. (1973). A catenated DNA molecule as an intermediate in the replication of the resistance transfer factor R6K in *Escherichia coli. Biochem. Biophys. Res. Commun.* **54,** 1451–1459.

Kuroiwa, T. (1982). Mitochondrial nuclei. *Int. Rev. Cytol.* **75,** 1–58.

Lagerkvist, U. (1981). Unorthodox codon reading and the evolution of the genetic code. *Cell* **23,** 305–306.

Lambowitz, A. M., Chua, N., and Luck, D. J. L. (1976). Mitochondrial ribosome assembly in *Neurospora.* Preparation of mitochondrial ribosomal precursor particles, site of synthesis of mitochondrial ribosomal proteins and studies on the poky mutant. *J. Mol. Biol.* **107,** 223–253.

Lansman, R. A., and Clayton, D. A. (1975). Mitochondrial protein synthesis in mouse L-cells: Effect of selective nicking on mitochondrial DNA. *J. Mol. Biol.* **99,** 777–793.

LaPolla, R. J., and Lambowitz, A. M. (1981). Mitochondrial ribosome assembly in *Neurospora crassa. J. Biol. Chem.* **256,** 7064–7067.

Laughnan, J. R., and Gabay, S. J. (1975a). Nuclear and cytoplasmic mutations to fertility in S male-sterile maize. *In* "International Maize Symposium: Genetics and Breeding" (D. B. Walden, ed.), p. 427. Wiley, New York.

Laughnan, J. R., and Gabay, S. J. (1975b). An episomal basis for instability of S male sterility in maize and some implications for plant breeding. *In* "Genetics and Biogenesis of Mitochondria and Chloroplasts" (C. W. Birky, P. S. Perlman, and T. J. Beyers, eds.), pp. 340–349. Ohio State Univ. Press, Columbus, Ohio.

Laughnan, J. R., and Gabay, S. J. (1978). Nuclear and cytoplasmic mutations to fertility in S male-sterile maize. *In* "Maize Breeding and Genetics" (D. B. Walden, ed.), pp. 427–446. Wiley, New York.

Laughnan, J. R., Gabay-Laughnan, S., and Carlson, J. E. (1981). Characteristics of cms-S reversion to male fertility in maize. *Stadler Symp.* **13,** 93–114.

Lazarus, C. M., and Küntzel, H. (1980). Amplification of a common mitochondrial DNA sequence in three new ragged mutants. *In* "The Organization and Expression of the Mitochondrial Genome" (A. M. Kroon and C. Saccone, eds.), pp. 87–90. Elsevier, Amsterdam.

Lazarus, C. M., and Küntzel, H. (1981). Anatomy of amplified mitochondrial DNA in "ragged" mutants of *Aspergillus amstelodami*: Excision points within protein genes and a common 215bp segment containing a possible origin of replication. *Curr. Genet.* **4,** 99–107.

Lazarus, C. M., Lünsdorf, H., Hahn, U., Stepien, P. P., and Küntzel, H. (1980a). Physical map of *Aspergillus nidulans* mitochondrial genes coding for ribosomal RNA: An intervening sequence in the large rRNA cistron. *Mol. Gen. Genet.* **177,** 389–397.

Lazarus, C. M., Earl, A. J., Turner, G., and Küntzel, H. (1980b). Amplification of a mitochondrial DNA sequence in the cytoplasmically inherited 'ragged' mutant of *Aspergillus amstelodami. Eur. J. Biochem.* **106,** 633–641.

Lazowska, J., and Slonimski, P. P. (1976). Electron microscopy analysis of circular repetitive mitochondrial DNA molecules from genetically characterized rho⁻ mutants of *Saccharomyces cerevisiae. Mol. Gen. Genet.* **146,** 61–78.

Lazowska, J., and Slonimski, P. P. (1977). Site-specific recombination in "petite colony" mutants of *Saccharomyces cerevisiae*: I. Electron microscopic analysis of the organization of recombinant DNA resulting from end to end joining of two mitochondrial segments. *Mol. Gen. Genet.* **156,** 163–175.

Lazowska, J., Jacq, C., and Slonimski, P. P. (1980). Sequence of introns and flanking exons in wild type and *box* 3 mutants of cytochrome b reveals an interlaced splicing protein coded by an intron. *Cell* **22,** 333–348.

Lazowska, J., Jacq, C., and Slonimski, P. P. (1981). Splice points of the third intron in the yeast mitochondrial cytochrome b gene. *Cell* **27,** 12–14.

Leaver, C. J., and Gray, M. W. (1982). Mitochondrial genome organization and expression in higher plants. *Annu. Rev. Plant Physiol.* **33,** 373–402.

Leaver, C. J., Forde, B. G., Dixon, L. K., and Fox, T. D. (1982). Mitochondrial genes and cytoplasmically inherited variation in higher plants. *In* "Mitochondrial Genes" (P. Slonimski, P. Borst, and G. Attardi, eds.), pp. 457–470. Cold Spring Harbor Laboratory, Cold Spring Harbor, New York.

Leon, W., Frasch, A. C. C., Hoeijmakers, J. H. J., Fase-Fowler, F., Borst, P., Brunel, F., and Davison, J. (1980). Maxi-circles and mini-circles in kinetoplast DNA from *Trypanosoma cruzi. Biochim. Biophys. Acta* **607,** 221–231.

Levings, C. S., III, and Pring, D. R. (1976). Restriction endonuclease analysis of mitochondrial DNA from normal and Texas cytoplasmic male-sterile maize. *Science* **193,** 158–160.

Levings, C. S., III, and Pring, D. R. (1978). The mitochondrial genome of higher plants. *Stadler Symp.* **10,** 77–94.

Levings, C. S., III, and Pring, D. R. (1979a). Molecular basis of cytoplasmic male sterility of maize. *In* "Physiological Genetics" (J. G. Scandalios, ed.), Vol. 5, p. 171. Academic Press, New York.

Levings, C. S., III, and Pring, D. R. (1979b). Mitochondrial DNA of higher plants and genetic engineering. *In* "Genetic Engineering" (J. K. Setlow and A. Hollaender, eds.), Vol. 1, pp. 205–222. Plenum, New York.

Levings, C. S., III, and Sederoff, R. R. (1981). Organization of the mitochondrial genome of maize. *In* "Levels of Genetic Control in Development" (S. Subtelny and U. K. Abbott, eds.), pp. 119–136. Liss, New York.

Levings, C. S., III, and Sederoff, R. R. (1983). Nucleotide sequence of the S-2 mitochondrial DNA from the S cytoplasm of maize. *Proc. Natl. Acad. Sci. U.S.A.* **80,** 4055–4059.

Levings, C. S., III, Shah, D. M., Hu, W. W. L., Pring, D. R., and Timothy, D. H. (1979). Molecular heterogeneity among mitochondrial DNAs from different maize cytoplasms. *In* "Extrachromosomal DNA" (D. J. Cummings, C. F. Fox, P. Borst, I. G. Dawid, and S. M. Weissman, eds.), pp. 63–73. ICN-UCLA Symposia on Molecular and Cellular Biology. Academic Press, New York.

Levings, C. S., III, Kim, B. D., Pring, D. L, Conde, M. F., Mans, R. J., Laughnan, J. R., and Gabay-Laughnan, S. J. (1980). Cytoplasmic reversion of csm-S in maize: Association with a transpositional event. *Science* **209,** 1021–1023.

Levings, C. S., III, Sederoff, R. R., Hu, W. W. L., and Timothy, D. H. (1982). Relationships among plasmid-like DNAs of the maize mitochondria. *In* "Structure and Function of Plant Genomes," pp. 363–372. Plenum, New York.

Levings, C. S., III, Sederoff, R. R., and Timothy, D. H. (1983). The molecular basis of cytoplasmic inheritance in plants. *In* "Cytogenetics of Crop Plants" (M. S. Swaminathan, P. K. Gupta, and U. Sinha, eds.), pp. 157–190. Macmillan, New York.

Lewin, R. (1982a). A baroque turn for intron processing. *Science* **218,** 1293–1295.

Lewin, R. (1982b). On the origin of introns. *Science* **217**, 921–922.

Lewin, R. (1983). Promiscuous DNA leaps all barriers. *Science* **219**, 478–479.

Locker, J., and Rabinowitz, M. (1976). Electron microscopic analysis of mitochondrial DNA sequences from petite and grande yeast. *In* "The Genetic Function of Mitochondrial DNA" (C. Saccone and A. M. Kroon, eds.), pp. 313–324. North-Holland Publ., Amsterdam.

Locker, J., and Rabinowitz, M. (1981). Transcription in yeast mitochondria: Analysis of the 21S rRNA region and its transcripts. *Plasmid* **6**, 302–314.

Locker, J., Rabinowitz, M., and Getz, G. S. (1974a). Electron microscopic and renaturation analysis of mitochondrial DNA of cytoplasmic petite mutants of *Saccharomyces cerevisiae*. *J. Mol. Biol.* **88**, 489–502.

Locker, J., Rabinowitz, M., and Getz, G. S. (1974b). Tandem inverted repeats in mitochondrial DNA of petite mutants of *Saccharomyces cerevisiae*. *Proc. Natl. Acad. Sci. U.S.A.* **71**, 1366–1370.

Locker, J., Lewin, A., and Rabinowitz, M. (1979). Review: The structure and organization of mitochondrial DNA from petite yeast. *Plasmid* **2**, 155–181.

Lonsdale, D. M., Thompson, R. D., and Hodge, T. P. (1981). The integrated forms of the S-1 and S-2 DNA elements of maize male sterile mitochondrial DNA are flanked by a large repeated sequence. *Nucleic Acids Res.* **9**, 3657.

Lumsden, W. H. R., and Evans, D. A., eds. (1976–1979). "Biology of the Kinetoplastida," Vols. 1–4. Academic Press, New York.

Lund, E., and Dahlberg, J. E. (1977). Spacer transfer RNAs in ribosomal RNA transcripts of *E. coli*: Processing of 30S ribosomal RNA *in vitro*. *Cell* **11**, 247–262.

Macino, G. (1980). Mapping of mitochondrial structural genes in *Neurospora crassa*. *J. Biol. Chem.* **255**, 10563–10565.

Macino, G., Coruzzi, G., Nobrega, F. G., Li, M., and Tzagoloff, A. (1979). Use of the UGA terminator as a tryptophan codon in yeast mitochondria. *Proc. Natl. Acad. Sci. U.S.A.* **76**, 3784–3785.

Macino, G., Scazzocchio, C., Waring, R. B., Berks, M. M., and Davies, R. W. (1980). Conservation and rearrangement of mitochondrial structural gene sequences. *Nature (London)* **288**, 404–406.

Mahler, H. R. (1981). Mitochondrial evolution: Organization and regulation of mitochondrial genes. *Ann. N.Y. Acad. Sci.* **361**, 53–75.

Mahler, H. R., Hanson, D. K., Lamb, M. R., Perlman, P. S., Anziano, P. Q., Glaus, K. R., and Haldi, M. L. (1982). Regulatory interactions between mitochondrial genes: Expressed introns—their function and regulation. *In* "Mitochondrial Genes" (P. Slonimski, P. Borst, and G. Attardi, eds.), pp. 185–199. Cold Spring Harbor Laboratory, Cold Spring Harbor, New York.

Malmstrom, B. G. (1979). Cytochrome c oxidase structure and catalytic activity. *Biochim. Biophys. Acta* **549**, 281–303.

Mannella, C. A., Goewert, R. R., and Lambowitz, A. M. (1979). Characterization of variant *Neurospora crassa* mitochondrial DNAs which contain tandem reiterations. *Cell* **18**, 1197–1207.

Manning, J. E., and Wolstenholme, D. R. (1976). Replication of kinetoplast DNA of *Crithidia acanthocephali*. I. Density shift experiments using deuterium oxide. *J. Cell Biol.* **70**, 406–418.

Manning, J. E., Wolstenholme, D. R., Ryan, R. S., Hunter, J. A., and Richards, O. C. (1971). Circular chloroplast DNA from *Euglena gracilis*. *Proc. Natl. Acad. Sci. U.S.A.* **68**, 1169–1173.

Marcou, D. (1961). Notion de longévité et nature cytoplasmique du déterminant de la senescence chez quelque champignons. *Ann. Sci. Nat. Bot.* **11**, 653–764.

Marini, J. C., Miller, K. G., and Englund, P. T. (1980). Decatenation of kinetoplast DNA by topoisomerases. *J. Biol. Chem.* **255**, 4976–4979.

Marini, J. C., Levene, S. D., Crothers, D. M., and Englund, P. T. (1982). Bent helical structure in kinetoplast DNA. *Proc. Natl. Acad. Sci. U.S.A.* **79**, 7664–7668.

Martens, P. A., and Clayton, D. A. (1979). Mechanism of mitochondrial DNA replication in mouse L-cells: Localization and sequence of the light-strand origin of replication. *J. Mol. Biol.* **135**, 327–351.

Merten, S., Synenki, R. M., Locker, J., Christianson, T., and Rabinowitz, M. (1980). Processing of precursors of 21S ribosomal RNA from yeast mitochondria. *Proc. Natl. Acad. Sci. U.S.A.* **77**, 1417–1421.

Merten, S. H., and Pardue, M. L. (1981). Mitochondrial DNA in *Drosophila.* An analysis of genome organization and transcription in *Drosophila melanogaster* and *Drosophila virilis. J. Mol. Biol.* **153**, 1–21.

Mery-Drugeon, E., Crouse, E. J., Schmitt, J. M., Bohnert, H.-J., and Bernardi, G. (1981). The mitochondrial genomes of *Ustilago cynodontis* and *Acanthamoeba castellanii. Eur. J. Biochem.* **114**, 577–583.

Messing, J. (1982). An integrative strategy of DNA sequencing and experiments beyond. *In* "Genetic Engineering—Principles and Methods" (J. K. Setlow and A. Hollaender, eds.), pp. 19–35. Plenum, New York.

Minuth, W., Tudzynski, P., and Esser, K. (1982). Extrachromosomal genetics of *Cephalosporium acremonium.* I. Characterization and mapping of mitochondrial DNA. *Curr. Genet.* **5**, 227–231.

Miyata, T., Hayashida, H., Kikuno, R., Hasegawa, M., Kobayashi, M., and Koike, K. (1982). Molecular clock of silent substitution: At least six-fold preponderance of silent changes in mitochondrial genes over those in nuclear genes. *J. Mol. Evol.* **19**, 28–35.

Montoya, J., Ojala, D., and Attardi, G. (1981). Distinctive features of the 5'-terminal sequences of the human mitochondrial mRNAs. *Nature (London)* **290**, 465–470.

Morel, C., and Simpson, L. (1980). Characterization of pathogenic trypanosomatidae by restriction endonuclease fingerprinting of kinetoplast DNA minicircles. *Am. J. Trop. Med. Hyg. Suppl.* **29**, 1070–1074.

Morel, C., Chiari, E., Camargo, E. P., Mattei, D. M., Romanha, A. J., and Simpson, L. (1980). Strains and clones of *Trypanosoma cruzi* can be characterized by pattern of restriction endonuclease products of kinetoplast DNA minicircles. *Proc. Natl. Acad. Sci. U.S.A.* **77**, 6810–6814.

Morimoto, R., and Rabinowitz, M. (1979). Physical mapping of the yeast mitochondrial genome. *Mol. Gen. Genet.* **170**, 25–48.

Morimoto, R., Lewin, A., and Rabinowitz, M. (1979). Physical mapping and characterization of the mitochondrial DNA and RNA sequences from mit⁻ mutants defective in cytochrome oxidase peptide 1 (OXI 3). *Mol. Gen. Genet.* **170**, 1–9.

Nagy, F., Török, I., and Maliga, P. (1981). Extensive rearrangements in the mitochondrial DNA in somatic hybrids of *Nicotiana tabacum* and *Nicotiana knightiana. Mol. Gen. Genet.* **183**, 437–439.

Nass, M. M. K. (1966). The circularity of mitochondrial DNA. *Proc. Natl. Acad. Sci. U.S.A.* **56**, 1215–1222.

Nass, S., Nass, M. M. K., and Hennix, U. (1965). Deoxyribonucleic acid in isolated rat-liver mitochondria. *Biochim. Biophys. Acta* **95**, 426–435.

Netter, P., Petrochilo, E., Slonimski, P. P., Bolotin-Fukuhara, M., Coen, D., Deutch, J., and Dujon, B. (1974). Mitochondrial genetics VII: Allelism and mapping studies of ribosomal mutants resistant to chloramphenicol, erythromycin and spiramycin in *Saccharomyces cerevisiae. Genetics* **78,** 1063–1100.

Netter, P., Jacq, C., Carignani, G., and Slonimski, P. P. (1982). Critical sequences within mitochondrial introns: Cis-dominant mutations of the "cytochrome-b-like" intron of the oxidase gene. *Cell* **28,** 733–738.

Netzker, R., Köchel, H. G., Basak, N., and Küntzel, H. (1982). Nucleotide sequence of *Aspergillus nidulans* mitochondrial genes coding for ATPase subunit 6, cytochrome oxidase subunit 3, seven unidentified proteins, four tRNAs and L-rRNA. *Nucleic Acids Res.* **10,** 4783–4794.

Newton, B. A. (1979). Biochemical approaches to the taxonomy of kinetoplastid flagellates. *Biol. Kinetoplast.* **5,** 406–434.

Nobrega, F. G., and Tzagoloff, A. (1980a). Assembly of the mitochondrial membrane system. DNA sequence and organization of the cytochrome b gene in *Saccharomyces cerevisiae* D273–10B. *J. Biol. Chem.* **255,** 9828–9837.

Nobrega, F. G., and Tzagoloff, A. (1980b). Assembly of the mitochondrial membrane system. Complete restriction map of the cytochrome b region of mitochondrial DNA in *Saccharomyces cerevisiae* D273-10B. *J. Biol. Chem.* **255,** 9821–9827.

Noller, H. F., and Woese, C. R. (1981). Secondary structure of 16S ribosomal RNA. *Science* **212,** 403–411.

Nomiyama, H., Sakaki, Y., and Takagi, Y. (1981a). Nucleotide sequence of a ribosomal RNA gene intron from slime mold *Physarum polycephalum. Proc. Nat. Acad. Sci. U.S.A.* **78,** 1376–1380.

Nomiyama, H., Kuhara, S., Kutika, T., Otsuka, T., and Sakaki, Y. (1981b). Nucleotide sequence of the ribosomal RNA gene of *Physarum polycephalum:* Intron 2 and its flanking regions of the 26S rRNA gene. *Nucleic Acids Res.* **9,** 5507–5520.

Novick, R. P., Smith, K., Sheehy, R. J., and Murphy, E. (1973). A catenated intermediate in plasmid replication. *Biochem. Biophys. Res. Commun.* **54,** 1460–1469.

Oakley, K. M., and Clark-Walker, G. D. (1978). Abnormal mitochondrial genomes in yeast restored to respiratory competence. *Genetics* **90,** 517–530.

O'Connor, R. M., McArthur, C. R., and Clark-Walker, G. D. (1975). Closed-circular DNA from mitochondrial-enriched fractions of four petite-negative yeasts. *Eur. J. Biochem.* **53,** 137–144.

Ojala, D., Montoya, J., and Attardi, G. (1981). tRNA punctuation model of RNA processing in human mitochondria. *Nature (London)* **290,** 470–474.

Osinga, K. A., DeHaan, M., Christianson, T., and Tabak, H. F. (1982). A nonanucleotide sequence involved in promotion of ribosomal RNA synthesis and RNA priming of DNA replication in yeast mitochondria. *Nucleic Acids Res.* **10,** 7993–8006.

Palmer, J. D., and Thompson, W. F. (1981). Rearrangements in the chloroplast genomes of mung bean and pea. *Proc. Natl. Acad. Sci. U.S.A.* **78,** 5533–5537.

Perlman, P. S., Douglas, M. G., Strausberg, R. L., and Butow, R. A. (1977). Localization of genes for variant forms of mitochondrial proteins on mitochondrial DNA of *Saccharomyces cerevisiae. J. Mol. Biol.* **115,** 675–694.

Powling, A. (1981). Species of small DNA molecules found in mitochondria from sugarbeet with normal and male sterile cytoplasms. *Mol. Gen. Genet.* **183,** 82–84.

Pring, D. R., and Levings, C. S., III. (1978). Heterogeneity of maize cytoplasmic genomes among male-sterile cytoplasms. *Genetics* **89,** 121–136.

Pring, D. R., Levings, C. S., III, Hu, W. W. L., and Timothy, D. H. (1977). Unique DNA associated with mitochondria in the "S" type cytoplasm of male-sterile maize. *Proc. Natl. Acad. Sci. U.S.A.* **74,** 2904–2908.

Pring, D. R., Conde, M. F., and Gengenbach, B. G. (1981). Cytoplasmic genome variability in tissue culture-derived plants. *Environ. Exp. Bot.* **21**, 369–377.

Pring, D. R., Conde, M. F., and Schertz, K. F. (1982). Organelle genome diversity in sorghum: Male-sterile cytoplasms. *Crop Sci.* **22**, 414–421.

Pritchard, A. E., and Cummings, D. J. (1981). Replication of linear mitochondrial DNA from *Paramecium*: Sequence and structure of the initiation-end crosslink. *Proc. Natl. Acad. Sci. U.S.A.* **78**, 7341–7345.

Prunell, A., and Bernardi, G. (1977). The mitochondrial genome of wild-type yeast cells. *J. Mol. Biol.* **110**, 53–74.

Prunell, A., Kopecka, H., Strauss, F., and Bernardi, G. (1977). The mitochondrial genome of wild-type yeast cells. *J. Mol. Biol.* **110**, 17–52.

Quetier, F., and Vedel, F. (1977). Heterogeneous population of mitochondrial DNA molecules in higher plants. *Nature (London)* **268**, 365–368.

Quetier, F., and Vedel, F. (1980). Physicochemical and restriction endonuclease analysis of mitochondrial DNA in higher plants. *In* "Genome Organization and Expression in Plants" (C. J. Leaver, ed.), pp. 401–406. Plenum, New York.

Ramirez, J. L., and Dawid, I. G. (1978). Mapping of mitochondrial DNA in *Xenopus laevis* and *X. borealis*: The positions of ribosomal genes and D-loops. *J. Mol. Biol.* **119**, 133–146.

Reilly, J. G., and Thomas, C. A., Jr. (1980). Length polymorphisms, restriction site variation, and maternal inheritance of mitochondrial DNA of *Drosophila melanogaster*. *Plasmid* **3**, 109–115.

Riou, G. (1976). The kinetoplast DNA from a *Trypanosoma cruzi* strain resistant to ethidium bromide. *In* "The Genetic Function of Mitochondrial DNA" (C. Saccone and A. M. Kroon, eds.), pp. 95–102. North-Holland Publ., Amsterdam.

Riou, G., and Barrois, M. (1979). Restriction cleavage map of kinetoplast DNA minicircles from *Trypanosoma equiperdum*. *Biochem. Biophys. Res. Commun.* **90**, 405–409.

Riou, G., and Saucier, J. (1979). Characterization of the molecular components in kinetoplast-mitochondrial DNA of *Trypanosoma equiperdum*. *J. Cell Biol.* **82**, 248–263.

Riou, G., and Yot, P. (1977). Heterogeneity of the kinetoplast DNA molecules of *Trypanosoma cruzi*. *Biochemistry* **16**, 2390–2396.

Riou, G., Yot, P., and Truhaut, M. R. (1975). Etude de l'ADN kinétoplastique de *Trypanosoma cruzi* à l'aide d'endonucléases de restriction. *C. R. Hebd. Seances Acad. Sci. Ser. D* **280**, 2701–2704.

Riou, G., Kayser, A., and Douc-Rasy, S. (1982). Decatenation of the kinetoplast DNA network of trypanosomes by a bacterial DNA-gyrase. *Biochimie* **64**, 285–288.

Robberson, D. L., and Clayton, D. A. (1973). Pulse-labeled components in the replication of mitochondrial DNA. *J. Biol. Chem.* **248**, 4512–4514.

Rodrick, G. E., Carter, C. E., Woodcock, C. L. F., and Fairbairn, D. (1977). *Ascaris suum*: Mitochondrial DNA in fertilized eggs and adult body muscle. *Exp. Parasitol.* **42**, 150–156.

Roe, B. A., Wong, J. F. H., Chen, E. Y., Armstrong, P. W., Stankiewicz, A., Ma, D.-P., and McDonough, J. (1982). A modified nucleotide 3' to the anticodon may modulate their codon response. *In* "Mitochondrial Genes" (P. Slonimski, P. Borst, and G. Attardi, eds.), pp. 45–49. Cold Spring Harbor Laboratory, Cold Spring Harbor, New York.

Rubenstein, J. L. R., Brutlag, D., and Clayton, D. A. (1977). The mitochondrial DNA of *Drosophila melanogaster* exists in two distinct and stable superhelical forms. *Cell* **12**, 471–482.

Ryan, R., Grant, D., Chiang, K., and Swift, H. (1978). Isolation and characterization of mitochondrial DNA from *Chlamydomonas reinhardtii*. *Proc. Natl. Acad. Sci. U.S.A.* **75**, 3268–3272.

Saccone, C., DeBenedetto, C., Gadaleta, G., Lanave, C., Pepe, G., Rainaldi, G., Sbisa, E., Cantatore, P., Gallerani, R., Quagliarello, C., Holtrop, M., and Kroon, A. M. (1982). Rat mitochondrial DNA: Evolutionary considerations based on nucleotide sequence analysis. *In* "Mitochondrial Genes" (P. Slonimski, P. Borst, and G. Attardi, eds.), pp. 121–128. Cold Spring Harbor Laboratory, Cold Spring Harbor, New York.

Sanders, J. P. M., Heyting, C., Verbeet, M. P., Meijlink, C. P. W., and Borst, P. (1977b). The organization of genes in yeast mitochondrial DNA. III. Comparison of the physical maps of the mitochondrial DNAs from three wild-type *Saccharomyces* strains. *Mol. Gen. Genet.* **157**, 239–261.

Sebald, W., Sebald-Althaus, M., and Wachter, E. (1977). Altered amino acid sequence of the DCCD binding protein of the nuclear oligomycin-resistant mutant AP-2 from *Neurospora crassa*. *In* "Mitochondria 1977—Genetics and Biogenesis of Mitochondria" (W. Bandlow, R. J. Schweyer, K. Wolf, and F. Kaudewitz, eds.), pp. 433–440. De Gruyter, Berlin.

Sebald, W., Hoppe, J., and Wachter, E. (1979). Amino acid sequence of the ATPase proteolipid from mitochondria, chloroplasts and bacteria (wild type and mutants). *In* "Function and Molecular Aspects of Biomembrane Transport" (E. Quagliarello, ed.), p. 63. Elsevier, Amsterdam.

Sederoff, R. R., Levings, C. S., III, Timothy, D. H., and Hu, W. W. L. (1981). Evolution of DNA sequence organization in mitochondrial genomes of *Zea*. *Proc. Natl. Acad. Sci. U.S.A.* **78**, 5953–5957.

Seilhamer, J. J., and Cummings, D. J. (1981). Structure and sequence of the mitochondrial 20S rRNA and tRNA tyr gene of *Paramecium primaurelia*. *Nucleic Acids Res.* **9**, 6391–6406.

Seitz-Mayr, G., and Wolf, K. (1982). Extrachromosomal mutator inducing point mutations and deletions in mitochondrial genome of fission yeast. *Proc. Natl. Acad. Sci. U.S.A.* **79**, 2618–2622.

Sekiya, T., Kobayashi, M., Seki, T., and Koike, K. (1980). Nucleotide sequence of a cloned fragment of rat mitochondrial DNA containing the replication origin. *Gene* **11**, 53–62.

Shah, D. M., and Langley, C. H. (1979). Inter- and intraspecific variation in restriction maps of *Drosophila* mitochondrial DNAs. *Nature (London)* **281**, 696–699.

Simpson, A., and Simpson, L. (1980). Kinetoplast DNA and RNA of *T. brucei*. *Mol. Biochem. Parasitol.* **2**, 93.

Simpson, L. (1972). The kinetoplast DNA of hemoflagellates. *Int. Rev. Cytol.* **32**, 139–207.

Simpson, L. (1979). Isolation of maxicircle component of kinetoplast DNA from hemoflagelate protozoa. *Proc. Natl. Acad. Sci. U.S.A.* **76**, 1585–1588.

Simpson, L., and Simpson, A. M. (1978). Kinetoplast RNA of *Leishmania tarentolae*. *Cell* **14**, 169–178.

Simpson, L., Simpson, A. M., Kidane, G., Livingston, L., and Spithill, T. W. (1980). The kinetoplast DNA of the hemoflagellate protozoa. *Am. J. Trop. Med. Hyg. Suppl.* **29**, 1053–1063.

Simpson, L., Simpson, A. M., Spithill, T. W., and Livingston, L. (1982). Sequence organization of maxicircle kinetoplast DNA from *Leishmania tarentolae*. *In* "Mitochondrial Genes" (P. Slonimski, P. Borst, and G. Attardi, eds.), pp. 435–439. Cold Spring Harbor Laboratory, Cold Spring Harbor, New York.

Slonimski, P. P., Claisse, M. L., Foucher, M., Jacq, C., Kochko, A., Lamouroux, A., Pajot, P., Perrodin, G., Spyridakis, A., and Wambier-Kluppel, M. L. (1978a). Mosaic organization and expression of the mitochondrial DNA region controlling cytochrome c reductase and oxidase. III. A model of structure and function. In "Biochemistry and Genetics of Yeast: Pure and Applied Aspects" (M. Bacila, B. L. Horecker, and A. O. M. Stoppani, eds.), pp. 391–401. Academic Press, New York.

Slonimski, P., Pajot, P., Jacq, C., Foucher, M., Perrodin, G., Kochko, A., and Lamouroux, A. (1978b). Mosaic organization and expression of the mitochondrial DNA region controlling cytochrome c reductase and oxidase. I. Genetic, physical and complementation maps of the *box* region. *In* "Biochemistry and Genetics of Yeasts: Pure and Applied Aspects" (M. Bacila, B. L. Horecker, and A. M. Stoppani, eds.), pp. 339–368. Academic Press, New York.

Slonimski, P., Borst, P., and Attardi, G., eds. (1982). "Mitochondrial Genes." Cold Spring Harbor Laboratory, Cold Spring Harbor, New York.

Smith, J. R., and Rubenstein, I. (1973). Cytoplasmic inheritance of the timing of senescence in *Podospora anserina*. *J. Gen. Microbiol.* **76**, 297–304.

Sparks, R. B., Jr., and Dale, R. M. K. (1980). Characterization of ^{3}H-labeled supercoiled mitochondrial DNA from tobacco suspension culture cells. *Mol. Gen. Genet.* **180**, 351–355.

Spencer, D. F., Bonen, L., and Gray, M. W. (1981). Primary sequence of wheat mitochondrial 5S ribosomal ribonucleic acid: Functional and evolutionary implications. *Biochemistry* **20**, 4022–4029.

Spruill, W. M., Levings, C. S., III, and Sederoff, R. R. (1980). Recombinant DNA analysis indicates that the multiple chromosomes of maize mitochondria contain different sequences. *Dev. Gen.* **1**, 363–378.

Spruill, W. M., Levings, C. S., III, and Sederoff, R. R. (1981). Organization of mitochondrial DNA in normal and Texas male sterile cytoplasms of maize. *Dev. Gen.* **2**, 319–336.

Stahl, U., Kück, U., Tudznyski, P., and Esser, K. (1978). Characterization and cloning of plasmid-like DNA of the ascomycete *Podospora anserina*. *Mol. Gen. Genet.* **178**, 639–646.

Stahl, U., Tudzynski, P., Kück, U., and Esser, K. (1982). Replication and expression of a bacterial-mitochondrial hybrid plasmid in the fungus *Podospora anserina*. *Proc. Natl. Acad. Sci. U.S.A.* **79**, 3641–3645.

Steinert, M., and Van Assel, S. (1980). Sequence heterogeneity in kinetoplast DNA: Reassociation kinetics. *Plasmid* **3**, 7–17.

Steinert, M., Van Assel, S., Borst, P., and Newton, B. A. (1976a). Evolution of kinetoplast DNA. *In* "The Genetic Function of Mitochondrial DNA" (C. Saccone and A. M. Kroon, eds.), pp. 71–81. North-Holland Publ., Amsterdam.

Steinert, M., Van Assel, S., and Steinert, G. (1976b). Mini-circular and non-mini-circular components of kinetoplast DNA. *In* "Biochemistry of Parasites and Host-Parasite Relationships" (H. Van den Bossche, ed.), pp. 193–202. North-Holland Publ., Amsterdam.

Stern, D. B., and Lonsdale, D. M. (1982). Mitochondrial and chloroplast genomes of maize have a 12kb DNA sequence in common. *Nature (London)* **299**, 698–702.

Stern, D. B., Dyer, T. A., and Lonsdale, D. M. (1982). Organization of the mitochondrial ribosomal RNA genes of maize. *Nucleic Acids Res.* **10**, 3333–3340.

Stinchcomb, D. T., Struhl, K., and Davis, R. W. (1979). Isolation and characterization of a yeast chromosomal replicator. *Nature (London)* **282**, 39–43.

Stohl, L. L., Collins, R. A., Cole, M. D., and Lambowitz, A. M. (1982). Characterization of the two new plasmids DNAs found in mitochondria of wild-type *Neurospora intermedia* strains. *Nucleic Acids Res.* **10**, 1439–1458.

Strausberg, R. L., and Butow, R. A. (1981). Gene conversion at the var 1 locus on yeast mitochondrial DNA. *Proc. Natl. Acad. Sci. U.S.A.* **78**, 494–498.

Strausberg, R. L., Vincent, R. D., Perlman, P. S., and Butow, R. A. (1978). Asymmetric gene conversion at inserted segments on yeast mitochondrial DNA. *Nature (London)* **276**, 577–583.

Stuart, K. (1979). Kinetoplast DNA of *Trypanosoma brucei*: Physical map of the maxicircle. *Plasmid* **2**, 520–528.

Stuart, K., and Gelvin, S. (1980). Kinetoplast DNA of normal and mutant *Trypanosoma brucei*. *Am. J. Trop. Med. Hyg. Suppl.* **29**, 1075–1081.

Stuart, K., and Gelvin, S. (1982). Localization of kinetoplast DNA maxicircle transcripts in bloodstream and procyclic form *Trypanosoma brucei*. *Mol. Cell. Biol.* **2**, 845–852.

Suyama, Y. (1982). Native and imported tRNAs in *Tetrahymena* mitochondria: Evidence for their involvement in intramitochondrial translation. *In* "Mitochondrial Genes" (P. Slonimski, P. Borst, and G. Attardi, eds.), pp. 449–455. Cold Spring Harbor Laboratory, Cold Spring Harbor, New York.

Suyama, Y., and Hamada, J. (1976). Imported tRNA: Its synthetase as a probable transport protein. *In* "Genetics and Biogenesis of Chloroplasts and Mitochondria" (T. Bucher, ed.), p. 763. Elsevier, Amsterdam.

Synenki, R. M., Levings, C. S., III, and Shah, D. M. (1978). Physicochemical characterization of mitochondrial DNA from soybean. *Plant Physiol.* **61**, 460–464.

Talen, J. L., Sanders, J. P. M., and Flavell, R. A. (1974). Genetic complexity of mitochondrial DNA from *Euglena gracilis*. *Biochim. Biophys. Acta* **374**, 129–135.

Tapper, D. P., and Clayton, D. A. (1981). Mechanism of replication of human mitochondrial DNA. Localization of the 5' ends of nascent daughter strands. *J. Biol. Chem.* **256**, 5109–5115.

Terpstra, P., and Butow, R. A. (1979). The role of var 1 in the assembly of yeast mitochondrial ribosomes. *J. Biol. Chem.* **254**, 12662–12669.

Terpstra, P., Zanders, E., and Butow, R. A. (1979). The association of var 1 with the 38 S mitochondrial ribosomal subunit in yeast. *J. Biol. Chem.* **254**, 12653–12661.

Thompson, R. D., Kemble, R. J., and Flavell, R. B. (1980). Variations in mitochondrial DNA organization between normal and male-sterile cytoplasms of maize. *Nucleic Acids Res.* **8**, 1999.

Timmis, J. N., and Scott, N. S. (1983). Sequence homology between spinach nuclear and chloroplast genomes. *Nature (London)* **305**, 65–67.

Timothy, D. H., Levings, C. S., III, Pring, D. R., Conde, M. F., and Kermicle, J. L. (1979). Organelle DNA variation and systematic relationships in the genus *Zea*: Teosinte. *Proc. Natl. Acad. Sci. U.S.A.* **76**, 4220–4224.

Timothy, D. H., Levings, C. S., III, Hu, W. W. L., and Goodman, M. M. (1982). *Zea diploperennis* may have plasmid-like mitochondria DNAs. *Maize Genet. Coop. News Lett.* **56**, 133.

Timothy, D. H., Levings, C. S., III, Hu, W. W. L., and Goodman, M. M. (1983). Plasmid-like mitochondrial DNAs in diploperennial teosinte. *Maydica* **28**, 139–149.

Tudzynski, P., and Esser, K. (1979). Chromosomal and extrachromosomal control of senescence in the ascomycete *Podospora anserina*. *Mol. Gen. Genet.* **173**, 71–84.

Tudzynski, P., Stahl, U., and Esser, K. (1980). Transformation to senescence with plasmid-like DNA in the ascomycete *Podospora anserina*. *Curr. Genet.* **2**, 181–190.

Turner, G., Earl, A. J., and Greaves, D. R. (1982). Interspecies variation and recombination of mitochondrial DNA in the *Aspergillus nidulans* species group and the selection of species-specific sequences by nuclear background. *In* "Mitochondrial Genes" (P. Slonimski, P. Borst, and G. Attardi, eds.), pp. 411–414. Cold Spring Harbor Laboratory, Cold Spring Harbor, New York.

Tzagoloff, A. (1982). "Mitochondria." Plenum, New York.

Tzagoloff, A., Macino, G., and Sebald, W. (1979). Mitochondrial genes and translation products. *Annu. Rev. Biochem.* **48**, 419–441.

Tzagoloff, A., Nobrega, M., Akai, A., and Macino, G. (1980). Assembly of the mitochondrial membrane system. Organization of yeast mitochondrial DNA in the oli 1 region. *Curr. Genet.* **2**, 149–157.

Upholt, W. B., and Dawid, I. G. (1977). Mapping of mitochondrial DNA of individual sheep and goats: Rapid evolution in the D loop region. *Cell* **11**, 571–583.

van den Boogaart, P., Samallo, J., and Agsteribbe, E. (1982a). Similar genes for a mitochondrial ATPase subunit in the nuclear and mitochondrial genomes of *Neurospora crassa. Nature (London)* **298**, 187–189.

van den Boogaart, P., Samallo, J., van Dijk, S., and Agsteribbe, E. (1982b). Structural and functional analyses of the genes for subunit II of cytochrome aa₃ and for a dicyclohexylcarbodiimide-binding protein in *Neurospora crassa* mitochondrial DNA. *In* "Mitochondrial Genes" (P. Slonimski, P. Borst, and G. Attardi, eds.), pp. 375–380. Cold Spring Harbor Laboratory, Cold Spring Harbor, New York.

Van Etten, R. A., Michael, N. L., Bibb, M. J., Brennicke, A., and Clayton, D. A. (1982). Expression of the mouse mitochondrial DNA genome. *In* "Mitochondrial Genes" (P. Slonimski, P. Borst, and G. Attardi, eds.), pp. 73–88. Cold Spring Harbor Laboratory, Cold Spring Harbor, New York.

Van Ommen, G. J. B., Groot, G. S. P., and Grivell, L. A. (1979). Transcription maps of mtDNAs of two strains of *Saccharomyces*: Transcription of strain-specific insertions; complex RNA maturation and splicing. *Cell* **18**, 511–523.

Van Ommen, G.-J. B., De Boer, P. H., Groot, G. S. P., de Haan, M., Roosendaal, E., Grivell, L. A., Haid, A., and Schweyen, R. J. (1980). Mutations affecting RNA splicing and the interaction of gene expression of the yeast mitochondrial loci cob and oxi-3. *Cell* **20**, 173–183.

Walberg, M. W., and Clayton, D. A. (1981). Sequence and properties of the human KB cell and mouse L cell D-loop regions of mitochondrial DNA. *Nucleic Acids Res.* **9**, 5411–5421.

Wallace, D. C. (1982). Structure and evolution of organelle genomes. *Microbiol. Rev.* **46**, 208–240.

Wallace, D. C., Oliver, N. A., Blanc, H., and Adams, C. W. (1982). A system to study human mitochondrial genes: Application to chloramphenicol resistance. *In* "Mitochondrial Genes" (P. Slonimski, P. Borst, and G. Attardi, eds.), pp. 105–116. Cold Spring Harbor Laboratory, Cold Spring Harbor, New York.

Wang, J. C. (1971). Interaction between DNA and an *Escherichia coli* protein. *J. Mol. Biol.* **55**, 523–533.

Ward, B. L., Anderson, R. S., and Bendich, A. J. (1981). The size of the mitochondrial genome is large and variable in a family of plants (Curcurbitaceae). *Cell* **25**, 793–803.

Waring, R. B., Davies, R. W., Lee, S., Grisi, E., Berks, M. M., and Scazzocchio, C. (1981). The mosaic organization of the apocytochrome b gene of *Aspergillus nidulans* revealed by DNA sequencing. *Cell* **27**, 4–11.

Waring, R. B., Davies, R. W., Scazzocchio, C., and Brown, T. A. (1982). Internal structure of a mitochondrial intron of *Aspergillus nidulans. Proc. Natl. Acad. Sci. U.S.A.* **79,** 6332–6336.

Weislogel, P. O., Hoeijmakers, J. H. J., Fairlamb, A. H., Kleisen, C. M., and Borst, P. (1977). Characterization of kinetoplast DNA networks from the insect trypanosome *Crithidia luciliae. Biochim. Biophys. Acta* **478,** 167–179.

Weiss-Brummer, B., Rödel, G., Schweyen, R. J., and Kaudewitz, F. (1982). Expression of the split gene cob in yeast: Evidence for a precursor of a "Maturase" protein translated from intron 4 and preceding exons. *Cell* **29,** 527–536.

Weissinger, A. K., Timothy, D. H., Levings, C. S., III, Hu, W. W. L., and Goodman, M. M. (1982). Unique plasmid-like mitochondrial DNAs from indigenous maize races of Latin America. *Proc. Natl. Acad. Sci. U.S.A.* **79,** 1.

Weissinger, A. K., Timothy, D. H., Levings, C. S., III, and Goodman, M. M. (1983). Patterns of mitochondrial DNA variation in indigenous maize races of Latin America. *Genetics* **104,** 365–379.

Weslowski, M., and Fukuhara, H. (1981). Linear mitochondrial deoxyribonucleic acid from the yeast *Hansenula mrakii. Mol. Cell. Biol.* **1,** 387–393.

Weslowski, M., Algeri, A., and Fukuhara, H. (1981). Gene organization of the mitochondrial DNA of yeasts: *Kluyveromyces lactis* and *Saccharomycopsis lipolytica. Curr. Genet.* **3,** 157–162.

Wheeler, M. R. (1981). The Drosophilidae: A taxonomic overview. *In* "The Genetics and Biology of Drosophila" (M. Ashburner, H. L. Carson, and J. N. Thompson, Jr., eds.), Vol. 3a, pp. 1–97. Academic Press, New York.

Wolf, K., Lang, B., Del Giudice, L., Anziano, P. Q., and Perlman, P. S. (1982). *Schizosaccharomyces pombe*: A short review of a short mitochondrial genome. *In* "Mitochondrial Genes" (P. Slonimski, P. Borst, and G. Attardi, eds.), pp. 355–360. Cold Spring Harbor Laboratory, Cold Spring Harbor, New York.

Wolstenholme, D. R., and Dawid, I. B. (1968). A size difference between mitochondrial DNA molecules of urodele and anuran amphibia. *J. Cell Biol.* **39,** 222–228.

Wolstenholme, D. R., and Gross, N. J. (1968). The form and size of mitochondrial DNA of the red bean, *Phaseolus vulgaris. Proc. Natl. Acad. Sci. U.S.A.* **61,** 245–252.

Wolstenholme, D. R., Renger, H. C., Manning, J. E., and Fouts, D. L. (1974). Kinetoplast DNA of *Crithidia. J. Protozool.* **21,** 622–631.

Wolstenholme, D. R., Goddard, J. M., and Fauron, C. M.-R. (1979). Structure and replication of mitochondrial DNA from the genus *Drosophila. In* "Extrachromosomal DNA: ICN-UCLA Symposia on Molecular and Cellular Biology" (D. J. Cummings, P. Borst, I. B. Dawid, S. M. Weissman, and C. F. Fox, eds.), pp. 410–425. Academic Press, New York.

Wolstenholme, D. R., Fauron, C. M.-R., and Goodard, J. M. (1982). Nucleotide sequence of *Rattus norvegicus* mitochondrial DNA that includes the genes for tRNA[ile], tRNA[gln], and tRNA[f-met]. *Gene* **20,** 63–69.

Wright, R. M., Horrum, M. A., and Cummings, D. J. (1982). Are mitochondrial structural genes selectively amplified during senescence in *Podospora anserina? Cell* **29,** 505–515.

Yin, S., Heckman, J., and RajBhandary, U. L. (1981). Highly conserved GC-rich palindromic DNA sequences flank tRNA genes in *Neurospora crassa* mitochondria. *Cell* **26,** 325–332.

Yin, S., Burke, J., Chang, D. D., Browning, K. S., Heckman, J. E., Alzner-DeWeerd, B., Potter, M. J., and RajBhandary, U. L. (1982). *Neurospora crassa* mitochondrial tRNAs and rRNAs: Structure, gene organization, and DNA sequences. *In* "Mitochondrial Genes" (P. Slonimski, P. Borst, and G. Attardi, eds.), pp. 361–373. Cold Spring Harbor Laboratory, Cold Spring Harbor, New York.

Yonekawa, H., Moriwaki, K., Gotoh, O., Hayashi, J., Watanabe, J., Miyashita, N., Petras, M. L., and Tagashira, Y. (1981). Evolutionary relationships among five subspecies of *Mus musculus* based on restriction enzyme cleavage patterns of mitochondrial DNA. *Genetics* **98,** 801–816.

Yoshikawa, H., and Ito, J. (1981). Terminal proteins and short inverted terminal repeats of the small *Bacillus* bacteriophage genomes. *Microbiology* **78,** 2596–2600.

Zakian, V. A. (1981). Origin of replication from *Xenopus laevis* mitochondrial DNA promotes high-frequency transformation of yeast. *Proc. Natl. Acad. Sci. U.S.A.* **78,** 3128–3132.

THE STRUCTURE AND EXPRESSION OF NUCLEAR GENES IN HIGHER PLANTS

John C. Sorenson

Experimental Agricultural Sciences, The Upjohn Company,
Kalamazoo, Michigan

I. Introduction

Information about the structure and organization of higher plant genes has been accumulating at a rapid rate over the past several years. Substantial progress has been made in the analysis of organelle genes and genomes (reviewed by Sederoff, this volume). Information

109

on nuclear gene organization has accumulated more slowly but, as described in the following pages, the body of information available on nuclear genes has become quite impressive. Although I have attempted to provide a thorough and critical review of the current literature, a body of new and exciting information exists that has not yet found its way into print. The result is that many of the systems described are considerably more advanced than this article might indicate.

To date, our best information is limited to a few specialized classes of genes, but information is beginning to accumulate on a variety of diverse systems. Here I have attempted to point out gene isolation and identification strategies where similar approaches could be applied to other systems, and I have included information on some systems that have not yet provided much specific information on gene structure and expression but that seem to hold considerable promise for doing so. In this article, my goal is to consolidate a rather diffuse literature and to suggest some areas that could use more attention.

II. Genome Structure in Higher Plants

The sequence organization of higher plant DNA has been reviewed most recently by Flavell (1980) and earlier by Walbot and Goldberg (1979) and Thompson and Murray (1979). These reviews were comprehensive and I will not reiterate that detail here. I will provide a brief overview of the subject and will discuss a few of the additional studies published since those reviews appeared.

Higher plant genomes are typically large, in some cases because many higher plant genomes are amphidiploid or polyploid. Several plants known to be highly polyploid have genome sizes in excess of 10^{11} nucleotide pairs (ntp). Most plant genomes range between 10^9 and 10^{10} ntp. The range of plant genome sizes is also large. The lower end of the range of plant genome sizes characterized to date is about 2×10^8 ntp. All of the variation in genome sizes cannot be accounted for by ploidy variations, however. The kinetic complexity, which reflects the unique informational content of the genome by factoring out the amount of DNA that is reiterated, varies by nearly two orders of magnitude.

Reassociation kinetic analyses have been conducted on a wide variety of plant species (Table 1). In general, they follow the classical short-period interspersion pattern typical of most animal genomes.

TABLE 1
Genome Size and Organization in Higher Plants

Plant	Haploid genome size (ntp × 10^9)	Percentage repetitive DNA	Percentage single-copy DNA in long period[a]	Percentage single-copy DNA in short period	References
Allium (7 sp.)	7–20	44–65	ND[b]	ND	Ranjekar et al. (1978)
Atriplex (8 sp.)	0.4–0.7	60–74	ND	ND	Belford and Thompson (1981a,b)
Barley	5	70	<9	>90	Rimpau et al. (1980)
Broadbean	44	80	ND	>65	Taylor and Bendich (1977)
Cotton	0.7	32	<20	>80	Walbot and Dure (1976)
Flax	0.4	56	<100[c]	ND	Cullis (1981)
Maize	5.7	60	<34	>66	Hake and Walbot (1980)
Mung bean	0.5	35	<46	>54	Murray et al. (1979)
Oats	13	70 -	<9	>90	Rimpau et al. (1980)
Oryza (8 sp.)	0.2–0.3	58–66	ND	ND	Iyengar et al. (1979)
Parsley	3	70	<17	>83	Kiper and Herzfeld (1978)
Pea	4.5	70	<3	>97	Murray et al. (1978)
Pearl millet	0.2	69	ND	>50	Wimpee and Rawson (1979)
Phaseolus aureus	0.6	45	ND	>40	Sheshardi and Ranjekar (1980)
Phaseolus vulgaris	0.3	45	ND	>40	Sheshardi and Ranjekar (1980)
Rye	7	75	ND	>80	Smith and Flavell (1977); Rimpau et al. (1978)
Soybean	1	41	<35	>65	Gurley et al. (1979)
Soybean	2	63	<32	>68	Goldberg (1978)
Spinach	0.8	74	ND	ND	Belford and Thompson (1981a,b)
Tobacco	1.5	55	<20	>80	Zimmerman and Goldberg (1977)
Wheat	5	75	ND	>80	Flavell and Smith (1976); Rimpau et al. (1978)

[a] Due to technical limitations these values are maximum values and the short-period values are minimum values.

[b] ND, Not determined in these experiments.

[c] These values were determined using interpretive methods not generally employed in similar analyses. Using more traditional interpretation, a substantial amount of the single-copy DNA appears to be interspersed with repetitive DNA at a mean interval of 800 ntp.

Some caution should be used in comparing the results from different laboratories because kinetic reassociation experiments are particularly sensitive to variation in method and technique. In many case, these may materially affect the conclusions derived from various experiments.

Most of the DNA in a "typical" plant consists of short lengths of single-copy DNA (200–4000 ntp) interspersed with short lengths of repeated sequence DNA (50–2000 ntp). The repeated sequence DNA consists primarily of families reiterated 10–100 times per sequence (low-frequency repetitive DNA) or 100–1000 times per sequence (moderately repeated DNA). There is also a (usually) minor fraction of highly repetitive DNA (greater than 10,000 copies per sequence). The amount of this highly repeated fraction is variable among plant species. Wheat and maize have relatively large amounts of highly repeated DNA. This sequence class composes nearly 10% of the total in wheat, in which it is preferentially associated with one of the three constituent genomes (the B genome; Gerlach and Peacock, 1980). In maize, as much as 20% of the DNA may be highly repetitive (Hake and Walbot, 1980), and in addition to a general short-period interspersion pattern it contains a substantial amount of long-period interspersion DNA. It is not clear, however, to what extent these results could be confounded by contamination with organelle DNA (cf. Murray et al., 1979). Pearl millet also shows a substantial amount of long-period interspersion (Wimpee and Rawson, 1979).

The best characterized exception to the short-period interspersion pattern is the mung bean. In contrast to most other plants, in which repetitive DNA sequences are separated by single-copy sequences ranging from 250 to 1500 ntp in length, nearly half of the single-copy sequences in the mung bean are in excess of 6700 ntp (Murray and Thompson, 1979). The mung bean genome is quite small (Table 1). The water mold *Achyla,* which has a very small genome (3.7×10^7 ntp), also exhibits long-period interspersion (Hudspeth *et al.,* 1977). In contrast, cotton has a relatively small genome, but over 80% of the single-copy DNA is found in the short-period interspersion pattern (Walbot and Dure, 1976). Maize, which has a relatively large genome, also has a substantial amount of long-period interspersion DNA. Therefore, the general trend toward small genomes having more long-period interspersion is not inviolate. In addition, the fact that three related legumes, pea, broadbean, and mung bean, have different interspersion patterns and very different genome sizes reflects our inability to ascribe any particular significance or function to genome organizational patterns. Pea has one of the shortest single-copy interspersion patterns ever reported for any species, and mung bean has one of the longest (Preisler and Thompson, 1981a,b).

The repeat units themselves occur in two basic types of patterns. The simplest, the so-called tandem repeat, consists of identical or similar repeating units arranged adjacently to each other in large arrays. "Satellite" DNAs are the classic example of this type of organization, although this pattern is by no means limited to satellites. The lengths of the repeat units in the arrays may vary from a few nucleotide pairs (as few as three in barley; Dennis *et al.*, 1980) to several thousand nucleotide pairs. The reiteration frequency of these repeat units sometimes exceeds a million copies per constituent genome. The second type of pattern consists of complex interspersions of repeat families. In contrast to the tandem arrays, in which the repeats of any sequence family tend to be localized in specific regions of chromosomes, the interspersed arrays tend to be scattered throughout the genome in various permutations. Repeat families of one reiteration frequency are frequently interspersed with other frequency classes.

There have been several studies on satellite DNA structures in higher plants. The best studied are the cereal genomes, although there have been several recent and useful studies concerning other species. The typical strategy for studying such structures has involved reassociation of repeated sequence DNA followed by an analysis of duplex fidelity and repeat length by either single-strand nuclease digestion or a thermal melting strategy. Similar techniques have been combined with heterologous DNA–DNA hybridizations to estimate the divergence of repeated sequence families among related species. The most important conclusion drawn from such studies is that the rate of evolution of such sequences is exceedingly rapid. Hexaploid wheat and relatives of its constituent species provide an elegant comparison of the rates of species divergence (Flavell *et al.*, 1979). Hexaploid wheat (actually an allohexaploid) has one *Triticum* (*monococcum*) and two *Aegilops* (*squarossa* and *speltoides* or a near relative) progenitor parents. *Triticum* and *Aeqilops* have unique, highly repeated DNA sequences. There has been sequence divergence between the *Aegilops* species and, in addition, there appears to have been some sequence divergence within the hexaploid wheat itself. There are repeated sequences associated with the A genome (from the *Triticum* progenitor) that are not found in *Triticum monococcum*. Gerlach and Peacock (1980) were able to use *in situ* hybridization to show that the most highly repeated sequences in hexaploid wheat are preferentially asso-

ciated with the B genome (*Aegilops speltoides*). In maize, highly repeated DNA is associated with knob heterochromatin (Peacock *et al.*,
1981). Additional comparisons have been made with wheat, rye, barley, and oats (Rimpau *et al.*, 1978, 1980; Ranjekar *et al.*, 1978b). Divergence of repetitive sequences among maize races and possible progenitor species was also detected in a similar study by Hake and Walbot
(1980).

Because the sensitivity of such experiments is relatively low, this
sort of result represents substantial divergence. In cases such as maize,
where archaeological and other lines of evidence allow a reasonable
estimate of species divergence, it is unlikely that such dramatic
changes can be explained by the traditional point-mutational model.

Using cloned DNA sequences, Bedbrook *et al.* (1980a,b) have been
able to more definitively study the details of repeated sequence DNA
arrangement in rye. Five of the seven highly repeated rye sequence
families studied were found to be compound, that is, composed of sub-
families with differing reiteration frequencies. Four major tandem arrays account for over half of the telomeric DNA.

Several recent studies have also addressed the nature of repeated
DNA structure in plants other than cereals. In soybeans the predominant repeated sequence DNA organization (except for that which is
interspersed with single-copy DNA) is long-tandem repeats (Goldberg,
1978; Pellegrini and Goldberg, 1979). This may reflect the fact that as
much as 25% of the soybean genome may occur as a satellite (Hepburn
et al., 1977). Gurley *et al.* (1979) also observed that complex networks
formed when large (>11,000 ntp) soybean DNA is reassociated to moderately low C_0t values, suggesting that long stretches of tandemly
repeated DNA may be interspersed with short stretches of DNA with
other reiteration frequencies. Parsley also shows a mixture of tandem
array (13% of the genome) and complex interspersion (20% of the genome; Kiper and Herzbeld, 1978).

An analysis of the satellite DNA in two *Cucumis* species by Hemleben *et al.* (1982) indicates a short repeat unit in the tandem array and a
surprisingly high rate of divergence (less than 10% homology among
related species).

Perhaps the most striking example of rapid divergence of repetitive
DNA to date was described by Preisler and Thompson (1981a,b) in a
comparison of the pea and mung bean genomes. Bendich and Andersen
(1977) proposed the concept of homogeneous and heterogeneous repeated sequence families, and described an experimental technique to distinguish between the two types. Homogeneous families contain se-

quences with a high degree of sequence homology and heterogeneous families have relatively less sequence homology. The generation of these two types of sequence structure would likely involve different evolutionary mechanisms. Based on this sort of analysis of the pea and mung bean genomes, Preisler and Thompson proposed a model for genome evolution that involves different rates of amplification of repeated DNA sequences in the two species. Such variable amplification rates could easily account for the wide range of genome sizes in higher plants as well as the variation in repetitive DNA structure. Verification of such variable rates could provide important new insights into the evolutionary process. A similar role for amplification in the evolution of cereals has also been discussed by Bedbrook *et al.* (1980b).

Two studies of the intrageneric evolution of higher plant genomes provide contrasting conclusions as to the degree of short-term divergence of DNA sequences among species of the same genus. Iyengar *et al.* (1979) found that among eight *Oryza* species very little divergence was observed in nonrepetitive DNA, whereas Ranjekar *et al.* (1978a) found substantial divergence in both the repetitive and nonrepetitive DNA in seven *Allium* species examined. The *Allium* genomes are approximately 10-fold larger than the *Oryza* genomes. These observations would be consistent with a model in which there is variation in the rate at which genomic segments are amplified in various species during evolution. Possible models of genome evolution are discussed in greater detail by Flavell (1980).

Belford and Thompson (1981a,b) have studied the divergence of single-copy DNA in *Atriplex* species. Only about 5×10^7 ntp of single-copy DNA are conserved among members of this genus, and this therefore sets a maximum limit on the amount of single-copy DNA necessary to specify the phenotypic characteristics common to *Atriplex* species. This amount of DNA would code for approximately the number of genes expressed in a mature tobacco plant (see below). These studies were conducted using solution hybridization and thermal denaturation of the heterologous DNA hybrids and, of course, do not directly assess the amount of point mutation at the levels necessary to alter gene function. The authors argue that deletion may be an important adjunct to amplification in the evolution of plant genomes.

In summary, plant genomes are large and possess a wide range of genome sizes. Although several extensive studies of genome organization have failed to provide any new insights into the functional significance of genome structure in general, new concepts regarding the mechanisms of plant genome evolution are emerging.

III. Genes Coding for Seed Storage Proteins

The most extensively studied genes in higher plants are those coding for seed storage proteins. In many plant species these proteins (and hence their mRNAs) are extremely abundant in easily manipulable tissues and are therefore attractive for the isolation and characterization of the gene sequences. In addition, these proteins are of considerable interest in agronomic species because they are the primary determinants of nutritional quality in edible seeds.

Major seed proteins have traditionally been classified on the basis of their solubility properties. The major storage proteins in cereals are the prolamins, which are soluble in aqueous alcohol. These proteins contain up to 30 mol% proline and up to 40 mol% glutamine. They tend to be poor in several essential amino acids, most notably lysine. There is limited immunological cross-reactivity among the prolamins of maize, barley, and wheat (zein, hordein, and gliadin, respectively; Dierks-Ventiling and Cozens, 1982), suggesting some homology among the proteins. Determination of the extent of this homology by nucleic acid sequence comparisons is, at this time, not available.

The other classes of abundant seed proteins are albumins and globulins, which are soluble in salt solutions, and the glutelins, which are soluble in alkalai. The genes for seed storage proteins have been isolated from several species, and the information on their structure and expression is summarized below.

A. MAIZE

Perhaps the best characterized nuclear gene family in higher plants is the zein gene family. Zein is the major storage protein in maize. Zeins (and glutelins) have been postulated to play a role as a nitrogen sink in maize and to regulate photosynthate flow into kernels (Tsai *et al.*, 1980). Together these two proteins compose nearly 80% of the protein in the maize endosperm. Zein polypeptides account for 60% of the protein in the endosperm. Zeins compose a heterogeneous group of proteins that can be grouped into apparent size classes of 22, 19, 15, and 10 kd (Misra *et al.*, 1975; Wall and Paulis, 1978). Recent DNA sequence data indicate that these sizes are considerable underestimates (see below). I shall continue to refer to these proteins as 19- and 22-kd proteins to avoid confusion with the current literature. The 22- and 19-kd size classes are prominent and the other two are relatively minor. There is considerable heterogeneity within each of the two major size classes. Between 10 and 20 zein polypeptides can be identified in these size classes by isoelectric focusing or two-dimensional gel electrophoresis (Righetti *et al.*, 1977; Hagen and Rubenstein, 1980) and by amino-terminal sequence analysis (Bietz *et al.*, 1979; Larkins and Pederson, 1982). The complexity of the zein protein patterns has made genetic analysis of the zein system difficult. In an ambitious set of experiments summarized by Salamini and Soave (1982), the isoelectric focusing patterns in various maize inbreds have been used to map several of the zein genes to specific chromosomes. There appears to be some clustering of genes coding for the 19-kd protein on chromosome 7 near the *opaque-2* locus (Soave *et al.*, 1981a). The significance of this finding is disucssed below. Zein genes have also been mapped on several other chromosomes. These mapping results are largely consistent with the gene locations determined by *in situ* hybridization experiments (Viotti *et al.*, 1980).

Larkins and Dalby (1975) and Larkins *et al.* (1976) demonstrated the synthesis of zein polypeptides on isolated membrane polyribosomes in proportions equivalent to those found in the cell, and were later able to isolate zein mRNAs from this fraction (Larkins *et al.*, 1976). Zein is found sequestered in the maize cell in membrane-bound protein bodies (Duvick, 1961). In a clever approach to isolating zein mRNAs, Burr and Burr (1976) were able to show that zein-synthesizing polyribosomes were associated with the external surface of the protein body membranes, and that isolated protein bodies could serve as a source of highly enriched zein mRNAs (Burr *et al.*, 1978).

The 22- and 19-kd polypeptides are coded by separate but similarly sized mRNAs approximately 1100 nucleotides in length (Larkins and Hurkman, 1978; Wienand and Feix, 1978). Viotti *et al.* (1979a) showed that zein products synthesized *in vitro* from isolated mRNAs showed a similar charge heterogeneity to native zein polypeptides. Park *et al.* (1980) used cloned zein cDNA sequences to demonstrate that although the zein mRNA sequences were indeed heterogeneous, they could be assigned to one of three major groups, based on sequence homology. Zein mRNAs, selected by hybridization to cDNA clones representing the three sequence-homology groups, direct the synthesis of most of the zein polypeptides observed in the *in vitro* translation products of total zein mRNA. One of the homology groups contains sequences coding for the larger 22-kd polypeptide and two groups contain sequences coding for the 19-kd polypeptide. Burr *et al.* (1982), using some of the same cDNA clones as well as several additional ones, found at least two homology groups for the 22-kd polypeptide, and three homology groups for the 19-kd polypeptide. Marks and Larkins (1982), using an independent set of cDNA clones, found evidence for two or three homology groups for the 22-kd polypeptide and two groups for the 19-kd polypeptide. In addition, they were able to use clones for the 15-kd polypeptide to establish a single homology group for those sequences. This apparent homology in the mRNAs coding for the 15-kd protein is consistent with the lack of heterogeneity of this protein on two-dimensional gels of the native zein polypeptides (Hurkman *et al.*, 1981). No clones coding for the 10-kd polypeptide were recovered.

Zein polypeptides synthesized *in vitro* are larger than the corresponding native polypeptides. This added size is due to the presence of leader sequences that are removed in a cotranslational processing step. Burr and Burr (1981a) used a cell-free translation system enriched with nuclease-treated endoplasmic reticulum membranes to demonstrate that mature processed zein polypeptides could be produced *in vitro*. Adding these membranes to fully synthesized but unprocessed zein does not result in efficient cleavage, suggesting that the removal of the leader sequence is a cotranslational event. The investigators were unable to find a protein fragment corresponding to the processed leader sequence. When zein mRNA is microinjected into *Xenopus* oocytes, the proteins are appropriately cleaved (Larkins *et al.*, 1979) and can even be found in association with membraneous structures similar to endosperm protein bodies (Hurkman *et al.*, 1981).

The structure and organization of the zein genes are being investigated by several groups. Maize genomic clones have been isolated by Lewis *et al.* (1981), Pederson *et al.* (1982), Wienand *et al.* (1981), and Pintor-Toro *et al.* (1982). When maize DNA is digested to apparent completion with any of several restriction endonucleases and analyzed on Southern blots, a wide range of fragment sizes hybridize to zein cDNA clones (Wienand and Feix, 1980; Hagen and Rubenstein, 1981). It is not clear to what extent such experiments could be confounded by incomplete digestion of the DNA (caused by internal methylation, for instance) and it is difficult to ascertain the significance of this high degree of variation. The size range of the genomic clones obtained in three labs does not coincide with the range of sizes observed in genomic DNA blots. The vectors used in cloning the genomic fragments would discriminate against some sizes of inserts, however, and the sample size is small.

The zein genes studied to date do not contain intervening sequences detectable by any of three techniques. Neither electron microscopy (Pintor-Toro *et al.*, 1982; Wienand and Feix, 1981), single-strand-specific nuclease digestion of hybrid molecules formed between genomic cloned DNA and either zein mRNA or zein cDNA (Langridge *et al.*, 1982; Pederson *et al.*, 1982), nor direct sequence comparisons of cDNA clones and genomic clones (Pederson *et al.*, 1982; Rubenstein, 1982) revealed evidence of any introns. The possibility cannot be ruled out that other zein genes may contain such sequences, although at this point that is beginning to seem unlikely.

The first zein genomic clone to be sequenced appears to have the classical recognition sequences common to most eukaryotic genes. Larkins and Pederson (1982) reported a TATAAATAT (TATA box) sequence 33 nucleotides upstream from the probable transcription initiation site and a potential polyadenylation signal (AATAAA) 24 nucleotides downstream from the termination codon. A cDNA clone sequenced by Geraghty *et al.* (1981) contains this sequence at the same location. This sequence is located considerably farther from the actual beginning of the poly(A) tail than is typical of eukaryotic messages. A variant of the canonical sequence, AATAAG, appears nearer the 3' terminus of the transcription unit. It is not known if either (or both) of these sequences are in fact polyadenylation signals in maize.

An interesting observation from two separate laboratories has come from the sequence comparisons of several zein cDNA clones. There are short amino acid repeats in the zein genes that are variable in length among clones (Larkins and Pederson, 1982; Geraghty *et al.*, 1981; Rubenstein, 1982). This type of variation could be responsible for some of the heterogeneity observed in the zein proteins. Zein appears to be a highly ordered protein (Argos *et al.*, 1982), and the variation in the number of repeating units could conceivably be more easily accommodated in a helical structure than random amino acid substitutions.

There is general consensus among investigators as to most aspects of zein gene structure and organization. The primary issue that remains unresolved concerns the number of zein genes in the maize genome. Pederson *et al.* (1980) estimated a reiteration frequency of 1–5 copies per haploid genome, using solution hybridization. Assuming the number of homology groups (described above) to be 5–7, one can estimate the number of zein genes to be 5–35 per haploid genome. Viotti *et al.* (1979b) conducted a similar analysis and concluded that the reiteration frequency of zein genes was 8–10. Using the number of homology groups described above, one can estimate the number of zein genes to be 40–70 per haploid genome. This magnitude of difference is well within the expected variation in this type of experiment between laboratories. However, Viotti *et al.* estimated the number of homology groups to be 15, based on the hybridization kinetics of cDNA to an excess of mRNA. Using this number, one would estimate up to 150 zein genes per haploid genome. This method of determining homology groups would seem to be inherently less reliable than the one obtained by the cross-hybridization of cloned cDNA sequences, as described above (Park *et al.*, 1980; Burr *et al.*, 1982; Marks and Larkins, 1982). Although the number of cDNA clones examined is relatively small, and the number of homology groups could be expected to increase somewhat (no clones for the 10-kd polypeptide have been found, for instance), it seems unlikely that the number of major homology groups will exceed 10. Viotti *et al.* (1979b) also conducted a hybridization–saturation experiment that they claimed indicated 110–130 gene copies. This determination relies on accurate values for the genome complexity and driver cDNA size uniformity, as well as assumptions about cDNA purity and DNA reactivity. These values cannot generally be determined to greater than a twofold reliability.

Additional evidence for zein gene copy numbers in the neighborhood of 100 comes from blot hybridization experiments (discussed by Rubenstein, 1982), in which reconstruction experiments were used to estimate reiteration frequency within a band of hybridization intensity.

The evidence on zein gene numbers may be summarized as follows:

1. Approximately 20–25 zein polypeptides can be observed on two-dimensional gels. This number represents a probable underestimate of the zein gene number, because not every gene variant would be detected and there is little evidence for zein heterogeneity caused by posttranslational modifications.

2. Solution hybridization studies from 2 labs suggest a reiteration frequency of 5–10 copies per homology group.

3. Cross-hybridization studies suggest 5–8 homology groups. Kinetic analyses suggest 15 homology groups; however, this number depends on necessary assumptions in the analysis.

4. Saturation–hybridization studies suggest 110–130 gene copies, although this value should only be considered reliable within a two- to threefold range.

5. Blotting experiments suggest approximately 100 gene copies. This represents a probable overestimate of gene number due to potential cross-hybridization with nonzein sequences or zein pseudogenes and possible partial digestion of genomic DNA.

Much of the controversy regarding zein gene numbers seems to have arisen from attempts to make precise estimates of gene numbers using techniques that are inherently imprecise. If one considers the probable direction and magnitude of the biases in each of these techniques and weights them appropriately, it seems that the actual gene numbers will fall somewhere intermediate to the extremes that have been proposed. The solution hybridization experiments would seem to indicate a gene copy number of 25–80. It seems futile to attempt to refine that range with the data currently available. The important conclusion— that zein polypeptides are encoded by a large and complex gene family—remains unchanged.

Certainly the elucidation of the details of zein transcription and the additional characterization of zein gene structure will be of interest in future studies. Such experiments will no doubt be challenging, given the complexity of the system. An additional subject that should prove interesting involves the regulation of zein synthesis. A number of loci have been described that control levels of zein in the maize endosperm (Mertz *et al.*, 1964; Nelson, 1969; McWhirter, 1971; Ma and Nelson, 1975; Salamini *et al.*, 1979). These genes do not appear to code directly for zein polypeptides. The best studied of these is the *opaque-2* locus. The accumulation of zein in the endosperm is reduced by an allele at this locus, several other seed proteins are enhanced (Dierks-Ventling, 1981), and the overall nutritional quality of the seed is improved. The mechanism of this action is unknown. Unfortunately, these high-lysine varieties do not yield well and have not been commercially accepted. It has been observed that several of these loci interact with each other at an as yet undefined level (DiFonzo *et al.*, 1980). Soave *et al.* (1981b) have identified a protein that appears to be associated with the action of *opaque-2* and *opaque-6* on the depression of zein synthesis. The prospect of elucidating gene interactions in a system as well defined as this one is exciting indeed.

B. SOYBEAN

The mature soybean seed contains 40–50% of its dry weight as protein. About 70% of this protein is in the form of two major storage proteins, glycinin (11 S storage protein, legumin) and β-conglycinin (7 S storage protein, vicilin).

β-Conglycinin is a glycoprotein containing mannose and glucosamine, and is composed of three major subunits (α, α', and β) and one minor subunit (Tanh and Shibasaki, 1977). The complex has a molecular weight of approximately 150 kd and sediments in a sucrose gradient at approximately 7 S. Six major isomers of β-conglycinin are composed of various combinations of α, α', and β subunits in trimeric combinations. The α and α' subunits are immunologically related to each other and have similar mass (57 kd), but are immunologically distinct from β subunits. The β subunit has a mass of 42 kd. The contribution of the minor subunit to the complex has not been defined. The β-conglycinin subunits are themselves heterogeneous.

Beachy *et al.* (1980) were able to demonstrate the synthesis of α and α' *in vitro* using mRNAs sedimenting between 21 and 25 S. Analysis of the tryptic peptides on HPLC demonstrated the identity of the α and α' subunits and suggested substantial homology between them. Tumer *et al.* (1981) isolated a 20 S mRNA fraction that also directs the synthesis of the α and α' subunits *in vitro*. The proteins are synthesized as slightly larger precursor molecules. Gayler and Sikes (1981) studied the appearance of α, α', and β subunits in developing soybean seeds and showed that their appearance was not synchronous. The α and α' subunits can be detected 15–17 days after flowering but the β subunits were undetectable until about 22 days after flowering. A cDNA clone for the β-conglycinin subunits prepared by Beachy *et al.* (1981) was used to study the expression of the β-conglycinin mRNAs during development (Meinke *et al.*, 1981). The pattern of appearance of the mRNAs precedes the pattern of appearance of the proteins by several days, and the staggered expression of the subunits was observed at the mRNA level as well. This differential temporal expression could explain the shift in isomers. At midmaturation (approximately 75 days after flowering) the α and α' subunits are represented at approximately 23,000 copies per cell. This amounts to nearly half of the mRNA in the cell (Goldberg *et al.*, 1981). In tissues in which β-conglycinin is not expressed, the mRNAs are present in only a few copies per cell.

Most of the molecular characterization of the soybean storage protein genes has been focused on the glycinin complex. The storage globulin glycinin is the predominant storage protein in the soybean seed. The protein has a mass of approximately 350 kd and is composed of approximately 12 subunits in a cylindrical array (Badley *et al.*, 1975). Half of the subunits are basic and half are acidic. The acidic and basic subunits are synthesized as a single precursor of approximately 60 kd and then cleaved (Barton *et al.*, 1982; Tumer *et al.*, 1981, 1982). The cleaved polypeptides are covalently attached to each other in the mature glycinin protein by a disulfide linkage. When glycinin mRNAs are microinjected into *Xenopus* oocytes, a signal peptide is cleaved from the amino terminus but the cleavage of the protein into the acidic and basic subunits does not occur (Tumer *et al.*, 1982).

Glycinin mRNA can first be detected in the cotyledon toward the end of the cotyledon stage (Meinke *et al.*, 1981; Goldberg *et al.*, 1981). After reaching a peak at midmaturation there is a sharp decline in mRNA levels. The embryonic axes contain approximately 500 copies of the mRNAs per cell at midmaturation, whereas the whole embryos contain 27,000 copies per cell. As is the case with the β-conglycinin mRNAs, tissues that do not synthesize the protein contain a few copies or less per cell (Goldberg *et al.*, 1981).

In contrast to the zein genes, both the β-conglycinin and the glycinin genes appear to be present in five copies or less per haploid genome, as determined by both solution hybridization and filter blot hybridization (Goldberg *et al.*, 1981; Fisher and Goldberg, 1982). There is no indication of any selective amplification or rearrangement of the glycinin genes during development. Two glycinin genomic clones were analyzed and found to contain at least one intervening sequence. The genes are not tandem nor are they adjacent to other abundant gene sequences (Fisher and Goldberg, 1982).

C. FRENCH BEAN

The G1 globulin (phaseolin) storage protein of the French bean (*Phaseolus vulgaris*) composes 40–50% of the total seed protein (Ma and Bliss, 1978). In the 'Tendergreen' variety this protein is composed of three nonidentical subunits, α, β, and γ (Hall *et al.*, 1978). These subunits have masses of 51, 48, and 45.5 kd, respectively (Hall *et al.*, 1977; McLeester *et al.*, 1973). There is genetically determined charge heterogeneity within each of the size classes (Brown *et al.*, 1980, 1981). Peptide mapping studies indicated substantial sequence homology among the phaseolin subunits (Ma *et al.*, 1980).

Phaseolin is synthesized on membrane-bound polysomes (Bolini and Chrispeels, 1979). Phaseolin mRNAs sediment at 16 S in sucrose gradients and yield only two bands when the *in vitro* translation products are analyzed on denaturing gel electrophoresis (Hall *et al.*, 1978). These two bands migrated slightly faster than the native β and γ subunits. It has been determined subsequently that all of the subunits are synthesized, but that the migration of the bands is anomalous due to the lack of glycosylation of the polypeptides when synthesized *in vitro*. This was established primarily by the translation of the mRNAs in *Xenopus* oocytes. The translation products made in this system co-migrate with native α and β subunits, suggesting that they are appropriately glycosylated (Matthews *et al.*, 1981). The migration of the γ subunits is slightly slower than that of the native polypeptide, suggesting aberrant processing of this protein in the *Xenopus* system.

A DNA sequence comparison between a phaseolin genomic clone and a cDNA clone representing about 40% of the coding sequence indicated the presence of three intervening sequences in the region covered by the clone (Sun *et al.*, 1981). All three of the intervening sequences begin with the GT consensus sequence and end with AG.

D. PEA

The major storage protein in peas, legumin, appears to be analogous to that of soybean (glycinin). Legumin is a complex protein with a mass of 360–400 kd. It is composed of six monomers, each of which contains two subunits covalently joined by a disulfide bond (Croy *et al.*, 1979; Derbyshire *et al.*, 1976). These subunits, which have masses of 20 and 40 kd when isolated from native legumin, are synthesized *in vitro* as a single polypeptide of about 60 kd that is subsequently cleaved (Evans *et al.*, 1979; Croy *et al.*, 1980). Legumin appears to be synthesized on membrane-bound polysomes and sequestered in the rough endoplasmic reticulum (Hurkman and Beevers, 1982). A developmental shift in the expression of the pea storage protein genes occurs during the development of the seed (Spencer *et al.*, 1980).

Two cDNA clones that cover approximately 35% of the coding region have been sequenced (Croy *et al.*, 1982). The coding sequence for the basic subunit is covered by the clones and extends for some 90 nucleotides past the amino terminus of the basic subunit. This presumably extends into the coding sequence of the acidic subunit and, as would be expected from the *in vitro* translation data, there is no termination codon between the sequences. The basic subunit region is at the 3' end of the gene. These clones hybridize to a single 2.2-kb mRNA that is sufficiently large to code for a 60-kd protein precursor. These clones hybridize to only four bands in an *Eco*RI restriction digest of total pea DNA, suggesting that the number of legumin genes may be low. It is also possible that there are multiple genes in each fragment class that are highly conserved. By analogy to the soybean system, however, this would seem unlikely.

The synthesis of another pea storage protein, vicillin, has also been demonstrated *in vitro* and information on vicillin gene structure is under investigation (Higgins and Spencer, 1981).

E. BARLEY

The alcohol-soluble prolamins of barley (hordeins) make up about 40% of the seed protein. They can be separated by electrophoresis into two molecular weight classes (B and C) that are themselves heterogeneous (Faulks *et al.*, 1981). Eight to 16 polypeptides can be identified in the B fraction in any 1 variety, and a total of 47 electromorphs have been identified in total. Based on cyanogen bromide cleavage patterns, at least three subgroups of the B class can be detected. Two loci have been mapped by classical genetic techniques, one of which codes for the B class and one of which codes for the C class (Shewry *et al.*, 1980). The fact that such a heterogeneous population of proteins maps to a single location raises the possibility that, unlike zein, the hordein genes may be tightly clustered.

Fifty hordein cDNA clones have been identified in a library of barley endosperm cDNAs (Forde *et al.*, 1981). They represent about 20% of the clones examined. Eleven of these were examined further, and all were found to hybridize to a 1.3-kb mRNA. This mRNA directs the synthesis of the hordein B-class polypeptide *in vitro*. Extensive cross-hybridization studies indicate considerable heterogeneity in the hordein B mRNAs and also suggest some homology between hordein B genes and other unidentified genes. Although the sample is small, it appears that the complexity of storage proteins in cereals is great and in direct contrast to the legumes, in which relatively simple gene families appear to be involved in storage protein synthesis. The mRNAs coding for the major storage prolamins of wheat (gliadins) have been isolated (Okita and Greene, 1982), and when clones are available it will be of interest to compare their structure to those of maize and barley.

IV. Genes Coding for Other Proteins

A growing number of genes coding for proteins other than storage proteins are being studied at the molecular level. The best characterized examples are still high-abundance proteins, but progress is being made on more minor genes as well.

A. LEGHEMOGLOBIN

Leghemoglobin is found exclusively in the root nodules of nitrogen-fixing legumes, in which it serves the dual function of scavenging free oxygen (protecting the oxygen-sensitive nitrogenase enzyme from denaturation) and acting as a carrier of oxygen to sites of oxidative metabolism. In soybean nodules leghemoglobin molecules may account for 20% or more of the total protein synthesis (Auger *et al.*, 1979). Even though leghemoglobin is found only in cells infected with *Rhizobium*, it is encoded by a nuclear gene in the plant host (Baulcombe and Verma, 1978; Sidloi-Lumbroso *et al.*, 1978).

There are five leghemoglobin polypeptides: a, b, c1, c2, and d. In infected soybean roots one finds primarily leghemoglobins a, c1, and c2. Leghemoglobins c1 and c2 differ from each other by only one amino acid and from leghemoglobin a by only six amino acids (Sievers *et al.*, 1978).

Leghemoglobin mRNA is about the same size as mammalian globin mRNA (Sidloi-Lumbroso and Schulman, 1977). Titration of soybean DNA with leghemoglobin cDNA suggests a reiteration frequency of approximately 40 copies per haploid genome (Baulcombe and Verma, 1978). Leghemoglobin cDNA clones (Truelsen *et al.*, 1979; Sullivan *et al.*, 1981) have been used to identify seven leghemoglobin genes representing all of the major classes (Sullivan *et al.*, 1981; Jensen *et al.*, 1981; Hyldig-Nielsen *et al.*, 1982; Wiborg *et al.*, 1982). All have intervening sequences. In several cases two of the intervening sequences are located in the same positions as the intervening sequences found in genes coding for mammalian globins.

In addition to the obvious interest in leghemoglobin for evolutionary studies and genetic engineering, the regulation of a eukaryotic structural gene by a bacterium should provide a unique opportunity for studying gene regulation and the interaction between simple and complex genomes.

B. RuBP CARBOXYLASE (SMALL SUBUNIT)

Ribulose-1,5-biphosphate carboxylase (fraction-1 protein) is a highly abundant chloroplast protein. In higher plants, this enzyme is an oligomer composed of eight large subunits (55 kd) and eight small subunits (14 kd). The large subunit is encoded in the chloroplast genome and the small subunit is encoded in the nuclear genome. Bedbrook *et al.* (1980c) were able to utilize the fact that the small subunit is light inducible in pea to identify the mRNA and, in turn, to identify a cDNA clone that codes for the protein. This clone has been sequenced, and contains the entire coding sequence for the mature subunit as well as a portion of the amino-terminal leader sequence. Although the mRNA is polyadenylated, no sequence resembling the canonical polyadenylation signal was observed. Because the location of the putative polyadenylation sequence in zein is in an unusual location, perhaps plants utilize some signal other than AATAAA for polyadenylation.

C. LECTINS

Lectins compose a group of moderately abundant seed proteins. Their physiological function is unknown, but their impact on cell-surface studies has been substantial indeed. Ironically, lectins are known to have numerous biological effects in animal cells. The structure of lectin proteins has been studied in some detail, and studies on lectin biosynthesis are beginning to appear (Roberts and Lord, 1981; Hemperly et al., 1982). The most advanced system for the study of lectin gene structure is the soybean lectin. Soybean lectin is a tetramer composed of four 30-kd polypeptides (Loten et al., 1974). The inheritance of the soybean lectins has been studied (Orf et al., 1978), and soybeans that lack detectable lectin protein are available (Pull et al., 1978). The amount of soybean lectin is variable in different lines, but typically represents up to 5% of the protein mass in the cell (Pueppke et al., 1978). The mRNA(s) coding for soybean trypsin inhibitor have been isolated by a polysome immunoprecipitation procedure (Vodkin, 1981). This mRNA has been shown to direct the synthesis of lectin in the rabbit reticulocyte system and has subsequently been used to isolate a cDNA clone from a soybean seed cDNA library (Goldberg et al., cited in Vodkin, 1981). In a parallel study, the soybean trypsin inhibitor mRNA has been isolated (Vodkin, 1981) and a cDNA clone for this gene has been isolated as well (Goldberg et al., 1981). Hybridization studies suggest a reiteration frequency of five copies per haploid genome.

D. SUCROSE SYNTHETASE

Sucrose synthetase is a moderately abundant enzyme found in the endosperm of maize which is involved in starch biosynthesis. It is encoded by a genetically well-characterized gene, *Shrunken-1*, and a second gene that is expressed at very low levels (Chourey and Nelson, 1976; Chourey, 1981). Controlling element events have been well studied at this locus, and the abundance of the product of the *Shrunken-1* locus has made this an attractive target system for the elucidation of the molecular basis of controlling element action in maize.

Sucrose synthetase cDNA clones have been identified by hybrid-selected translation (Geiser *et al.*, 1980; Burr and Burr, 1981a), and these have been used to identify a genomic clone (Burr and Burr, 1982). To date, these clones have been used primarily to study the details of controlling element structure (Geiser *et al.*, 1980; Doring *et al.*, 1981; Burr and Burr, 1981a,b, 1982), although sequence information on the *Shrunken* gene itself should be available soon.

E. ACTIN

Actins compose a ubiquitous group of proteins in eukaryotic cells. They are highly conserved proteins, and are therefore of some interest in evolutionary studies. Soybean actin has been shown to be immunologically cross-reactive with mammalian actin (Metcalf *et al.*, 1980). The nucleotide sequence of a soybean actin gene (identified with a heterologous probe) has been determined (Shah *et al.*, 1982). The gene contains three intervening sequences, the borders of which are similar to the canonical border junctions and those of other plants that have been studied. A probable transcription initiation site and a possible promoter have been identified.

F. OTHER GENES TO WATCH

There are a number of other gene systems about which structural information should be attainable in the relatively near future. I shall very briefly describe a few of these systems in this section.

Maize alcohol dehydrogenase (ADH) is a genetically defined system that involves at least three structural genes (reviewed by Scandalios, 1977; Freeling and Birchler, 1981). The regulation of ADH is of interest in that both dosage compensation (Birchler, 1979) and a temporal regulatory gene (Lai and Scandalios, 1980) have been observed. In addition, maize ADH is one of a small group of proteins that are induced in response to anaerobiosis in roots (Sachs and Freeling, 1978; Ferl *et al.*, 1980; Sachs *et al.*, 1980).

The anaerobic induction of the ADH mRNAs was utilized in conjunction with hybrid-selected translation in the isolation of cDNA clones of the major ADH gene (Sachs *et al.*, 1982). A partial nucleotide sequence has been published (Gehrlach *et al.*, 1982).

Maize catalase is a genetically and biochemically well-defined gene system that is of particular interest as a developmental model system (Scandalios *et al.*, 1980). The catalase mRNAs have been isolated by a polysome immunoprecipitation technique and have been utilized to identify catalase cDNA clones (Sorenson, 1982). These, in turn, are being utilized to isolate genomic clones and to characterize the structure and expression of the genes.

Barley α-amylase is one of the best studied inducible systems in higher plants (see review by Jacobsen, 1977). It is synthesized in the aleurone layer of the barley seed in response to gibberellic acid. α-Amylase accounts for more than 50% of the protein synthesized *in vitro* in response to poly(A)⁺ RNA from gibberellic acid-stimulated aleurone layers. This has allowed the isolation of the mRNA coding for this protein (Mozer, 1980), and cDNA clones should be available in due course. Isolation of wheat α-amylase may also be feasible (Boston *et al.*, 1982).

Thaumatin is a highly sweet protein that is found in the fruits of the west African shrub *Thaumatococcus daniellii*. It is an extremely abundant protein, and this has allowed the isolation of cDNA clones by hybrid-selected translation using a cDNA library (Edens *et al.*, 1982). The nucleotide sequence has been published and the protein successfully expressed in *Escherichia coli*.

V. Unidentified Genes in Regulated Sets

The work described in Sections III and IV has approached the question of gene structure and regulation through the characterization of specific and well-studied genes. A number of groups have taken a different approach to the problem, namely, to study a spectrum of unidentified genes that share certain regulatory features, typically abundance, developmental expression, or inducibility by some treatment.

Studies in tobacco (Goldberg *et al.*, 1978; Kamalay and Goldberg, 1980) suggested that less than 5% of the tobacco genome (about 60,000 structural genes) is expressed in the mature tobacco plant. The mRNAs fall into three general abundance classes with average frequencies of 4500, 340, and 17 copies per cell. Approximately half of the mRNA mass is found in the intermediate abundance class. About 8000 diverse mRNA species were common to all organs examined and would appear to represent "housekeeping" genes. An interesting finding was that mRNAs not found in the cytoplasm of a given organ were nonetheless present in the nuclear RNA, suggesting a substantial degree of posttranscriptional regulation of gene expression. This does not appear to be the case in soybean (see below). It would be of interest to examine the tobacco nuclear RNA with specific cloned DNA sequences.

Studies on embryogenesis in tobacco are not feasible, due to extremely small seed size. Soybean, however, has a well-defined embryogenic pattern. Goldberg et al. (1981) have investigated mRNA expression during soybean embryogenesis, using cloned probes as well as mRNA excess hybridizations to single-copy DNA. They determined that the number of diverse mRNAs is similar throughout the period examined (14,000–18,000 structural genes represented), but that the relative abundance of various mRNAs within that group changes dramatically. A small number of superabundant mRNAs (which code for the superabundant seed proteins) compose approximately 50% of the mRNA mass at midmaturation (75 days after flowering).

A series of experiments in cotton embryogenesis have allowed the identification of 11 subsets of mRNAs in which the concentration of the constituent mRNAs change coordinately with other members of the subset (Galau and Dure, 1981). These subsets contain mRNAs of varying abundance, and would seem to be prime sets of genes in which to look for common regulatory sequences. Similar sets of expression groups can be identified at the protein level (Dure et al., 1981). It has also been shown in this series of studies that polyadenylated and non-polyadenylated mRNAs contain essentially the same sequences (Galau et al., 1981). A compilation of the number of mRNAs found in a number of higher plants can be found in Galau and Dure (1980).

There are a number of treatments that can dramatically alter the expression of genes in higher plants. Perhaps two of the more interesting developments in this area should be highlighted. The availability of cloned probes for several as yet unidentified genes that are regulated by auxin should prove invaluable in elucidating the mechanisms of the hormonal control of gene expression (Baulcombe et al., 1981). One of the major problems in previous studies in this area has been the inability of investigators to look at specific gene sequences. It now appears that studies can be conducted on specific (albeit unidentified) genes.

Another system that has proven useful as a model of gene regulation in other organisms, and which is beginning to do so in plants as well, is the heat shock system (Key *et al.*, 1981; Scharf and Nover, 1982). The relatively small groups of proteins that are induced by a brief elevation of the temperature seem to be generally analogous to the well-characterized heat shock phenomenon in *Drosophila* and will hopefully add to our knowledge of coordinate gene regulation.

VI. Genes Coding for Ribosomal RNA

An important group of genes that is relatively easy to study is the one coding for ribosomal RNA (hereafter referred to as rDNA). The subject has been reviewed previously by Leaver (1979) and Ingle (1979).

A "typical" plant contains approximately 10 million ribosomes (Leaver, 1979). Mitochondral ribosomes account for less than 1% of the total, although in leaves, chloroplast ribosomes may account for 25–40%. More than half of the ribosome is rRNA, and the rest is protein. rRNA composes more than 90% of the RNA in a plant cell. The large ribosomal subunit contains the large (25 S) rRNA, 1 or 2 small RNAs (5.8 S and 5 S), and 30–40 ribosomal proteins. The small subunit contains a single small rRNA (18 S) and 20–30 proteins.

The 25 S rRNA is approximately 3.7 kb in length, and has undergone considerable evolution. The mass of the large rRNA has increased from a mass of 1.3×10^6 d (about 3700 nucleotides) in plants and protozoa to 1.75×10^6 d (nearly 5000 nucleotides) in mammals (Loening, 1968). In contrast, the small rRNA has remained constant at about 700,000 d (about 2100 nucleotides). The 5.8 S rRNA (about 160 nucleotides) is hydrogen bonded to the 25 S RNA in the large ribosome subunit. This RNA is found in cytoplasmic ribosomes, but not in organelle ribosomes. The 5 S RNA (about 120 nucleotides) is found in the large subunit of both cytoplasmic and organellar ribosomes, and its sequence has been evolutionarily conserved (Dyer *et al.*, 1976).

The 26, 18, and 5.8 S genes (with appropriate spacer regions) are contiguous in the chromosome and, as in animals, are synthesized as a single transcript (hereafter referred to as the major transcriptional unit) that is subsequently cleaved. The 5 S RNA is encoded in one or more clusters of genes located at distinct chromosomal site(s). The major transcriptional unit is repeated, generally in a tandem fashion. The number of such repeats in a sample of 45 higher plants ranges from 1250 copies per telophase nucleus in the orange to 31,900 copies per telophase nucleus in the hyacinth (Ingle, 1979).

In situ hybridization is capable of detecting genes that are reiterated 50 times or more in a chromosomal site (Phillips and Wang, 1982), and this technique has been utilized to map ribosomal gene clusters in several higher plant species. In maize, the major transcriptional unit is associated with the nucleolar organizer region on chromosome 6 (Phillips *et al.,* 1979), and the 5 S gene cluster is located on the long arm of chromosome 2 (Wimber *et al.,* 1974; Mascia *et al.,* 1981). The major transcriptional unit is associated with chromosomes 1B, 6B, and 5D in hexaploid wheat, chromosome 1R in rye, and chromosomes 6 and 7 in barley (Appels *et al.,* 1980). In all cases, the gene clusters are associated with a nucleolus organizer. It is curious that no rDNA clusters were observed in association with the A genome in hexaploid wheat. It is possible that the genes are present, but in low copy numbers. In *Triticum monococcum,* a proposed donor of the A genome, the clusters are readily observable. The 5 S gene clusters have also been localized. In wheat and rye they are found in proximity to the sites of the major transcriptional unit gene clusters. In barley, as in maize, the 5 S genes are localized on a chromosome not related to the other rDNA sequences.

There are two classic examples of the regulation of rDNA copy number. In *Xenopus* oocytes, the rDNA of the major transcriptional unit is amplified to a reiteration frequency of well over a million copies per nucleus and can represent as much as 70% of the total DNA in those cells (Lewin, 1980). The other classic example is the compensatory regulation of rDNA in the *Drosophila "bobbed"* mutation (Henderson and Ritossa, 1970). In general, plants do not seem to utilize selective amplification or compensation of rDNA sequences, perhaps because the number of rDNA genes is already in excess of what is needed for maximum growth (Ingle, 1979). One interesting example of the regulation of rDNA copy number is found in flax. Heritable changes can be induced in some flax varieties when they are grown for one generation in certain environments (Durrant, 1962). Flax can be induced to form stable lines (genotrophs) by treatment with fertilizer. The original intermediate genotroph can be "induced" to form the large genotroph with high nitrogen or a small genotroph with high phosphate. The genotrophs are stable and heritable, provided that the seed of each generation are germinated in a warm environment. If large genotroph plants are germinated in the cold they will, over the course of several generations, revert to the small genotroph. These physical changes are associated with an overall amplification of the genome such that the large genotroph contains about 16% more DNA than the small genotroph. The reiteration of the rDNA is increased by about 70%, however. The amplification does not appear to involve (extensive) rearrangement of the rDNA sequences (Goldsbrough and Cullis, 1981). During reversion of the large genotroph there is a progressive loss of amplified DNA such that full revertants have an amount of DNA similar to that of the small genotroph. The loss of rDNA also occurs, but at a slower rate (Cullis, 1976; Cullis and Charlton, 1981). There does not appear to be any change in the reiteration of the 5 S genes (Goldsbrough et al., 1981). The mechanisms controlling this phenomenon should be of considerable importance to those interested in gene regulation, evolution, and the nature of complex traits in plants.

Information on the molecular organization of rRNA genes in plants has been obtained in a number of species. Sizes of the repeat units in various plants are listed in Table 2. Heterogeneity in the sizes of the repeating units has been observed in some plants. Although in some instances this heterogeneity has been carefully established, in others suitable controls for variable methylation of the sequences have not been conducted.

The genes coding for 5 S rRNA have been studied in several species.

TABLE 2

Repeat Unit Structures in the rDNA Coding for the Major Transcriptional Unit (25 S, 18 S, and 5.8 S Genes)

Plant	Size of repeat unit (kbp)	Size heterogeneity	Methylation	References
Barley	7.8	No	Yes	Appels et al. (1980)
Flax	8.6	No	Yes	Goldsbrough and Cullis (1981)
Lemon	9.5, 10	Yes	Yes	Fodor and Beridze (1980)
Onion	12.7	Yes	ND[a]	Maggini and Carmona (1981)
Radish	11.1	Yes	ND	Delseny et al. (1979)
Rice	7.7, 8.0	Yes	ND	Oono and Sugiura (1980)
Rye	9.0	No	ND	Appels et al. (1980)
Soybean	7.8	No	Yes	Varsanyi-Breiner et al. (1979)
Soybean	9.0	No	ND	Friedrich et al. (1979)
Wheat	9.5	No	ND	Appels et al. (1980)

[a]ND, Not determined.

In maize these genes are organized in a heterogeneous 320-base pair (bp) repeating unit (Mascia *et al.,* 1981). A similar heterogeneous 350–370-bp repeating unit was observed in flax (Goldsbrough *et al.,* 1981). The 5 S repeat unit in wheat and rye is somewhat larger (500 bp) and in barley the repeat unit appears to be heterogeneous and no dominant repeat unit was observed (Appels *et al.,* 1980).

In summary, the ribosomal gene clusters in higher plants appear to be essentially similar to those in animal systems with the exception that they are generally more highly reiterated. The wide variability in the reiteration frequency among species suggests that the rDNA in higher plants should be thought of as being evolutionarily dynamic. At least in the case of flax, there appears to be a plasticity that is remarkable indeed.

VII. Conclusions and Thoughts about the Future

Tremendous progress has been made in understanding gene structure and expression over the past few years. I have attempted to summarize our current knowledge about the subject, and hope that by organizing this substantial body of knowledge some new perspectives and insights may be gained. Although much remains to be done and much remains beyond our present technical capabilities, some clear and important principles are emerging. In the next few years our data base should expand substantially.

Perhaps the greatest single impediment to understanding gene expression in plants is the lack of a reliable genetic transformation system for plants that would allow the biological testing of theories on important gene structures. Work is progressing on such systems, and when they are developed there will be a veritable explosion of new information and the firm elucidation of the principles of gene regulation in higher plants. I look forward to that day with great anticipation!

References

Appels, R., Gerlach, W. L., Swift, D. H., and Peacock, W. J. (1980). Molecular and chromosomal organization of DNA sequences coding for the ribosomal RNAs in cereals. *Chromosoma* **78,** 293–311.

Argos, P., Pederson, K., Marks, M. D., and Larkins, B. A. (1982). A structural model for zein proteins. *J. Biol. Chem.* **257,** 9984–9990.

Auger, S., Baulcombe, D., and Verma, D. P. S. (1979). Sequence complexities of the poly(A) containing mRNA in uninfected soybean root and the nodule tissue developed due to the infection by *Rhizobium. Biochim. Biophys. Acta* **563**, 496–507.

Badley, R. A., Atkinson, D., Hauser, H., Oldani, D., Green, J. P., and Stubbs, J. M. (1975). The structure, physical and chemical properties of the soybean protein glycinin. *Biochim. Biophys. Acta* **412**, 214–218.

Barton, K. A., Thompson, J. F., Madison, J. T., Rosenthal, R., Jarvis, N. P., and Beachy, R. N. (1982). The biosynthesis and processing of high molecular weight precursors of soybean glycinin subunits. *J. Biol. Chem.* **257**, 6089–6095.

Baulcombe, D., and Verma, D. P. S. (1978). Preparation of a complementary DNA for leghemoglobin and direct demonstration that leghemoglobin is encoded by the soybean genome. *Nucleic Acids Res.* **5**, 4141–4153.

Baulcombe, D. C., Kroner, P. A., and Key, J. L. (1981). Auxin and gene regulation. *In* "Levels of Genetic Control in Development" (S. Subtelny and U. K. Abbott, eds.), pp. 83–97. Liss, New York.

Beachy, R. N., Barton, K. A., Thompson, J. F., and Madison, J. T. (1980). *In vitro* synthesis of the α and α' subunits of the 7s storage proteins (conglycinin) of soybean seeds. *Plant Physiol.* **65**, 990–994.

Beachy, R. N., Jarvis, N. P., and Barton, K. A. (1981). *In vivo* and *in vitro* biosynthesis of subunits of the soybean 7s storage protein. *J. Mol. Appl. Genet.* **1**, 19–27.

Bedbrook, J. R., Jones, J., O'Dell, M., Thompson, R., and Flavell, R. B. (1980a). A molecular description of telomeric heterochromatin in *Secale* species. *Cell* **19**, 545–560.

Bedbrook, J. R., O'Dell, M., and Flavell, R. B. (1980b). Amplication of rearranged repeated DNA sequences in cereal plants. *Nature (London)* **288**, 133–137.

Bedbrook, J. R., Smith, S. M., and Ellis, R. J. (1980c). Molecular cloning and seqencing of cDNA encoding the precursor to the small subunit of chloroplast ribulose-1, 5-bisphosphate carboxylase. *Nature (London)* **287**, 692–697.

Belford, H. S., and Thompson, W. F. (1981a). Single copy DNA homologies in *Atriplex*. I. Cross reactivity estimates and the role of deletions in genome evolution. *Heredity* **46**, 91–108.

Belford, H. S., and Thompson, W. F. (1981b). Single copy DNA homologies in *Atriplex*. II. Hybrid thermal stabilities and molecular phylogeny. *Heredity* **46**, 109–122.

Bendich, A. J., and Anderson, R. S. (1977). Characterization of families of repeated DNA sequences from four vascular plants. *Biochemistry* **16**, 4655–4663.

Bietz, J. A., Paulis, J. W., and Wall, J. S. (1979). Zein subunit homology revealed through amino-terminal sequence analysis. *Cereal Chem.* **56**, 327–332.

Birchler, J. A. (1979). A study of enzyme activities in a dosage series of the long arm of chromosome one in maize. *Genetics* **92**, 1211–1229.

Bollini, R., and Chrispeels, M. J. (1979). The rough endoplasmic reticulum is the site of reserve protein synthesis in developing *Phaseolus vulgaris* cotyledons. *Planta* **146**, 487–501.

Boston, R. S., Miller, T. J., Mertz, J. E., and Burgess, R. R. (1982). *In vitro* synthesis and processing of wheat α-amylase. *Plant Physiol.* **69**, 150–154.

Brown, J. W. S, Bliss, F. A., and Hall, T. C. (1980). Microheterogeneity of globulin-1 storage protein from french bean with isoelectric focusing. *Plant Physiol.* **66**, 838–840.

Brown, J. W. S., Ma, Y., Bliss, F. A., and Hall, T. C. (1981). Genetic variation in the subunits of globulin-1 storage protein of french bean. *Theor. Appl. Genet.* **59**, 83–88.

Burr, B., and Burr, F. A. (1976). Zein synthesis in maize by polyribosomes attached to proetin bodies. *Proc. Natl. Acad. Sci. U.S.A.* **73**, 515–519.

Burr, F. A., and Burr, B. B. (1981a). *In vitro* uptake and processing of prezein and other maize preproteins by maize membranes. *J. Cell Biol.* **90**, 427–434.

Burr, B., and Burr, F. A. (1981b). Controlling element events at the *shrunken* locus in maize. *Genetics* **98**, 143–156.

Burr, B. B., and Burr, F. A. (1982). Ds controlling elements of maize at the *Shrunken* locus are large and dissimilar insertions. *Cell* **29**, 977–986.

Burr, B., Burr, F. A., Rubenstein, I., and Simon, M. N. (1978). Purification and translation of zein mRNA from maize endosperm protein bodies. *Proc. Natl. Acad. Sci. U.S.A.* **75**, 696–700.

Burr, B., Burr, F. A., St. John, T. P., Thomas, M., and Davis, R. W. (1982). Zein storage protein gene family of maize. An assessment of heterogeneity with cloned messenger RNA sequences. *J. Mol. Biol.* **154**, 33–49.

Chourey, P. S. (1981). Genetic control of sucrose synthetase in maize endosperm. *Mol. Gen. Genet.* **184**, 372–376.

Chourey, P. S., and Nelson, O. E. (1976). The enzymatic deficiency conditioned by the *shrunken-1* mutations in maize. *Biochem. Genet.* **14**, 1041–1055.

Croy, R. R. D., Derbyshire, E., Krishna, T. G., and Boulter, D. (1979). Legumin of *Pisum sativum* and *Vicia faba. New Phytol.* **83**, 29–35.

Croy, R. R. D., Gatehouse, J. A., Evans, I. M., and Boulter, D. (1980). Characterization of the storage protein subunits synthesized *in vitro* by polyribosomes and RNA from developing pea (*Pisum sativum* L.) I. Legumin. *Planta* **148**, 49–56.

Croy, R. R. D., Lycett, G. W., Gatehouse, J. W., Yarwood, J. N., and Boulter, D. (1982). Cloning and analysis of cDNAs encoding plant storage protein precursors. *Nature (London)* **295**, 76–78.

Cullis, C. A. (1976). Environmentally induced changes in rRNA cistron number in flax. *Heredity* **36**, 73–79.

Cullis, C. A. (1981). DNA sequence organization in the flax genome. *Biochim. Biophys. Acta* **652**, 1–15.

Cullis, C. A., and Charlton, L. (1981). The induction of ribosomal DNA changes in flax. *Plant Sci. Lett.* **20**, 213–217.

Delseny, M., Aspart, L., Cooke, R., Grellet, F., and Penon, P. (1979). Restriction analysis of radish nuclear genes coding for rRNA: Evidence for heterogeneity. *Biochem. Biophys. Res. Commun.* **91**, 540–547.

Dennis, E. S., Gerlach, W. L., and Peacock, W. J. (1980). Identical polypyrimidine-polypurine satellite DNA's in wheat and barley. *Heredity* **44**, 349–366.

Derbyshire, E., Wright, D. J., and Boulter, D. (1976). Legumin and vicilin storage proteins of legume seeds. *Phytochemistry* **15**, 3–24.

Dierks-Ventling, C. (1981). Storage proteins in *Zea mays* (L.): Interrelationship of albumins, globulins, and zeins in the *opaque-2* mutation. *Eur. J. Biochem.* **120**, 177–182.

Dierks-Ventling, C., and Cozens, K. (1982). Immunochemical crossreactivity between zein, hordein and gliadin. *FEBS Lett.* **142**, 147–150.

DiFonzo, N., Fornasari, E., Salamini, F., Reggiani, R., and Soave, C. (1980). Interaction of maize mutants *floury-2* and *opaque-7* with *opaque-2* in the synthesis of endosperm proteins. *J. Hered.* **71**, 397–402.

Doring, H. P., Geiser, M., and Starlinger, P. (1981). Transposable element Ds at the *shrunken* locus in *Zea mays. Mol. Gen. Genet.* **184**, 377–380.

Dure, L., Greenway, S. C., and Galau, G. A. (1981). Developmental biochemistry of cottonseed embryogenesis and germination: Changing messenger ribonucleic acid populations as shown by *in vitro* and *in vivo* protein synthesis. *Biochemistry* **20**, 4162–4168.

Durrant, A. (1982). The environmental induction of heritable change in *Linum. Heredity* **17**, 27–61.

Duvick, D. N. (1961). Protein granules of maize endosperm cells. *Cereal Chem.* **38**, 373–385.

Dyer, T. A., Bowman, C. M., and Payne, P. I. (1976). The low molecular weight rRNAs of plant ribosomes: Their structure, function, and evolution. *In* "Nucleic Acids and Protein Synthesis in Plants" (L. Bogorad and J. H. Weil, eds.), p. 121. Plenum, New York.

Edens, L., Heslinga, L., Klok, R., Ledeboer, A. M., Maat, J., Toonen, M. Y., Visser, C., and Verrips, C. T. (1982). Cloning of cDNA encoding the sweet tasting plant protein thaumatin and its expression in *E. coli. Gene* **18**, 1–12.

Evans, I. M., Croy, R. R. D., Hutchinson, P., Boulter, D., Payne, P. I., and Gordon, M. E. (1979). Cell free synthesis of some storage protein subunits by polyribosomes and RNA isolated from developing seeds of pea (*Pisum sativum* L.). *Planta* **144**, 57–63.

Faulks, A. J., Shewry, P. R., and Miflin, B. J. (1981). The polymorphism and structural homology of storage polypeptides (hordein) coded by the *Hor-2* locus in barley (*Hordeum vulgare* L.). *Biochem. Genet.* **19**, 841–858.

Ferl, R. J., Brennan, M. D., and Schwartz, D. (1980). *In vitro* translation of maize ADH: Evidence for the anaerobic induction of mRNA. *Biochem Genet.* **18**, 681–691.

Fisher, R. L., and Goldberg, R. B. (1982). Structure and flanking regions of soybean seed protein genes. *Cell* **29**, 651–660.

Flavell, R. (1980). The molecular characterization and organization of plant chromosomal DNA sequences. *Annu. Rev. Plant Physiol.* **31**, 569–596.

Flavell, R., and Smith, D. (1976). Nucleotide sequence organization in the wheat genome. *Heredity* **37**, 231.

Flavell, R., O'Dell, M., and Smith, D. (1979). Repeated sequence DNA comparisons between *Triticum* and *Aegilops* species. *Heredity* **42**, 309–322.

Fodor, I., and Beridze, T. (1980). Structural organization of plant ribosomal DNA. *Biochem. Int.* **1**, 493–501.

Forde, B. G., Kreis, M., Bahramian, M. B., Matthews, J. A., and Miflin, B. J. (1981). Molecular cloning and analysis of cDNA sequences derived from polyA(+) RNA from barley endosperm: Identification of B hordein related clones. *Nucleic Acids Res.* **9**, 6689–6707.

Freeling, M., and Birchler, J. A. (1981). Mutants and variants of the *alcohol dehydrogenase-1* gene in maize. *In* "Genetic Engineering: Principles and Methods" (J. K. Setlow and A. Hollaender, eds.), pp. 223–264. Plenum, New York.

Friedrich, H., Hemleben, V., Meagher, R B., and Key, J. L. (1979). Purification and restriction endonuclease mapping of soybean 18s and 25s ribosomal RNA genes. *Planta* **146**, 467–473.

Galau, G. A., and Dure, L. S. (1980) RNA complexity during seed formation and germination. *Isr. J. Bot.* **29**, 281–292.

Galau, G. A., and Dure, L. S. (1981). Developmental biochemistry of cottonseed embryogenesis and germination: Changing messenger ribonucleic acid populations as shown by reciprocal heterologous complementary deoxyribonucleic acid-messenger ribonucleic acid hybridization. *Biochemistry* **20**, 4169–4178.

Galau, G. A., Legocki, A. B., Greenway, S. C., and Dure, L. S. (1981). Cotton messenger RNA sequences exist in both polyadenylated and nonpolyadenylated forms. *J. Biol. Chem.* **256**, 2551–2560.

Gayler, K. R., and Sykes, G. E. (1981). B-conglycinins in developing soybean seeds. *Plant Physiol.* **67**, 958–961.

Geiser, M., Doring, H. P., Wostemeyer, J., Behrens, U., Tillman, E., and Starlinger, P. (1980). A cDNA clone from *Zea mays* endosperm sucrose synthetase mRNA. *Nucleic Acids Res.* **8**, 6175–6188.

Geraghty, D., Peifer, M. A., Rubenstein, I., and Messing, J. (1981). The primary structure of a plant storage protein: Zein. *Nucleic Acids Res.* **9**, 5163–5174.

Gerlach, W. L., and Peacock, W. J. (1980). Chromosomal locations of highly repeated DNA sequences in wheat. *Heredity* **44**, 269–276.

Gerlach, W. L., Pryor, A. J., Dennis, E. S., Ferl, R. J., Sachs, M. M., and Peacock, W. J. (1982). cDNA cloning and induction of the alcohol dehydrogenase gene (ADH1) of maize. *Proc. Natl. Acad. Sci. U.S.A.* **79**, 2981–2985.

Goldberg, R. B. (1978). DNA sequence organization in the soybean plant. *Biochem. Genet.* **16**, 45–68.

Goldberg, R. B., Hoschek, G., Kamalay, J., and Timberlake, W. E. (1978). Sequence complexity of nuclear and polysomal RNA in leaves of the tobacco plant. *Cell* **14**, 123–131.

Goldberg, R. B., Hoscheck, G., Ditta, G. S., and Breidenbach, R. W. (1981). Developmental regulation of cloned superabundant embryo mRNAs in soybean. *Dev. Biol.* **83**, 218–231.

Goldsbrough, P. B., and Cullis, C. A. (1981). Characterization of the genes for ribosomal RNA in flax. *Nucleic Acids Res.* **9**, 1301–1309.

Goldsbrough, P. B., Ellis, T. H. N., and Cullis, C. A. (1981). Organization of the 5s RNA genes in flax. *Nucleic Acids Res.* **9**, 5895–5904.

Gurley, W. B., Hepburn, A. G., and Key, J. L. (1979). Sequence organization of the soybean genome. *Biochim. Biophys. Acta* **561**, 167–183.

Hagen, G., and Rubenstein, I. (1980). Two-dimensional gel analysis of the zein proteins in maize. *Plant Sci. Lett.* **19**, 217–223.

Hagen, G., and Rubenstein, I. (1981). Complex organization of zein genes in maize. *Gene* **13**, 239–249.

Hake, S., and Walbot, V. (1980). The genome of *Zea mays,* its organization and homology to related grasses. *Chromosoma* **79**, 251–270.

Hall, T. C., McLeester, R. C., and Bliss, F. A. (1977). Equal expression of the maternal and paternal alleles for the polypeptide subunits of the major storage protein of the bean *Phaseolus vulgaris* L. *Plant Physiol.* **59**, 1122–1124.

Hall, T. C., Ma, Y., Buchbinder, B. U., Pyne, J. W., Sun, S. M., and Bliss, F. A. (1978). Messenger RNA for G1 protein of french bean seeds: Cell free translation and product characterization. *Proc. Natl. Acad. Sci. U.S.A.* **75**, 3196–3200.

Hemleben, V., Leweke, B., Roth, A., and Stadler, J. (1982). Organization of highly repetitive satellite DNA of two *Curcurbitaceae* species (*Cucumis melo* and *Cucumis sativus*). *Nucleic Acids Res.* **10**, 631–644.

Hemperly, J. J., Mostov, K. E., and Cunningham, B. A. (1982). *In vitro* translation and processing of a precursor form of favin, a lectin from *Vicia faba*. *J. Biol. Chem.* **257**, 7903–7909.

Henderson, A., and Ritossa, R. M. (1970). On the inheritance of rDNA in magnified bobbed loci in *D. melanogaster*. *Genetics* **66**, 463–473.

Hepburn, A. G., Gurley, W. B., and Key, J. L. (1977). Organization of the soybean genome. *In* "Nucleic Acids and Protein Synthesis in Plants" (J. H. Weil and L. Bogorad, eds.), pp. 113–116. Plenum, Amsterdam.

Higgins, T. J. V., and Spencer, D. (1981). Precursor forms of pea vicilin subunits. Modification by microsomal membranes during cell free translation. *Plant Physiol.* **67**, 205–211.

Hudspeth, M. E. S., Timberlake, W. E., and Goldberg, R. B. (1977). DNA sequence organization in the water mold *Achlya*. *Proc. Natl. Acad. Sci. U.S.A.* **74**, 4332–4335.

Hurkman, W. J., and Beevers, L. (1982). Sequestration of pea reserve proteins by rough microscomes. *Plant Physiol.* **69**, 1414–1417.

Hurkman, W. J., Smith, L. D., Richer, J., and Larkins, B. A. (1981). Subcellular compartmentalization of maize storage proteins in *Xenopus* oocytes injected with zein mRNAs. *J. Cell Biol.* **89**, 292–299.

Hyldig-Nielsen, J. J., Jensen, E. O., Paludan, K., Wiborg, O., Garrett, R., Jorgensen, P., and Marcker, K. A. (1982). The primary structures of two leghemoglobin genes from soybean. *Nucleic Acids Res.* **10**, 689–701.

Ingle, J. (1979). Plant ribisomal RNA genes—a dynamic system? *In* "Molecular Biology of Plants" (I. Rubenstein, R. L. Phillips, C. E. Green, and B. G. Gegenbach, eds.), pp. 139–164. Academic Press, New York.

Iyengar, G. A. S., Gaddipati, J. P., and Sen, S. K. (1979). Characteristics of nuclear DNA in the genus *Oryza*. *Theor. Appl. Genet.* **54**, 219–224.

Jacobsen, J. V. (1977). Regulation of ribonucleic acid metabolism by plant hormones. *Annu. Rev. Plant Physiol.* **28**, 537–564.

Jensen, E. O., Paludan, K., Hyldig-Nielsen, J. J., Jorgensen, P., and Marcker, K. A. (1981). The structure of a chromosomal leghaemoglobin gene from soybean. *Nature (London)* **291**, 677–670.

Kamalay, J. C., and Goldberg, R. B. (1980). Regulation of structural gene expression in tobacco. *Cell* **19**, 935–946.

Key, J. L., Lin, C. Y., and Chen, Y. M. (1981). Heat shock proteins of higher plants. *Proc. Natl. Acad. Sci. U.S.A.* **78**, 3526–3530.

Kiper, M., and Herzfeld, F. (1978). DNA sequence organization in the genome of *Petroselinum sativum* (Umbelliferae). *Chromosoma* **65**, 335–351.

Lai, Y. K., and Scandalios, J. G. (1980). Genetic determination of the developmental program for maize scutellar alcohol dehydrogenase: Involvement of a recessive, trans-acting, temporal regulatory gene. *Dev. Genet.* **1**, 311–324.

Langridge, P., Pintor-Toro, J. A., and Feix, G. (1982). Zein genomic clones from maize. *In* "Maize for Biological Research" (W. F. Sheridan, ed.), pp. 183–187. U. N. D. Press, Grand Forks, North Dakota.

Larkins, B. A., and Hurkman, W. J. (1978). Synthesis and deposition of zein in protein bodies of maize. *Plant Physiol.* **62**, 256–263.

Larkins, B. A., and Dalby, A. (1975). *In vitro* synthesis of zein-like protein by maize polyribosomes. *Biochem. Biophys. Res. Commun.* **66**, 1048–1054.

Larkins, B. A., and Pedersen, K. (1982). Cloning of maize zein genes. *In* "Maize for Biological Research" (W. F. Sheridan, ed.), pp. 177–182. U. N. D. Press, Grand Fords, North Dakota.

Larkins, B. A., Jones, R. A., and Tsai, C. Y. (1976). Isolation and *in vitro* translation of zein messenger RNA. *Biochemistry* **15**, 5506–5511.

Larkins, B. A., Pedersen, K., Handa, A. K., Hurkman, W. J., and Smith, L. D. (1979).

Synthesis and processing of maize storage proteins in *Xenopus laevis* oocytes. *Proc. Natl. Acad. Sci. U.S.A.* **76,** 6448–6452.

Leaver, C. J. (1979). Ribosomal RNA of plants. *In* "Nucleic Acids in Plants" (T. C. Hall and J. W. Davies, eds.) pp. 193–215. CRC Press, Boca Raton, Florida.

Lewin, B. (1980). "Gene Expression," Vol. 1. Wiley (Interscience), New York.

Lewis, E. D., Hagen, G., Mullins, J. I., Mascia, P. N., Park, W. D., Benton, W. D., and Rubenstein, I. (1981). Cloned genomic segments of *Zea mays* homologous to zein mRNAs. *Gene* **14,** 205–215.

Loening, U. E. (1968). Molecular weights of ribosomal RNA in relation to evolution. *J. Mol. Biol.* **38,** 355.

Lotan, R., Siegelman, H. W., Lis, H., and Sharon, N. (1974). Subunit structure of soybean agglutinin. *J. Biol. Chem.* **249,** 1219–1224.

Ma, Y., and Bliss, F. A. (1978). Seed proteins of common bean. *Crop Sci.* **17,** 431–437.

Ma, Y., and Nelson, O. E. (1975). Amino acid composition of storage proteins in two high lysine mutants in maize. *Cereal Chem.* **52,** 412–419.

Ma, Y., Bliss, F. A., and Hall, T. C. (1980). Peptide mapping reveals considerable sequence homology among the three polypeptide subunits of G1 storage protein from french bean seed. *Plant Physiol.* **66,** 897–902.

McLeester, R. C., Hall, T. C., Sun, S. M., and Bliss, F. A. (1973). Comparison of globulin proteins from *Phaseolus vulgaris* with those of *Vicia faba. Phytochemistry* **12,** 85–93.

McWhirter, K. S. (1971). A floury endosperm, high lysine locus on chromosome 10. *Maize Genet. Coop. Newslett.* **45,** 184.

Maggini, F., and Carmona, M. J. (1981). Sequence heterogeneity of the ribosomal DNA in *Allium cepa* (Liliaceae). *Protoplasma* **108,** 163–171.

Marks, M. D., and Larkins, B. A. (1982). Analysis of sequence microheterogeneity among zein messenger RNAs. *J. Biol. Chem.* **257,** 9976–9983.

Mascia, P. N., Rubenstein, I., Phillips, R. L., Wang, A. S., and Xiang, L. Z. (1981). Localization of the 5s rRNA genes and evidence for diversity in the 5s rDNA region of maize. *Gene* **15,** 7–20.

Matthews, J. A., Brown, J. W. S., and Hall, T. C. (1981). Phaseolin mRNA is translated to yield glycosylated polypeptides in *Xenopus* oocytes. *Nature (London)* **294,** 175–176.

Meinke, D. W., Chen, J., and Beachy, R. N. (1981). Expression of storage protein genes during soybean seed development. *Planta* **153,** 130–139.

Mertz, E. T., Bates, L. S., and Nelson, O. E. (1964). Mutant gene that changes protein composition and increases lysine content of maize endosperm. *Science* **145,** 279–280.

Metcalf, T. N., Szabo, L. J., Schubert, K. R., and Wang, J. L. (1980). Immunochemical identification of an actin-like protein from soybean seedlings. *Nature (London)* **285,** 171–172.

Misra, P. S., Mertz, E. T., and Glover, D. V. (1975). Characteristics of proteins in single and double endosperm mutants of maize. *In* "High Quality Protein Maize," pp. 291–305. CYMMIT-Purdue Univ. Press, Lafayette, Indiana.

Mozer, T. J. (1980). Partial purification and characterization of the mRNA for α-amylase from barley. *Plant Physiol.* **65,** 834–837.

Murray, M. G., and Thompson, W. F. (1979). Long period interspersion in the mung bean genome. *Carnegie Inst. Year Book* **78,** 204–207.

Murray, M. G., Cuellar, R. E., and Thompson, W. F. (1978). DNA sequence organization in the pea genome. *Biochemistry* **17,** 5781–5790.

Murray, M. G., Palmer, J. D., Cuellar, R. E., and Thompson, W. F. (1979). Deoxyribonucleic acid sequence organization in the mung bean genome. *Biochemistry* **18**, 5259–5266.

Nelson, O. E. (1969). Genetic modification of protein quality in plants. *Adv. Agron.* **21**, 171–194.

Okita, T. W., and Greene, F. C. (1982). Wheat storage proteins. Isolation and characterization of the gliadin mRNAs. *Plant Physiol.* **69**, 834–839.

Oono, K., and Sugiura, M. (1980). Heterogeneity of the ribosomal gene clusters in rice. *Chromosoma* **76**, 85–89.

Orf, J. F., Hymowitz, T., Pull, S. P., and Pueppke, S. G. (1978). Inheritance of a soybean seed lectin. *Crop Sci.* **18**, 899–900.

Park, W. D., Lewis, E. D., and Rubenstein, I. (1980). Heterogeneity of zein mRNA and protein in maize. *Plant Physiol.* **65**, 98–106.

Peacock, W. J., Dennis, E. S., Rhoads, M. M., and Pryor, A. J. (1981). Highly repeated DNA sequence limited to knob heterochromatin in maize. *Proc. Natl. Acad. Sci. U.S.A.* **78**, 4490–4494.

Pedersen, K., Bloom, K. S., Anderson, J. N., Glover, D. V., and Larkins, B. A. (1980). Analysis of the complexity and frequency of zein genes in the maize genome. *Biochemistry* **19**, 1644–1650.

Pedersen, K., Devereaux, J., Wilson, D. R., Sheldon, E., and Larkins, B. A. (1982). Cloning and sequence analysis reveal structural variation among related zein genes in maize. *Cell* **29**, 1015–1026.

Pellegrini, M., and Goldberg, R. B. (1979). DNA sequence organization in soybean investigated by electron microscopy. *Chromosoma* **75**, 309–326.

Phillips, R. L., and Wang, A. S. (1982). *In situ* hybridization with maize meiotic cells. *In* "Maize for Biological Research" (W. F. Sheridan, ed.), pp. 121–122. U. N. D. Press, Grand Forks, North Dakota.

Phillips, R. L., Wang, A. S., Rubenstein, I., and Park, W. D. (1979). Hybridization of ribosomal RNA to maize chromosomes. *Maydica* **24**, 7–21.

Pintor-Toro, J. A., Langridge, P., and Feix, G. (1982). Isolation and characterization of maize genes coding for zein proteins of the 2100 dalton size class. *Nucleic Acids Res.* **13**, 3845–3860.

Preisler, R. S., and Thompson, W. F. (1981a). Evolutionary sequence divergence within repeated DNA families of higher plant genomes. I. Analysis by reassociation kinetics. *J. Mol. Evol.* **17**, 78–84.

Preisler, R. S., and Thompson, W. F. (1981b). Evolutionary sequence divergence within repeated DNA families of higher plant genomes. II. Analysis of thermal denaturation. *J. Mol. Evol.* **17**, 85–93.

Pueppke, S. G., Bauer, W. D., Keegstra, K., and Ferguson, A. L. (1978). Role of lectins in plant microorganism interactions. II. Distribution of soybean lectin in tissues of *Glycine max*. *Plant Physiol.* **61**, 779–784.

Pull, S. P., Pueppke, S. G., Hymowitz, T., and Orf, J. H. (1978). Soybean lines lacking the 120,000 dalton seed lectin. *Science* **200**, 1277–1279.

Ranjekar, P. K., Pallotta, D., and Lafontaine, J. G. (1978a). Analysis of plant genomes V. Comparative study of molecular properties of seven *Allium* species. *Biochem. Genet.* **16**, 957–970.

Ranjekar, P. K., Pallotta, D., and Lafontaine, J. G. (1978b). Analysis of plant genomes III. Denaturation and reassociation properties of cryptic satellite DNAs in barley

(*Hordeum vulgare*) and wheat (*Triticum aestivum*). *Biochim. Biophys. Acta* **520,** 103–110.

Righetti, P. G., Gianazza, E., Viotti, A., and Soave, C. (1977). Heterogeneity of storage proteins in maize. *Planta* **136,** 115–123.

Rimpau, J., Smith, D., and Flavell, R. (1978). Sequence organization analysis of the wheat and rye genomes by interspecies DNA/DNA hybridization. *J. Mol. Biol.* **123,** 327–359.

Rimpau, J., Smith, D. B., and Flavell, R. B. (1980). Sequence organization in barley and oats chromosomes revealed by interspecies DNA/DNA hybridization. *Heredity* **44,** 131–149.

Roberts, L. M., and Lord, J. M. (1981). The synthesis of *Ricinnus communis* agglutinin. Cotranslational and posttranslational modification of agglutinin polypeptides. *Eur. J. Biochem.* **119,** 31–41.

Rubenstein, I. (1982). The zein multigene family. *In* "Maize for Biological Research" (W. F. Sheridan, ed.), pp. 189–195. U. N. D. Press, Grand Forks, North Dakota.

Sachs, M. M., and Freeling, M. (1978). Selective synthesis of alcohol dehydrogenase during anaerobic treatment of maize. *Mol. Gen. Genet.* **161,** 111–115.

Sachs, M. M., Freeling, M., and Okimoto, R. (1980). The anaerobic proteins in maize. *Cell* **20,** 761–767.

Sachs, M. M., Lorz, H., Dennis, E. S., Elizur, A., Ferl, R. J., Gerlach, W. L., Pryor, A. J., and Peacock, W. J. (1982). Molecular genetic analysis of the maize anaerobic response. *In* "Maize for Biological Research" (W. F. Sheridan, ed.), pp. 139–144. U. N. D. Press, Grand Forks, North Dakota.

Salamini, F., and Soave, C. (1982). Zein: Genetics and biochemistry. *In* "Maize for Biological Research" (W. F. Sheridan, ed.), pp. 155–160. U. N. D. Press, Grand Forks, North Dakota.

Salamini, F., DiFonzo, N., Gentinetta, E., and Soave, C. (1979). A dominant mutation interfering with protein accumulation in maize seeds. *In* "Seed Protein Improvement in Cereals and Grain Legumes" (W. Neuherberg, ed.), pp. 97–108. FAC-IAEA, Germany.

Scandalios, J. G. (1977). Isozymes—genetic and biochemical regulation of alcohol dehydrogenase. *Proc. Phytochem. Soc.* **14,** 129–155.

Scandalios, J. G., Chang, D. Y., McMillin, D. E., Tsaftaris, A. S., and Moll, R. H. (1980). Genetic regulation of the catalase developmental program in maize scutellum: Identification of a temporal regulatory gene. *Proc. Natl. Acad. Sci. U.S.A.* **77,** 5360–5364.

Scharf, H. D., and Nover, L. (1982). Heat-shock induced alterations of ribosomal protein phosphorylation in plant cell cultures. *Cell* **30,** 427–437.

Seshadri, M., and Ranjekar, P. K. (1980). An unusual pattern of genome organization in two *Phaseolus* plant species. *Biochim. Biophys. Acta* **610,** 211–220.

Shah, D. M., Hightower, R. C., and Meagher, R. B. (1982). Complete nucleotide sequence of a soybean actin gene. *Proc. Natl. Acad. Sci. U.S.A.* **79,** 1022–1026.

Shewry, P. R., Faulks, A. J., Pickering, R. A., Jones, I. T., Finch, R. A., and Miflin, B. J. (1980). The genetic analysis of barley storage proteins. *Heredity* **44,** 483.

Sidloi-Lumbroso, R., and Schulman, H. M. (1977). Purification and properties of soybean leghemoglobin messenger RNA. *Biochim. Biophys. Acta* **476,** 295–302.

Sidloi-Lumbroso, R., Kleinman, L., and Schulman, H. M. (1978). Biochemical evidence that the leghemoglobin genes are present in the soybean but not in the *Rhizobium* genome. *Nature (London)* **273,** 558–560.

Sievers, G., Huhtala, M. L., and Ellfolk, N. (1978). The primary structure of soybean (*Glycine max*) leghemoglobin c. *Acta Chem. Scand. Ser. B* **32**, 380–386.

Smith, D. B., and Flavell, R. B. (1977). Nucleotide sequence organization in the rye genome. *Biochim. Biophys. Acta* **474**, 82.

Soave, C., Reggiani, R., DiFonzo, N., and Salamini, F. (1981a). Clustering of genes for 20 kd zein subunit in the short arm of maize chromosome 7. *Genetics* **97**, 363–377.

Soave, C., Tardani, L., DiFonzo, N., and Salamini, F. (1981b). Zein level in maize endosperm depends on a protein under control of the *opaque-2* and *opaque-6* loci. *Cell* **27**, 403–410.

Sorenson, J. C. (1982). Catalase: A system for studying the molecular basis of developmental gene regulation. *In* "Maize for Biological Research" (W. F. Sheridan, ed.), pp. 135–138. U. N. D. Press, Grand Forks, North Dakota.

Spencer, D., Higgins, T. J. V., Button, S. C., and Davey, R. A. (1980). Pulse-labeling studies on protein synthesis in developing pea seeds and evidence of a precursor form of legumin small subunit. *Plant Physiol.* **66**, 510–515.

Sullivan, D., Brisson, N., Goodchild, B., and Verma, D. P. S. (1981). Molecular cloning and organization of two leghemoglobin genomic sequences of soybean. *Nature (London)* **289**, 516–518.

Sun, S. M., Slightom, J. L., and Hall, T. C. (1981). Intervening sequences in a plant gene—comparison of the partial sequence of cDNA and genomic DNA of french bean phaseolin. *Nature (London)* **289**, 37–41.

Taylor, W. C., and Bendich, A. J. (1977). Personal communication cited in Walbot and Goldberg (1979).

Thanh, V. H., and Shibasaki, K. (1977). Beta-conglycinin from soybean proteins. Isolation and immunological and physicochemical properties of monomeric forms. *Biochim. Biophys. Acta* **490**, 370–384.

Thompson, W. F., and Murray, M. G. (1979). The nuclear genome: Structure and function. *In* "The Biochemistry of Plants: A Comprehensive Treatise" (A. Marcus, ed.), Vol. 6, pp. 1–81. Academic Press, New York.

Truelsen, E., Gausing, K., Jochimsen, B., Jorgensen, P., Marcker, K. A. (1979). Cloning of soybean leghemoglobin structural gene sequences synthesized *in vitro*. *Nucleic Acids Res.* **9**, 3061–3072.

Tsai, C. Y., Huber, D. M., and Warren, H. L. (1980). A proposed role of zein and glutelin as N sinks in maize. *Plant Physiol.* **66**, 330–333.

Tumer, N. E., Thanh, V. H., and Nielsen, N. C. (1981). Purification and characterization of mRNA from soybean seeds. *J. Biol. Chem.* **256**, 8756–8760.

Tumer, N. E., Richter, J. D., and Nielsen, N. C. (1982). Structural characterization of the glycinin precursors. *J. Biol. Chem.* **257**, 4016–4018.

Varsanyi-Breiner, A., Gusella, J. F., Keys, C., and Housman, D. E. (1979). The organization of a nuclear DNA sequence from a higher plant: Molecular cloning and characterization of soybean ribosomal DNA. *Gene* **7**, 317–334.

Viotti, A., Sala, E., Alberi, P., and Soave, C. (1979a). Heterogeneity of zein synthesized *in vitro*. *Plant Sci. Lett.* **13**, 365–375.

Viotti, A., Sala, E., Marotta, E., Alberi, P., Balducci, C., and Soave, C. (1979b). Genes and mRNA coding for zein polypeptides in *Zea mays*. *Eur. J. Biochem.* **102**, 211–222.

Viotti, A., Pogna, N. E., Balducci, C., and Durante, M. (1980). Chromosomal localization of zein genes by *in situ* hybridization in *Zea mays*. *Mol. Gen. Genet.* **178**, 35–41.

Vitale, A., Soave, C., and Galante, E. (1980). Peptide mapping of IEF zein components from maize. *Plant Sci. Lett.* **18**, 57–64.

Vodkin, L. O. (1981). Isolation and characterization of messenger RNAs for seed lectin and Kunitz trypsin inhibitor in soybeans. *Plant Physiol.* **68,** 766–771.

Walbot, V., and Dure, L. (1976). Developmental biochemistry of cotton seed embryogenesis and germination. VII. Characterization of the cotton genome. *J. Mol. Biol.* **101,** 503.

Walbot, V., and Goldberg, R. (1979). Plant genome organization and its relationship to classical plant genetics. *In* "Nucleic Acids in Plants" (T. C. Hall and J. W. Davies, eds.), pp. 3–40. CRC Press, Boca Raton, Florida.

Wall, J. S., and Paulis, J. W. (1978). Corn and sorghum grain proteins. *Adv. Cereal Sci. Technol.* **1978,** 135–219.

Wiborg, O., Hyldig-Nielsen, J. J., Jensen, E. O., Paludan, K., and Marcker, K. A. (1982). The nucleotide sequences of two leghemoglobin genes from soybean. *Nucleic Acids Res.* **10,** 3487–3494.

Wienand, U., and Feix, G. (1978). Electrophoretic fractionation and translation *in vitro* of poly(rA)-containing RNA from maize endosperm. Evidence for two mRNAs coding for zein protein. *Eur. J. Biochem.* **92,** 605–611.

Wienand, U., and Feix, G. (1980). Zein specific restriction enzyme fragments of maize DNA. *FEBS Lett.* **116,** 14–16.

Wienand, U., Langridge, P., and Feix, G. (1981). Isolation and characterization of a genomic sequence of maize coding for a zein gene. *Mol. Gen. Genet.* **182,** 440–444.

Wimber, D. E., Duffey, P. A., Stefensen, D. M., and Prensky, W. (1974) Localization of the 5 S RNA genes in *Zea Mays* by RNA-DNA hybridization *in situ. Chromosoma* **47,** 353–359.

Wimpee, C. F., and Rawson, J. R. Y. (1979). Characterization of the nuclear genome of pearl millet. *Biochim. Biophys. Acta* **562,** 192–206.

Zimmerman, J. L., and Goldberg, R. B. (1977). DNA sequence organizations in the genome of *Nicotiana tabacum. Chromosoma* **59,** 227–252.

CHROMATIN STRUCTURE AND GENE REGULATION IN HIGHER PLANTS

Steven Spiker

Department of Genetics, North Carolina State University,
Raleigh, North Carolina

I. Introduction: The Concept of Gene Regulation through Chromatin Structure

One of the major questions in modern biology is whether cellular differentiation and the development of eukaryotic organisms are due to differential gene expression. There are countless examples of

ADVANCES IN GENETICS, Vol. 22

changes in gene expression that accompany development, and the assumption is widely made that selective gene expression is at least partially responsible for differentiation. The question then becomes, what controls the selective expression of genes? Gene expression can be potentially controlled at many levels in eukaryotes, e.g., protein stability, rate of translation of messenger RNA, transport, processing, stability of the primary transcript, and rate of transcription. The evidence for the occurrence of such controls has recently been reviewed by Darnell (1982), who concluded that transcription, assessed by the rate of synthesis of specific nuclear RNA, is the primary level of control. Nearly all the examples cited in his review were from animal systems. Comparable studies that show developmentally correlated changes in nuclear RNA have not been reported in plants (Kamalay and Goldberg, 1980). However, changes in levels of specific, cytoplasmic messenger RNAs have been demonstrated and at least some of these changes are likely to be due to transcriptional control. See, for example, Dure *et al.* (1981), Galau and Dure (1981), Baulcombe and Key (1980), Bedbrook *et al.* (1980), Goldberg *et al.* (1981a,b), Key *et al.* (1981), and Zurfluh and Guilfoyle (1982). In his review, Darnell mentioned only briefly the biochemical mechanisms that have been proposed to regulate the rate of transcription of specific genes throughout development. A number of mechanisms have, of course, been proposed. Among them are mechanisms analogous to well-studied prokaryotic systems in which specific DNA sequences are directly involved in binding allosteric proteins. The demonstrations of the requirement of sequences, such as the Goldberg–Hogness or TATA box and the CCAAT box for the transcription of eukaryotic genes *in vivo* (e.g., Grosveld *et al.*, 1982; Grosschedl and Birnstiel, 1982), are evidence that control of transcription is operative at this level in eukaryotic organisms. However, it is becoming increasingly obvious that control of transcription in eukaryotes also occurs at other levels. One of these is the chromatin structure level.

Chromatin was originally defined as the part of the cell nucleus that stains densely with nuclear stains. It is now taken to mean the material of which chromosomes (both metaphase and interphase) are composed. When chromatin is removed from the cell, fixed, or spread in order to study it biochemically or microscopically, undoubtedly some of its inherent properties are altered. For this reason some people have preferred to call such preparations "nucleoprotein." Fully realizing that some perturbations in structure must occur upon manipulation, the term "chromatin" will nevertheless be used in this article for the chromosomal material we study both under the microscope and in the test tube.

The exact composition of chromatin then depends not only upon the organism and cell type from which it is isolated but also upon the method of isolation. In general terms, chromatin consists mainly of (1) about equal parts by weight of DNA and a group of low-molecular-weight basic proteins, the histones, and (2) varying proportions of a diverse group of other proteins that for convenience are called "non-histone chromatin proteins."

The idea that the structure of chromatin might play a role in the control of gene expression is an old one, going back at least as far as 1928 to Heitz (see Brown, 1966, for discussion). At that time Heitz first realized the relationship between chromosomes and the darkly staining aggregates that had been seen in the nuclei of plants and animals since the time of the nineteenth century cytologists. He called the aggregates "heterochromatin" and proposed that they had special genetic properties that translate in modern terms to be "nontranscribed portions of the genome."

In more recent history, support for a role of chromatin structure in the control of gene expression has come from many lines of research. Among these are early microscopic studies, e.g., the inactivation of the X chromosome in mammalian females, in which a change in structure visible by light microscopy accompanies a loss in function of an entire chromosome (Lyon, 1961). Classic electron microscope autoradiography studies indicate that in mammals most RNA synthesis occurs in diffuse chromatin rather than in condensed chromatin (Hsu, 1962; Littau *et al.*, 1964).

Ideas for a mechanism of how chromatin structure might influence gene expression date at least to the 1950 hypothesis of Stedman and Stedman, in which they stated that the tissue-specific differences in histones that they observed might be responsible for tissue-specific patterns of gene repression. We now know that many of the differences in histones that the Stedmans observed were artifacts. Nevertheless, the idea of a role of chromatin proteins in gene regulation has persisted. In the 1960s histones were shown to be general repressors of gene activity *in vitro* (Bonner *et al.*, 1968), but improved technology for handling the proteins led to the conclusion that histone heterogeneity was limited and the probability that histones acted as specific gene repressors was considered highly unlikely (Panyim and Chalkley, 1969a). It was also obvious that developmentally important differences in chromatin structure were much more subtle than the differences between condensed and diffuse chromatin. Progress on the character-ization of histones was rapid at this time, but chromatin did not begin to yield the secrets of its structure until the discovery of the nu-cleosome.

In 1974 several lines of experiments carried out in several laborato-ries crystallized into the concept of the "nucleosome" as the elemen-tary structural unit of chromatin (Kornberg, 1974). The nucleosome was a radical departure from earlier models in which chromatin was visualized as a uniform supercoil of DNA covered with proteins. We now think of chromatin as being composed of subunits, globular cores of eight histone molecules with the DNA wrapped around the outside (Kornberg, 1977). The nucleosomes are connected by "spacer" DNA so that the original electron micrographs give the appearance of "beads on a string" (Olins and Olins, 1974). The "spacer" DNA is presumably more accessible to nucleases, such as micrococcal nuclease, so that when chromatin is treated with this enzyme a series of discrete sizes of double-stranded DNA results. These discrete lengths correspond to the amount of DNA in 1, 2, 3, etc., nucleosomes—about 200, 400, 600, etc., base pairs in most organisms (see Fig. 1).

The higher order structure of chromatin is not yet clear. However, it is thought that the 100-Å nucleosome fiber can undergo supercoiling in order to form the circa 300-Å fiber commonly observed by electron microscopy in interphase nuclei. This fiber can then undergo further coiling and condensation and, with the involvement of a protein scaf-fold, result in the structure observable by light microscopy as the mito-tic chromosome (Igo-Kemenes et al., 1982).

With the "toehold" of the nucleosome and the increasing use of re-striction enzymes and other endonucleases as tools, progress toward understanding the structure of chromatin underwent a rapid increase after 1974. The progress was dramatic enough that Pierre Chambon titled his summary of the 1977 Cold Spring Harbor Symposium on Chromatin, "The molecular biology of the eukaryotic genome is com-ing of age" (Chambon, 1978). Since 1977, the ever-expanding knowl-edge of chromatin structure in general and of the distinction between "active" and "inactive" chromatin has been reviewed many times. For example, see Weisbrod (1982a), McGhee and Felsenfeld (1980), Mathis et al. (1980), Kornberg (1977), Lilly and Pardon (1979), Elgin (1981), Isenberg (1979), Felsenfeld (1978), Thomas (1978), and Igo-Kemenes et al. (1982). A review by Thompson and Murray (1981) included mate-rial on the structure and function of plant chromatin published up to 1978.

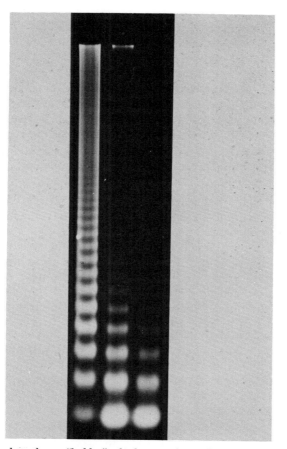

FIG. 1. Micrococcal nuclease "ladder" of wheat embryo chromatin. Nuclei isolated from wheat embryos were adjusted to 1 mg DNA/ml and digested at 37°C with 50 units of micrococcal nuclease/ml for (left to right) 1, 5, and 10 minutes. DNA fragments were separated on nondenaturing agarose gels and visualized under UV after staining with ethidium bromide as described by Spiker *et al.* (1983).

The idea that gene expression is regulated through chromatin structure is obviously valid only if we view chromatin structure as only one of the levels of gene regulation. In most cases thus far investigated, the idea has arisen that an active chromatin structure is necessary for transcription, but in order for transcription to actually occur, a number of other criteria must be met. Thus, the dichotomy is more accurately that of "inactive" chromatin versus "potentially active" chromatin.

With this reservation in mind, the term "active" chromatin will be used in this article for convenience even though "potentially active" would be more accurate.

In order for the DNA in chromatin to be transcribable it is obvious that it must be accessible to RNA polymerases and possibly other regulatory molecules. We shall consider two levels at which polymerases may be blocked from potential DNA templates by chromatin structure: (1) higher order coiling of the chromatin fibers and (2) by the mere presence of the canonical nucleosome. Evidence for the involvement of higher order structure in gene repression includes classical demonstrations of enhanced RNA synthesis in loops of lampbrush chromosomes and puffs of polytene chromosomes. Additional evidence that higher order structure, as well as the structure of individual nucleosomes, are involved in gene regulation has come from studies using relatively nonspecific endonucleases as tools.

After the general features of the nucleosome structure of chromatin were known, one of the first questions that arose was, are transcriptionally active genes organized into nucleosomes? Active chromatin does have a nucleosome structure (at least for certain periods) as evidenced by the appearance of several genes in circa 200 base-pair repeats generated by micrococcal nuclease (Mathis et al., 1980). Among the genes demonstrated to have a nucleosomal organization by this method is the single copy sequence of ovalbumin in chicken oviduct (Bellard et al., 1978). Despite the apparent location of active genes within nucleosomes, steric considerations make it difficult to envision RNA polymerase reading DNA through the canonical nucleosome. Thus, proposals have been made that in order for transcription to occur through the site of a nucleosome, some sort of perturbation must occur. The nucleosome must transiently disassemble, then reassemble. The histone core must split into two heterotypic tetramers, or perhaps the core becomes associated with only one DNA strand (Gould et al., 1980; Weintraub et al., 1976; Palter et al., 1979). Is there something special about the nucleosomes in active chromatin that would enhance such perturbations? Are nucleosomes of active chromatin somehow different?

Evidence that there is indeed something unique about the nucleosomes of active genes came from the observation of Weintraub and Groudine (1976) that the nonspecific endonuclease DNase I preferentially digests the globin-gene chromatin of chicken erythrocyte nuclei. The same globin sequences are not preferentially digested in other tissues, e.g., brain. The feature of globin active chromatin recognized by DNase I is apparently a function of the structure of individual nucleosomes and not of higher order structure because in chromatin digested to the mononucleosome level by micrococcal nuclease, the globin genes retain their enhanced sensitivity to DNase I. Furthermore, Weintraub and Groudine noted that the enhanced sensitivity was probably not due to the presence of transcriptional complexes on the genes because mature erythrocytes, which are transcriptionally silent, retain the preferential sensitivity of the globin genes to DNase I. Thus, even though transcription was not actually occurring, the chromatin structure of the gene was maintained in a "potentially active" conformation.

Garel *et al.* (1977) also presented evidence that the DNase I sensitivity was not due to the presence of transcriptional complexes. They studied the DNase I sensitivity of genes transcribed at low rates by using cDNAs to rare messenger RNAs as probes. The genes transcribed at low rates were also preferentially digested by DNase I. Earlier, the same group (Garel and Axel, 1976) had investigated the DNase I sensitivity of the ovalbumin genes in chicken. They observed that the gene was preferentially digested in oviduct chromatin but not in erythrocyte chromatin. In contrast to the results of Weintraub and Groudine on the globin gene, Garel and Axel found that when oviduct chromatin is digested to mononucleosomes with micrococcal nuclease, the enhanced sensitivity of the ovalbumin gene to DNase I is lost. Thus, the feature of ovalbumin chromatin recognized by DNase I appears to be a function of higher order structure and not inherent in the structure of individual nucleosomes.

Since these experiments were reported, the DNase I sensitivity of active chromatin has undergone intensive investigation (see Weisbrod, 1982a; Igo-Kemenes *et al.,* 1982; Elgin, 1981, for review). The coding regions of genes have a high sensitivity. In the region just 5' to the coding regions there are often relatively specific sites that have even greater sensitivity which have been termed the "hypersensitive" sites. Stalder *et al.* (1980) have demonstrated a 25–50 kb region adjacent to the globin-coding sequences in erythrocytes that has a DNase I sensitivity intermediate between that of an expressed gene and that of a totally inactive gene. On the basis of this work and the work that indicates that the nonhistone, high mobility group (HMG) proteins are a part of active genes (Weisbrod, 1982a,b; Mathis *et al.*, 1980), Stalder *et al.* proposed a model that is analogous to lampbrush chromosomes and relates DNase I sensitivity to higher order and nucleosome-based structure of active chromatin. Figure 2 is an adaptation of that model.

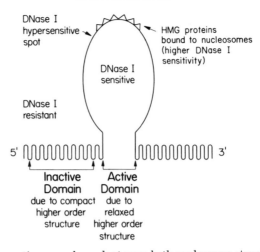

FIG. 2. Model of an active gene dependent upon both nucleosome structure and higher order structure. According to the model, RNA polymerases would not have access to inactive domains because of condensed higher order structure—folding to at least the level of the 250-Å fiber. The inactive domains would have a high resistance to endonucleases. An active domain (containing active genes and adjacent sequences) would be unfolded to the 100-Å fiber (nucleosome) level and thus be more accessible to endonucleases (Stalder *et al.*, 1980). The active genes are characterized by an altered nucleosome structure, including bound HMG proteins that render the genes more sensitive to DNase I and presumably accessible to RNA polymerases. Adjacent to the 5′ end of the coding regions are specific sites that are hypersensitive to DNase I (about an order of magnitude more sensitive than the coding regions). These sites may be nucleosome free and have been hypothesized to contain regulatory DNA sequences (Igo-Kemenes *et al.*, 1982).

From these studies and others reviewed by Mathis *et al.* (1980), Weisbrod (1982a), and Igo-Kemenes *et al.* (1982), it is obvious that there is more than one type of active chromatin; there is active chromatin based on higher order structure, and there is active chromatin based on the structure of individual nucleosomes. In addition, there are variations of the active structure within these two general categories, and upon occasion changes of chromatin structure from inactive to active involve both higher order changes and changes at the nucleosome level. We are only beginning to understand what constitutes an active chromatin structure. Beyond our grasp at present are the mechanisms that may be involved in programmed changes of chromatin structure during development. Current questions on the structure of active chromatin include the following: is initiation of transcription controlled by the position of the nucleosome relative to certain sequences of DNA (Weintraub, 1980; Zachau and Igo-Kemenes, 1981; Kornberg, 1981), and can such phasing be controlled by sequence-recognizing proteins? Are H1 histones (which bind to chromatin outside the nucleosomes) involved in active chromatin structure (Mathis *et al.*, 1980)? If HMG proteins do render a nucleosome active, what controls the binding of HMG proteins to specific sections of chromatin during development? It is obvious that although we have learned much about chromatin structure, a great number of fundamental biological questions remain to be answered.

Most of the work that has been reported on chromatin structure and gene regulation has been done with animal systems. It is likely that many of the principles that have been discovered are general and apply to all eukaryotic organisms. However, it is obvious that there are variations on basic themes. Different genes are regulated in different ways. Different organisms may have unique solutions for similar problems. The study of a different organism not only reveals information about that particular organism but also clarifies our understanding of what is of general biological significance and which are special cases. The study of chromatin structure and gene regulation in plants has pointed out several features in common with animal systems, but differences have also been observed. Even though chromatin structure at the nucleosome level seems to be similar in plants and animals, some aspects of the gross architecture of the nuclei are distinct (Nagl, 1976, 1979a,b; LaFontaine, 1974). The problem of cellular differentiation is completely different in plants, in which a perpetual supply of undifferentiated cells in the meristems must be continuously shunted down various developmental pathways throughout the life of the organism. The totipotency of plant cells is well known. A wide variety of nucleated cells (perhaps all?) from differentiated tissues can be grown in culture and regenerated into whole, normal plants. The fact that differentiated plant cells are totipotent argues strongly that cell differentiation does not proceed by loss of DNA sequences. It is also apparent that if plant cell differentiation is in part controlled by changes in chromatin structure, those changes must be reversible.

This article has been written from the point of view that differences in the way plants and animals develop may be reflected in differences in how chromatin structure is involved in gene regulation, and that this possibility should be considered when interpreting data concerning plant chromatin structure. The purposes of this article are (1) to bring together the work that has been done with chromatin of higher plants, (2) to point out the similarities and differences in what we know of chromatin structure in plants and other eukaryotes, and (3) to identify gaps in our knowledge and areas in which work might be expected to yield significant advances in the near future.

II. General Features of Chromatin in Higher Plants: Comparison to Other Eukaryotes

A. NUCLEOSOMES AND HIGHER ORDER STRUCTURE

1. Microscopic Investigations

The first published electron micrographs that showed a regular repeating subunit structure of chromatin were those of Olins and Olins (1974), who used material from rat thymus and liver and from chicken erythrocytes. The first similar micrographs showing nucleosome-like material in plants were those of Nicolaieff *et al.* (1976) and Gigot *et al.* (1976). These investigators estimated the average size of the plant nucelosomes to be 160 Å, which was slightly larger than the 125 Å reported by Oudet *et al.* (1975), who used a similar technique for various tissues of calf and chicken. These size differences were not considered to be significant and were attributed to variations in granularity of platinum shadowing. However, these workers did report a structure in the plant chromatin that was not seen in animal chromatin. When the plant chromatin was spread in 0.3 to 0.6 M NaCl, a 280-Å ring structure appeared. The suggestion was made that these rings might correspond to an alternate chromatin structure in plants, reflecting possible differences in plant histones. No further work has been presented concerning a ring structure in plant chromatin. Indeed, all work published since then indicates no differences between plant and animal nucleosomes at this level of observation (Greimers and Deltour, 1981; Lutz and Nagl, 1980; Moreno *et al.*, 1978). An electron micrograph of plant nucleosomes is shown in Fig. 3.

Differences do exist, however, in the gross ultrastructural organization of plant and animal chromatin (LaFontaine, 1974; Nagl, 1976, 1979a,b). In animals the amount of condensed euchromatin in nuclei varies from tissue to tissue and has often been taken to be an indication of mechanisms for massive repression and derepression of the genome. In plants the amount of condensed euchromatin is species specific rather than tissue specific (Nagl, 1979b). These differences in nuclear architecture must be due to differences in the higher order coiling of chromatin fibers because, at the electron microscope level, the nucleosomes of plants and animals appear to be the same.

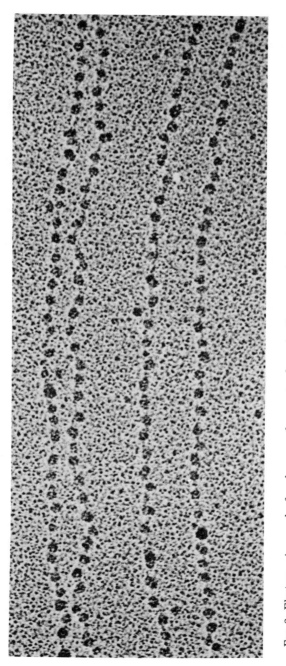

Fig. 3. Electron micrograph of nucleosomes from roots of germinating corn embryos showing typical "beads-on-a-string" appearance. The mean diameter of the nucleosomes is 125 ± 13 Å (Greimers and Deltour, 1981). (Reproduced with permission of Wissenschaftliche Verlagsgesellschaft mbH, Stuttgart.)

2. Biochemical Investigations

The principal biochemical evidence for a nucleosome structure in chromatin has been (1) the generation of "nucleosome ladders" upon digestion of chromatin with micrococcal nuclease, (2) a limit micrococcal nuclease digest nucleosome particle with a sedimentation coefficient of about 11 S that contains circa 146 base pairs of DNA and 2 each of the 4 "nucleosomal histones" H2A, H2B, H3, and H4, (3) generation of a ladder of single-stranded DNA fragments differing in length by multiples of about 10 bases upon digestion of nucleosomes by DNase I and indicating that DNA lies on the outside of the nucleosome, and (4) the demonstration that the nucleosomal histones can interact specifically with one another to form a globular, multisubunit core for the nucleosome and that this core and DNA can be used to reconstitute nucleosomes *in vitro* (see Kornberg, 1977, for review). Plants have been demonstrated to have canonical nucleosomes by all these criteria. An example of a micrococcal nuclease ladder generated from plant chromatin is shown in Fig. 1.

The first demonstration of a nucleosome ladder in plants was presented by McGhee and Engel (1975), who estimated a repeat size of 170 ± 30 base pairs of DNA in peas. Gigot *et al.* (1976) separated micrococcal nuclease-generated tobacco mono-, di-, and trinucleosomes on sucrose gradients and determined the sedimentation coefficient of the monomer to be 12.9 S. The monomer contained all the non-H1 histones in a 1/1 weight ratio with DNA. Frado *et al.* (1977) did similar work with pea chromatin and obtained a monomer sedimentation coefficient of 11 S. Philipps and Gigot (1977) demonstrated the existence of a circa 140 base-pair fragment after extensive micrococcal nuclease digestion of tobacco and barley chromatin. They also showed that DNA isolated from DNase I-digested tobacco nuclei can be separated on denaturing gels into a series of fragments that are multiples of 10 bases in size.

The plant histones H2A, H2B, H3, and H4 interact with each other to form nucleosomes in exactly the same way as do animal histones (Spiker and Isenberg, 1977; Martinson and True, 1979). Furthermore, plant and animal histones are interchangeable in the pairwise interactions that stabilize the histone core of the nucleosome (Spiker and Isenberg, 1978), indicating evolutionary conservation of histone– histone binding sites. Additional evidence for the importance of the evolutionary stability of the histone–histone contacts in nucleosome formation comes from reconstitution studies. Wilhelm *et al.* (1979) and Liberati-Langenbuch *et al.* (1980) reconstituted nucleosome cores, using combinations of histones from plants (corn or tobacco) and animals (chicken). The success of these reconstitutions demonstrates that the essential histone–histone contacts have been evolutionarily conserved even though there are distinct differences in plant and animal H2A and H2B histones (see Section II,B).

The general features of the nucleosome structure of plant chromatin have been established by numerous other investigators (see LaRue and Pallotta, 1976a,b; Cheah and Osborne, 1977; Yakura *et al.*, 1978; Leber and Hemleben, 1979a,b; Grellet *et al.*, 1980; Muller *et al.*, 1980; Willmitzer and Wagner, 1981). The work thus far points to no obvious differences in the structure of bulk chromatin between plants and animals. Whether there are subtle differences or differences that are important in constituting an active chromatin structure is yet to be determined.

B. Histones

The pioneering work of Fambrough and Bonner (1966, 1968, 1969; Fambrough *et al.*, 1968) laid the foundations for characterizing plant histones. They demonstrated the limited heterogeneity of plant histones, pointed out their similarity to animal histones, and showed that plant histones, like animal histones, could be divided into three groups according to content of basic amino acids: (1) the very lysine rich, now known as H1, (2) the slightly lysine rich, now called H2A and H2B, and (3) the arginine rich, now called H3 and H4. As reviewed by Isenberg (1979), the work of Fambrough and Bonner led to the finding of extreme evolutionary conservation of the arginine-rich histones. Both cow and pea histone H4 molecules have 102 amino acid residues. Their sequences are identical at 100 positions, and the 2 substitutions are conservative ones. Histone H3 is nearly as conserved, having a sequence identity in 131 of 135 positions. Based on the general similarities of all plant and animal histones and the demonstrated sequence conservation for the arginine-rich histones, the assumption was widely made (e.g., Nagl, 1976) that the slightly lysine-rich histones, H2A and H2B, must be nearly as conserved as H3 and H4. In fact, Fambrough and Bonner had suggested only that the slightly lysine-rich histones of plants were *similar* to those of animals. Their data predicted that these histones would not be as conserved as the arginine-rich histones. This point was further established by other laboratories.

Using both sodium dodecyl sulfate (SDS) and acetic acid–urea polyacrylamide gels, several investigators demonstrated large differences in electrophoretic mobilities and apparent molecular weights between plant and animal H2 histones (Panyim *et al.*, 1970; Spiker and Chalkley, 1971; Spiker and Krishnaswamy, 1973; Spiker, 1975, 1976a,b; Spiker *et al.*, 1976; Nadeau *et al.*, 1974; Stout and Phillips, 1973; Towill and Nooden, 1973; Burkhanova *et al.*, 1975) (see Fig. 4). These differences led Nadeau *et al.* (1974) to use the term "PH" for the plant histones that migrated just slower than H3 in both SDS– and acetic acid–urea–polyacrylamide gel electrophoresis. Such a designation was also used by Brandt and Von Holt (1975) and was considered warranted until the physiological correspondence between these histones and animal H2A and H2B could be established. In addition to the distinct electrophoretic mobilities and amino acid contents of the slightly lysine-rich plant histones, these proteins could not be defined as H2A or H2B if the standard operational criteria formulated by Johns (1964) were applied (Oliver *et al.*, 1972; Spiker *et al.*, 1976; LaRue and Pallotta, 1976a; Nadeau *et al.*, 1977; Fazal and Cole, 1977).

The reason for the uncertainty about what to call the slightly lysine-rich histones was, of course, the fact that the histones were originally categorized on the basis of operational criteria (Johns, 1964) and not on the basis of physiological function. The histones have no readily assayable biological function (such as enzyme activity) that could serve to identify H2A or H2B. The operational definitions served well for vertebrate histones and, in fact, for the histones of a wide variety of animals (Oliver *et al.*, 1972). They also served well for the plant arginine-rich histones but not for the slightly lysine-rich histones. In an attempt to refocus on the probable functional similarities of plant and animal slightly lysine-rich histones, an identification scheme was devised (Spiker *et al.*, 1976). The scheme used some of the classical criteria of Johns (1964) and additional criteria, including detergent-binding and staining characteristics. The criteria were general enough to allow the identification of H2A and H2B histones from yeast (Mardian and Isenberg, 1978) at a time when the existence of these histones in fungi was in doubt. A similar scheme for identifying and fractionating plant histones has recently appeared (Sidorova and Konarev, 1981).

Still, a definition of the histone groups based on an assayable biological function was lacking. It is still lacking today. However, the observation that the H2A, H2B, H3, and H4 histones formed the globular core of the nucleosome and the observation of a pattern of specific pairwise interactions between calf thymus histones (D'Anna and Isenberg, 1974) suggested a *quasi-functional* definition of the histone groups. D'Anna and Isenberg, using the techniques of fluorescence

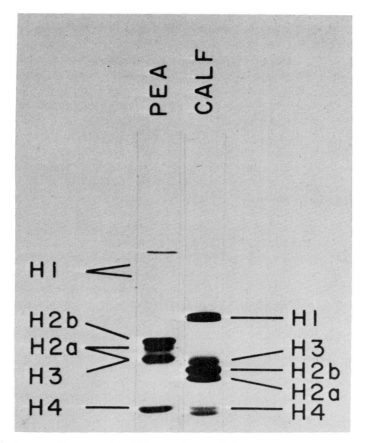

FIG. 4. Acetic acid–urea gels of histones from calf thymus and pea shoots. Histones were isolated and gels run according to Panyim *et al.* (1970). The evolutionarily conserved histones H3 and H4 have essentially identical electrophoretic mobilities. (The doublet in calf H4 is due to acetylation of lysine residues in about half the molecules.) Plant H2A and H2B histones have substantially lower electrophoretic mobilities than their animal counterparts.

anisotropy, light scattering, and circular dichroism, showed that in solution strong interactions occurred between the histone pairs H3–H4, H2B–H4, and H2A–H2B. Any interaction between H2A and H4 was too weak to measure. Thus, H2B could be defined as the slightly lysine-rich histone that interacts strongly with H4. Conversely, H2A could be defined as the histone that does not interact strongly with H4.

Experiments using these techniques with pea histones showed clearly that the same pairwise interactions that occurred with animal histones also occurred with plant histones (Spiker and Isenberg, 1977). The association constants measured for the plant histone complexes were the same as those measured for calf thymus histones. Furthermore, interkingdom hybrid histone complexes could be formed between calf and pea histones (Spiker and Isenberg, 1978). The interkingdom complexes displayed the same association constants as those observed when the histones of pea or calf alone were used. Thermodynamic arguments led to the prediction that the surfaces involved in the histone–histone interactions must have highly (if not completely) conserved amino acid sequences (Spiker and Isenberg, 1978). A further prediction of a large interaction surface was made, based on the observation that upon complexing there was a large increase in the number of amino acid residues assuming an α-helical conformation. Some of these predictions were confirmed by Martinson and True (1979), who showed by cross-linking experiments that the same histone–histone contacts demonstrated for histone pairs in solution actually occurred *in vivo* and the amino acids at the sites of contact were evolutionarily conserved.

Thus, a refined picture of histone evolution has emerged (Isenberg, 1979). Histones H3 and H4 are evolutionarily conserved to an extreme degree and throughout the entire length of the molecule. Histones H2A and H2B are evolutionary hybrids; one part of the molecule is involved in the histone–histone interactions that stabilize the core of the nucleosome. This portion is presumably highly conserved and there is additional evidence for this (see Section III,C,1). The remaining portions of the H2A and H2B molecules are not conserved based on amino acid content (Spiker and Isenberg, 1977), tryptic fingerprinting (Nadeau *et al.*, 1977), and partial sequence (Rodriguez *et al.*, 1979; Von Holt *et al.*, 1979; Hayashi *et al.*, 1977). The conserved portions of the H2A and H2B molecules are in the C-terminal regions. The N-termini are variable. If we assume, as the evidence strongly suggests, that histone–histone binding is the function for the sequence-invariant portion of these molecules, a question then arises; what is the function of the nonconserved, N-terminal portion of the molecules? Is it other than nonspecific interaction of DNA with positive charges? Why is it larger in plants than in animals? The difference between H2A and H2B in plants and animals does not appear to have anything to do with the basic nucleosome structure, as mentioned above. Might the differences in the nonconserved, N-terminal regions be concerned with the mechanism by which an active chromatin conformation is generated?

The variation in the N-terminal regions of the H2 histones is apparent not only upon comparison of plants and animals but also within the animal kingdom. For example, Elgin *et al.* (1979) sequenced an entire H2B molecule from *Drosophila*. Despite the overall similarity to calf thymus H2B in physical properties and amino acid content, the sequence was different. In fact, in the first 26 residues there were 6 sequence gaps (insertions or deletions) and 15 substitutions. Vertebrate H2 histones also vary from one another as demonstrated by electrophoretic mobility (Panyim *et al.*, 1971), as do a wide variety of vascular and lower plants (Spiker, 1975; Nadeau *et al.*, 1974; Stout and Phillips, 1973; Pitel and Durzan, 1978; Towill and Nooden, 1973; Nicklisch *et al.*, 1976; Tessier *et al.*, 1980). The possibility should not be overlooked that the variations in the N-terminal regions have functional significance. Perhaps they interact with species-specific regulatory molecules. The other extreme view is also possible—that their function is so nonspecific that there are limited constraints on mutations and only a general size and charge requirement. If only a general size requirement were necessary, the strict delimitation in size between plant and animal H2 histones would require explanation. These questions will be further considered in Section III.

Despite the great effort that has gone into sequencing histone molecules (Isenberg, 1979), to date no plant histones have been completely sequenced except for pea H3 and H4. Part of this may be due to the greater number of investigators who work with animal histones. However, this is not the only reason. The major difficulty concerns the character of the histone variants in plants. Histone variants are subtypes of the major groups of histones (e.g., H2A) that differ in primary structure. Although there are some exceptions (West and Bonner, 1980a), the general rule is that within an animal the histone variants are due to the substitution of one amino acid for another (Franklin and Zweidler, 1977; Urban *et al.*, 1979; Isenberg, 1979). Thus, complete purification of the variants is not necessary for sequencing. When a mixture of variants is split into peptides for sequencing, the peptides containing the amino acid substitutions can be picked out and sequenced separately (Franklin and Zweidler, 1977). Or, if sequencing is carried out directly from the N-terminal by automated Edman degradation, two or more amino acids will appear at a point, but before and after that point no ambiguities will exist. In a single plant there are variants of the H2A and H2B histones that display not only point substitutions but also insertions or deletions. This leads to a complicated peptide pattern that is difficult to interpret. Automated sequencing is similarly complicated (Von Holt *et al.*, 1979; Rodrigues *et al.*, 1979; Hayashi *et al.*, 1977). Thus, in order to sequence a plant H2A or H2B histone, the variant in question must be separated from the other variants in its class. This has proven a difficult task. Histone variants and their possible role in forming an active chromatin structure will be discussed in Section III.

Even though we have no complete sequence information concerning any plant histone other than H3 and H4, some information is available. Hayashi *et al.* (1977) have deduced a partial sequence of pea H2B based on amino acid content of several peptides. They projected that the amino acid sequence in the middle and carboxy terminal region of the histone is very similar to that of calf thymus H2B. The amino terminal is different from calf and has blocked N-terminal residues. It is larger and there are amino acid substitutions, insertions, and deletions. Little data were given on pea H2A, but the authors suggested that pea H2A differs from calf H2A in a similar manner.

Von Holt *et al.* (1979) reported the partial sequence of H2B from wheat and found that the N-terminal is blocked. None of the several known variants has been sequenced in the N-terminal region, but the number of residues seems to vary in this region. The sequence of the 89 C-terminal residues of a mixture of H2B variants has been determined and is identical to the calf thymus H2B sequence at 69 positions. Most of the substitutions are conservative ones.

Wheat H2A has also been partially sequenced (Rodrigues *et al.*, 1979). Wheat has two major groups of H2A, as previously reported (Spiker, 1976b). The same situation exists in peas (Sommer and Chalkley, 1974; Spiker *et al.*, 1976; Spiker and Isenberg, 1977) and in rye (LaRue and Pallotta, 1976a). The lower molecular weight form, termed $H2A_{(1)}$, was sequenced by automatic Edman degradation to residue 63 (by the numbering of Rodrigues and co-workers). When aligned for maximum homology (Fig. 5) to one of the calf thymus H2A variants, wheat $H2A_{(1)}$ was missing 2 N-terminal residues, had a deletion at residue 17, and had 14 other substitutions in the first 52 residues. A mixture of the higher molecular weight H2A variants, $H2A_{(2)}$ and $H2A_{(3)}$, was also sequenced. Both variants were 9 residues longer than $H2A_{(1)}$, and they differed from each other at 4 positions in the first 37 residues. At least 1 of the $H2A_{(2)}$ and $H2A_{(3)}$ variants differed from $H2A_{(1)}$ at seven positions in addition to an insertion and the additional 9 residues at the N-termini.

Sequence work has proceeded more slowly on the animal H1 histones than on the animal nucleosomal histones (Isenberg, 1979). The

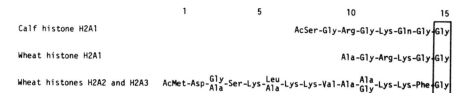

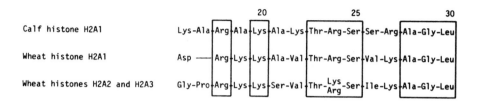

FIG. 5. Partial sequences of H2A variants from wheat embryos and calf thymus. Identical residues are framed. The symbol ——— is a gap in the alignment. The numbering system refers to alignment positions and not to sequence positions. (Data from Rodrigues *et al.*, 1979.) The number of the calf H2A variant shown is according to Franklin and Zweidler (1977).

major reason for this has been the variability of H1 histones in animals. Plant H1 histones are also variable (Stout and Phillips, 1973; Spiker, 1976b). Very little sequence work has been carried out with plant H1 histones. The only reported work is that of Hurley and Stout (1980), who sequenced about 25% of one of the H1 variants from maize. Maize H1 histones, like other plant H1 histones, appear to have higher molecular weights than their animal counterparts. Despite this difference, the partial characterization by Hurley and Stout demonstrated that the plant H1 is organized in generally the same way as is animal H1. That is, it appears to have three structural domains: (1) a random coil, basic N-terminal region, (2) a central globular region, and (3) a highly basic C-terminal region. The central globular section seems to have a high degree of sequence conservation in animals (Boulikas *et al.*, 1980). This maize region is not completely sequenced, but the residues that have been established are highly conserved. Cross-linking data, which show this region in contact with specific regions of H2A in animals (Boulikas *et al.*, 1980), have led these authors to conclude that the H1–H2A contacts are evolutionarily conserved and serve to close the ends of DNA around the nucleosome.

C. NONHISTONE CHROMATIN PROTEINS

As noted in the review by Thompson and Murray (1981), nonhistone chromatin (NHC) proteins have long presented a fascinating potential for regulation of gene activity in plants. But the study of these proteins presents some difficulties. For example, many proteins that might not properly be considered chromatin proteins may be nonspecifically absorbed to chromatin during isolation. Thus, determining whether a protein (especially one present in low quantity) should be considered a chromatin protein presents a real problem. It follows that observed changes in NHC proteins during development must be viewed with great caution. Nevertheless, the great apparent variability in NHC proteins, their higher turnover rate, and tissue specificity have inspired a great many studies of NHC proteins and their influence on gene expression in plants (Thompson and Murray, 1981). These studies have most often taken one of two routes: (1) developmentally correlated changes in the total acid-insoluble protein content of some chromatin preparation are described, or (2) the effect of NHC proteins added back to an *in vitro* transcription system is studied—usually with no characterization of the RNA produced. These studies have produced interesting and potentially useful data but so far have not contributed substantially to our understanding of gene regulation. In most cases the proteins that change during development have not been isolated and well characterized so that further experimentation could be carried out with them. The putative effects of an NHC preparation on *in vitro* transcription have most often not been attributed to any well-characterized protein so that the possible mechanisms of the effect could be tested. [Plant RNA polymerases are exceptions in that they have been well characterized (Guilfoyle and Jendrisak, 1978). However, these proteins will not be considered in this article because, even though they are obviously found associated with chromatin, they are not chromatin proteins in the usual sense. Plant polymerases have been recently reviewed (Guilfoyle, 1981).]

The studies of NHC proteins in animal systems have met with the same difficulties. However, a substantial number of these animal proteins have been well characterized—especially the high mobility group (HMG) proteins (Goodwin *et al.*, 1973, 1978; Johns, 1982). With these proteins the original approach to studying the role of NHC proteins in gene regulation was different from those approaches mentioned above. Instead of trying to track down and characterize an elusive protein with a putative function, Johns' group set out to systematically characterize a major group of NHC proteins from calf thymus for which no special function had been postulated at the time. There is now strong evidence that these proteins are involved in the structure of transcriptionally active chromatin (see Section III). Here a brief account will be given of the characterization of HMG proteins in mammals and of the evidence we have for the existence of such proteins in plants.

HMG proteins were first noticed as contaminants of H1 histone preparations in calf thymus (Goodwin *et al.*, 1973). They are extracted from chromatin with 0.35 M NaCl (lower than required for histone extraction), are soluble in 2% trichloroacetic acid, and are about 10% as abundant as any one of the histone fractions. These proteins were observed to have higher electrophoretic mobility on polyacrylamide gels than the 2% trichloroacetic acid-insoluble proteins. Thus, HMG proteins are operationally defined as the proteins that can be extracted from purified chromatin with 0.35 M NaCl and which are soluble in 2% trichloroacetic acid. In addition to high electrophoretic mobilities, corresponding to molecular weights of 26,500 for calf thymus HMG 1 and 9250 for HMG 17, the HMG proteins have very unusual amino acid compositions. All are highly charged with about 25% acidic and 25% basic residues. For example, calf thymus HMG 17 contains 12.0 mol% aspartic acid, 10.5 mol% glutamic acid, 24.3 mol% lysine, and 4.1 mol%

TABLE 1

Amino Acid Content of HMG Proteins from Calf Thymus and Wheat Embryos (Mol%)

Residue	Calf thymus				Wheat embryo		
	HMG 1[a]	HMG 2[a]	HMG 14[a]	HMG 17[a]	HMGa[b]	HMGb[c]	HMGd[b]
Asx	10.7	9.3	8.1	12.0	7.2	11.7	10.7
Thr	2.5	2.7	4.2	1.2	3.7	1.4	5.2
Ser	5.0	7.4	7.8	2.3	7.0	8.3	7.6
Glx	18.1	17.5	17.1	10.5	5.3	12.7	12.0
Pro	7.0	8.9	8.5	12.9	12.6	4.7	5.8
Gly	5.3	6.5	6.5	11.2	8.6	13.7	11.8
Ala	9.0	8.1	14.5	18.4	20.3	15.9	14.8
Val	1.9	2.3	4.2	2.0	3.2	3.6	5.4
Cys	Trace	Trace	0.7	—	—	—	—
Met	1.5	0.4	—	—	1.7	—	1.8
Ile	1.8	1.3	0.5	—	2.7	1.5	0.4
Leu	2.2	2.0	2.0	1.0	4.5	2.2	1.4
Tyr	2.9	2.0	0.4	—	1.9	0.8	1.2
Phe	3.6	3.0	0.6	—	0.7	2.7	3.1
Lys	21.3	19.4	19.0	24.3	13.4	17.0	15.9
His	1.7	2.0	0.3	—	0.3	1.3	0.2
Arg	3.9	4.7	5.6	4.1	7.1	2.5	2.7
Lys + Arg	25.2	24.1	24.6	28.4	20.5	19.5	18.6
Asx + Glx	28.2	26.8	25.2	22.5	12.5	24.4	22.7

[a]Goodwin et al. (1978).
[b]Spiker (unpublished data).
[c]Spiker et al. (1978).

arginine (see Table 1). This protein also has no cysteine, methionine, isoleucine, tyrosine, tryptophan, phenylalanine, or histidine.

In the initial preparations of HMG proteins, a large number of bands appeared (Goodwin et al., 1973). Some of these bands were later found to be degradation products of other HMG proteins or of H1 proteins (Goodwin et al., 1978). Of the originally described HMG proteins, only HMG 1, 2, 14, and 17 are considered authentic. Thus, when describing HMG proteins from new species or tissues, care must be taken that the proteins are not degradation artifacts.

Calf thymus HMG proteins 1, 2, 14, and 17 have been shown to be bound to mononucleosomes (Goodwin et al., 1977). Other proteins have been found bound to mononucleosomes that satisfy the operational criteria for classification as HMG proteins (Walker et al., 1978). One of these, ubiquitin, does not have the characteristic amino acid composition found in other HMG proteins. Thus, a philosophical question arises. What really is an HMG protein? The difficulty is, of course, the same as previously described with the histone fractions; when proteins are classified by operational criteria, functional distinctions are tacitly implied. But without assayable biological functions, the significance of the operational criteria is difficult to evaluate. For example, Weber and Isenberg (1980) have pointed out another difficulty in identifying HMG proteins. They characterized one yeast chromosomal protein that meets amino acid composition criteria for being an HMG protein but which is not extractable from chromatin with 0.35 M NaCl. This protein can be extracted with 0.25 N HCl, and has been classified as HMGa. Sommer (1978) studied a protein that by amino acid composition and electrophoretic mobility on SDS gels is probably the same protein as HMGa. Sommer, however, considered the protein to be the elusive H1 of yeast. Watson *et al.* (1979) have sequenced a protein from trout testis, H6, which has a striking degree of homology with calf thymus HMG 17 yet has only 13% acidic residues.

In addition to the HMG proteins of calf thymus, yeast, and trout, proteins that have been called HMG proteins because of one or more criteria have been reported in several other organisms, including (1) vertebrates (Goodwin *et al.*, 1978), (2) *Drosophila* (Alfageme *et al.*, 1976; Bassuk and Mayfield, 1982) and *Ceratitis* (Franco *et al.*, 1977; Marquez *et al.*, 1982), and (3) *Tetrahymena pyriformis* (Hamana and Iwai, 1979).

Plants also appear to have HMG proteins (Spiker *et al.*, 1978; Mayes, 1982). Four major proteins are extractable from purified wheat chromatin with 0.35 M NaCl and are soluble in 2% trichloroacetic acid. They have been called wheat HMGa,b,c, and d to discourage any premature comparison with calf thymus HMG proteins 1, 2, 14, and 17. The wheat HMG proteins have the same general electrophoretic mobilities as calf thymus HMG proteins, although no specific protein comigrates with any calf thymus HMG. Apparent molecular weights of the wheat HMG proteins estimated by SDS–polyacrylamide gel electrophoresis are the following: HMGa, 24,300; HMGb, 20,400; HMGc, 17,400; HMGd, 14,500 (Spiker, unpublished data). In Fig. 6, SDS gels of pig and wheat HMG proteins are shown.

The amino acid contents of wheat HMGb and HMGd support their identification as HMG proteins (see Table 1). Their compositions resemble calf HMG 17 in that (1) Glx and Asx are abundant and present in about equal quantities, (2) lysine is the most abundant amino acid, and (3) alanine is the second most abundant amino acid. However, unlike calf thymus HMG 17, these wheat HMG proteins have his-

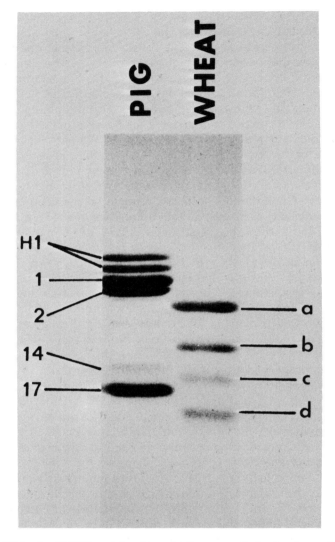

FIG. 6. SDS gels of HMG proteins from pig thymus and wheat embryos. Pig thymus HMG proteins were isolated by perchloric acid extraction according to Goodwin *et al.* (1975). Wheat HMG proteins (a–d) were isolated according to Spiker *et al.* (1978) and analyzed by SDS–polyacrylamide gel electrophoresis (Spiker, 1982).

tidine, phenylalanine, tyrosine, and isoleucine. Their content of basic amino acids is somewhat lower than that of calf HMG proteins. The amino acid composition of wheat HMGa is remarkable in that it has only 12.5 mol% acidic amino acids. It is reminiscent of histone H1 in that it has a high alanine content. However, its content of acidic amino acids and peptide maps argues against it being an H1 degradation product. Peptide mapping by the method of Cleveland *et al.* (1977) has shown that none of the four wheat HMGs is a degradation product of either other HMG proteins or of the H1 histones (Spiker, unpublished data).

Vertebrate HMG proteins can be fractionated by solubility in trichloroacetic acid (Goodwin *et al.*, 1975; Weisbrod and Weintraub, 1979). HMG proteins 1 and 2 are insoluble in 10% trichloroacetic acid whereas HMG proteins 14 and 17 are soluble. Such a fractionation does not work with wheat HMG proteins because all are insoluble in 10% trichloroacetic acid, and no lower concentration has been found that will efficiently serve to fractionate the wheat HMG proteins (Spiker, unpublished data). The wheat HMG proteins are associated with mononucleosomes, as indicated by the fact that they cosediment with mononucleosomes on sucrose gradients. Further evidence for the authenticity of the wheat HMG proteins, and evidence that they are associated with transcriptionally active chromatin, will be discussed (see Section III).

Among the chromosomal proteins that have been observed to change in plants during development are two wheat NHC proteins that diminish in quantity as germination proceeds (Sugita *et al.*, 1979). These proteins have been termed histone-like proteins (HLP) 1 and 2 with molecular weights 24,000 and 21,000, respectively (Motojima and Sakaguchi, 1981). On SDS–polyacrylamide gel electrophoresis, the proteins run in approximately the same positions as wheat HMG proteins a and b. Motojima and Sakaguchi (1981) consider these as HLP because they are only poorly extracted with 0.35 M NaCl whereas they are well extracted with 0.4 N H_2SO_4. However, it must be remembered that yeast has a protein with the same extraction characteristics, and this protein has been considered an HMG protein on the basis of amino acid content (Spiker *et al.*, 1978; Weber and Isenberg, 1980). Motojima and Sakaguchi (1981) did determine amino acid content of these proteins but unfortunately reported only that the mol% lysine was 14.3 for HLP 1 and 10.0 for HLP 2. They also reported a high degree of N-ϵ-methyl lysine in these two proteins. N-ϵ-methyl lysine has long been known to occur in histones (see Byvoet *et al.*, 1978), and the HMG proteins are also known to be methylated (Boffa *et al.*, 1979).

Preliminary studies on NHC proteins of plants have been carried out by various investigators (Simon and Becker, 1976; Lin *et al.*, 1973; Pipkin and Larson, 1973; Pitel and Durzan, 1978; Towill and Nooden, 1975).

III. Characterization of Transcriptionally Active Chromatin

A. Visualization by Electron Microscopy

Data obtained by biochemical means (see below) suggest that actively transcribed genes are organized into nucleosomes. The obvious difficulty with such data is that only the average of a very large number of genes may be monitored. At any precise moment some of these genes may be transcribing while some may not. The techniques of electron microscopy allow individual transcription units to be monitored, but results of electron microscope studies have not been straightforward (see Mathis *et al.*, 1980; Labhart and Koller, 1982). Rapidly transcribed genes, in particular ribosomal genes, appear to be devoid of nucleosomes. Other genes that are transcribed at lower rates have nucleosome-like particles between the RNA polymerase II molecules. The question may be one of what happens to the nucleosome at the very instant of read-through by RNA polymerase and what are the kinetics of establishing "pre-read-through" conformation.

Greimers and Deltour (1981) have presented an impressive set of electron micrographs of chromatin from nuclei of corn root cells. In the micrographs, nucleolar transcription units, characterized by "Christmas tree-like" figures, are devoid of nucleosomes. Other stretches of chromatin, which contain structures that the authors interpret as ribonucleoprotein particles of transcriptional complexes, contain nucleosomes immediately before and after the site of attachment of the presumed ribonucleoprotein particles.

Our best information suggests that active chromatin is at least uncoiled to the 100-Å fiber (nucleosome) level (Mathis *et al.*, 1980). Because coiling to a higher level of structure may be one of the mechanisms for preventing transcription, understanding the mechanisms for higher order coiling is of obvious importance. Several models for higher order structure based on solenoids of nucleosomes or superbeads have been presented (McGhee and Felsenfeld, 1980). None of the work reported thus far has been with plant chromatin.

B. FRACTIONATION OF CHROMATIN AND NUCLEASE SENSITIVITY OF TRANSCRIPTIONALLY ACTIVE CHROMATIN

In order to determine just what structural properties of chromatin render a gene active or inactive, we would ideally like to isolate a regulated gene with its native chromatin structure intact so that it could be studied. So far no one has succeeded in such a task, but work toward that end has been proceeding for some time (Gottesfeld, 1978). The best that has been done so far is to isolate a fraction of chromatin which, compared to bulk chromatin, is enriched several-fold in transcribing sequences.

The approaches to fractionation have depended on presumed structural differences between active and inactive chromatin. Thus, properties of density, susceptibility to shear, thermal stability, solubility, susceptibility to nucleases, etc., have been used. The most successful attempts have been those that used endonucleases to lightly digest chromatin and then took advantage of differential solubility of active and inactive chromatin in various salt concentrations (Mathis et al., 1980). Gottesfeld (1978) has mentioned five criteria to test the validity of any fractionation scheme:

1. The percentage of the DNA in the active fraction should be low because we assume that only a small portion of the genome is transcribed at any one time. Exactly what "low" means cannot be stated precisely. In fact, from a chromatin structure viewpoint, as discussed in Section I, a larger portion of the genome probably has an active structure than is actually being transcribed.

2. The sequence complexity of the active fraction should be a subset of the whole genome.

3. The sequences in the active fraction must be complementary to cellular messenger RNA.

4. Some of the sequences in the active fraction should be specific for certain cell types or stages of development. Gottesfeld gives the example that globin sequences should be found in the active fraction of reticulocyte chromatin but not in brain cell chromatin.

5. Nascent RNA chains should be associated with active chromatin. This last criterion assumes, of course, that the chromatin is actually being transcribed. As discussed in Section I, a gene may have a chro-

matin structure that is active (conducive to transcription) yet for other reasons the gene may not actually be undergoing transcription in a certain tissue.

One way of fractionating chromatin into active and inactive is by relatively extensive digestion with DNase I, as done by Weintraub and Groudine (1976). DNase I preferentially attacks the active genes, as shown by the depletion of transcribed sequences in the DNase I-resistant DNA. Therefore, the study of the active chromatin is indirect. It is logical to assume that as the DNA of active chromatin is digested away, the proteins found in the active structure would be released into free solution. Vidali *et al.* (1977) observed that among the proteins released from DNase I-treated chromatin, the HMG proteins were prominent. This was the first clue that the HMG proteins might be structural proteins of active genes. This observation was only suggestive, however. More extensive evidence for the involvement of HMG proteins in active chromatin will be discussed in Section III,C, on NHC proteins.

Treatment of chromatin with DNase II offers the possibility of directly studying chromatin enriched in actively transcribed sequences (Gottesfeld *et al.*, 1974, 1975; Gottesfeld and Butler, 1977; Gottesfeld and Partington, 1977). Chromatin is digested lightly with DNase II and separated by centrifugation into supernatant and pellet fractions. The supernatant is further separated into magnesium-soluble and magnesium-insoluble fractions. The DNA/histone ratios of all fractions are similar and all are organized into nucleosomes. However, the magnesium-soluble fraction, which consists of about 10% of the DNA, has a high content of NHC proteins, contains nascent RNA chains, and is enriched in sequences complementary to messenger RNA. This method appears to have great potential for studying active chromatin structure. However, Weisbrod (1982a) has noted that the method has been difficult to reproduce and thus has not been widely used. The reasons for the lack of reproducibility may be, as warned by Gottesfeld (1978), that the commercial preparations of DNase II contain both RNase and protease activities.

Micrococcal nuclease was originally supposed to have no preference for active sections of the genome (Mathis *et al.*, 1980). However, several workers have shown that chromatin fractions enriched in transcribing sequences can be obtained with light micrococcal-nuclease digestion (Sanders, 1978; Bloom and Anderson, 1978; Georgieva *et al.*, 1981; Levy *et al.*, 1979a; Mathis *et al.*, 1980). In cases where HMG proteins have been studied specifically, the transcriptionally active fraction has been found to be enriched in these proteins. Thus, HMG proteins are suspected of being structural proteins of active chromatin for three reasons: (1) when chromatin is treated with DNase I, the transcribed genes are preferentially digested and HMG proteins are released; (2) when chromatin is separated into active and inactive fractions with micrococcal nuclease, the HMG proteins are found with the active fraction; and (3) the reconstitution experiments of Weintraub's group (see Section III,D,1) show that HMG proteins are necessary to confer DNase I sensitivity upon transcriptionally active genes. Because such a structure can be reconstituted, it is obvious that HMG proteins must recognize some DNA sequence or some other structural features of chromatin for faithful reconstitution. So far no sequence specificity has been found. The histones in the nucleosomes do seem to be highly acetylated and the DNA undermethylated (see Section III,D,1), but how these could serve as specific binding signals for the HMG proteins is unclear. Histone variants have the potential to act as specific HMG binding sites, but so far the only difference in primary structures of histones that has been found is an unequal distribution of some histone H3 variants in some of the chromatin fractions generated by Sanders (1978).

Thus far, most of the attempts to separate chromatin into active and inactive fractions have been carried out with systems other than higher plants. However, it does appear as though some of the procedures will be successful with higher plants. Sugita and Sasaki (1982) have shown that the general susceptibility of wheat embryo chromatin to DNase II increases upon germination. We have recently shown (unpublished) that when wheat embryo chromatin is fractionated, using the micrococcal nuclease procedure of Jackson et al. (1979), the proteins appearing in the putative active and inactive fractions are analogous to those of mammalian systems. That is, the putative inactive fraction contains only histones. The putative active fraction contains histones and an almost equal amount of HMG proteins.

We have also examined the DNase I sensitivity of active chromatin in wheat embryos (Spiker et al., 1983). Using methods similar to those of Garel et al. (1977), Levy and Dixon (1977), and Lohr and Hereford (1979), isolated nuclei were treated with DNase I and the concentration of transcribed genes in the nuclease-resistant fraction was assayed by solution reassociation kinetics. The probe representing messenger RNA was a cDNA population complementary to polysomal poly(A)$^+$ RNA. The DNase I-resistant DNA drove the reassociation of the probe more slowly than did control DNA sheared to the same size.

This indicates that the transcribed sequences were preferentially digested with DNase I. The data obtained for wheat embryos were very similar to those obtained by Garel *et al.* (1977) and Levy and Dixon (1977) using similar probes in animal systems. Lohr and Hereford (1979) used a similar, general probe with yeast and found there was no preferential digestion of transcriptionally active genes with DNase I. They postulated that the small yeast genome may exist entirely in a transcriptionally active state comparable to that of only a restricted portion of the genome in multicellular eukaryotes. Sledziewski and Young (1982) have shown, however, that when a specific probe is used, differences in chromatin structure between a repressed and an active gene can be demonstrated using DNase I in yeast.

Thus, the transcriptionally active genes of plants display a general DNase I sensitivity comparable to that demonstrated in animal systems. Probes for specific genes have not as yet been investigated in plants. We would expect the presence of DNase I-hypersensitive sites to be associated with transcriptionally active genes in plants, but as of yet we have no data on this point.

From the above discussion it appears that the endonucleases recognize some feature of active chromatin that allows them to attack it more readily. The assumption is that the chromatin structure which the nuclease recognizes is involved in causing the gene to be active. Until the mechanisms for recognition and activation are known, we cannot be sure of this. However, such an assumption is justified in the early stages of investigation as a working hypothesis. Thus, a common approach to the problem of finding what constitutes an active gene structure is to ask what constitutes a nuclease-sensitive gene structure.

C. Do Histones Have a Role in Determining Active Chromatin Structure?

1. Histone Variants

a. A Background of Stability in Nucleosomal Histones throughout Development. In the early years of chromatin research histones were suspected of being specific gene repressors. There were many reports of tissue specificity of histones and changes in histones during development. Most of the reported changes were subsequently shown to be artifacts, and the relative amounts of the nucleosomal histones have been demonstrated to be constant throughout development and to have no tissue specificity in plants (Spiker and Chalkley, 1971; Spiker and Krishnaswamy, 1973; Burkhanova *et al.*, 1975; Towill and Nooden, 1973; Nicklisch *et al.*, 1976; Gregor *et al.*, 1974; Sugita *et al.*, 1979).

Recently, an increase in the H2 + H3/H4 ratio has been reported upon embryo induction of carrot suspension cells by Fujimura *et al.* (1981). In view of what we now know about the nucleosome structure of chromatin, such gross changes must be viewed with caution until they can be verified and the reasons for the changes explained.

b. Variants of the H1 Histones.   In contrast to the stability of the nucleosomal histones, the H1 histones have long been known to be variable. H1 in both plants and animals consists of several subfractions (variants) that differ in primary sequence. There are differences in H1 subfractions between tissues in an organism, and the subfractions can vary with the developmental stage of the organism (Isenberg, 1979). The most spectacular and well-documented change in H1 histones is the partial replacement of histone H1 with histone H5 (an H1-like histone) in the course of maturation of nucleated avian red blood cells. A corresponding change in the H1 content of slowly dividing tissue is the increase in the histone called $H1^0$ in mammals (Panyim and Chalkley, 1969b). This histone has a high degree of homology with avian H5 (Smith *et al.*, 1980).

Cole and co-workers have been early and consistent contributors to the establishment of primary sequence variants of H1 (Bustin and Cole, 1968; Isenberg, 1979). They have studied the differences in the physical properties of chromatin that are imparted by various H1 subfractions (e.g., Liao and Cole, 1981). Indeed, they have found differences in the behavior of H1 subfractions in their ability to condense chromatin. Because an open chromatin structure is assumed to be a prerequisite for transcription, the H1 subfractions could well play roles in gene expression by controlling higher order structure. Such an idea has existed for several years (e.g., DeLange and Smith, 1971), but as yet no convincing mechanism has appeared.

One of the problems in investigating a role for different H1 variants is that these histones are notoriously susceptible to proteolytic degradation (Isenberg, 1979). Thus, when tissue or developmental stage specificity is claimed for H1 histones, especially on a quantitative basis, extreme care must be taken to preclude artifactual sources for the differences.

Developmentally associated changes have been reported in the H1 histones of plants. Fambrough *et al.* (1968) reported a change in the total H1 quantity and relative quantities of each subfraction during development in peas. Pipkin and Larson (1972) reported the appearance of new H1 subfractions during development of anthers of *Hippeastrum belladonna*. Nicklisch *et al.* (1976) reported changes in H1 subfractions in various regions of corn root tips. Sugita *et al.* (1979) reported a decrease in the amount of H1 (especially H1d) upon germination of wheat embryos. These changes may well be real, but as yet no in-depth study has proven that changes in expression of the genes for histone H1 variants occur during higher plant development.

    c. Variants of the Nucleosomal Histones. Although the existence of variants of H1 histones has been widely recognized for many years, until recently the nucleosomal histones have been considered to be invariant in primary sequence. Variants of the H3 histones in both calf and pea were known (Patthy *et al.*, 1973), but the small differences in structure were usually ignored in the context of histone function. The existence of variations in the nucleosomal histones suddenly became appreciated when the work of Cohen *et al.* (1975) appeared. They demonstrated that variants of the nucleosomal histones existed and that specific variants were synthesized at specific stages of sea urchin development. It was suspected that the nucleosomal histone variants were primary structure variants (not due to posttranslational modifications), and this was later proven (Brandt *et al.*, 1979). Newrock *et al.* (1978a) showed that the appearance of the histone variants correlated with the appearance of their specific messenger RNAs. There is now a large body of evidence for the existence of primary structure variants in animals, fungi, and protozoans (Urban *et al.*, 1979; Ajiro *et al.*, 1981; Strickland *et al.*, 1981; West and Bonner, 1980a; Brandt *et al.*, 1980; Allis *et al.*, 1980).

    Weintraub *et al.* (1978) and Newrock (1978b) have suggested that these histone variants may function in the control of gene expression by being involved in the structuring of chromatin into inactive and active forms. All the sea urchin histone variants are found in nucleosomes (Shaw *et al.*, 1981), and the change in histone variant content of the nucleosomes imparts different physical properties to the nucleosomes (Savic *et al.*, 1981; Cognetti and Shaw, 1981; Simpson, 1981). Among the changes is the susceptibility to DNase I digestion. The rate of digestion of pluteus stage core particles is about half that of particles from blastula. Blankstein and Levy (1976) have demonstrated a dramatic shift in the relative quantities of two H2A variants in Friend leukaemic cells in response to subculturing and acquisition of the capacity of the cells to be induced to differentiate by treatment with dimethyl sulfoxide. Russanova *et al.* (1980) found tissue-specific quantitative differences in the variants of H2A and H3 in rats. Allis *et al.* (1980) have shown that in *Tetrahymena* a specific histone variant is found in the transcriptionally active macronucleus but not in the micronucleus. Halleck and Gurley (1980, 1982) have shown that in cultured deer mouse cells the constitutive heterochromatin is enriched in the more hydrophobic variant of histone H2A.

The observations mentioned in the preceding paragraphs indicate that the variants of the nucleosomal histones may have special physiological functions. Evidence against any special functions for the histone variants also has been presented. Zalenskaya *et al.* (1981) have argued that because genetically inactive sea urchin sperm contain histone variants, the variants cannot be involved in structural distinctions between active and inactive chromatin. Rykowski *et al.* (1981) have shown that yeast can proceed normally through the cell cycle with only one of its two histone H2B variants. However, it should be kept in mind that yeast has a very small genome, which Lohr and Hereford (1979) have suggested is essentially all in an active conformation. Thus, it is still possible that histone variants may play a role in determining an active chromatin structure in eukaryotes with larger genomes.

Higher plants also have variants of the nucleosomal histones. Patthy *et al.* (1973) showed that 60% of pea H3 molecules have alanine at position 96 and 40% have serine at this position. Data of Sommer and Chalkley (1974) indicated that peas have two subfractions of H2A that differ in electrophoretic mobility. Based on amino acid content of peptides from pea H2A and H2B, Hayashi *et al.* (1977) showed that these molecules must have amino acid point substitutions. Rodrigues *et al.* (1979) have presented partial sequences of H2A in wheat. Their findings (discussed with sequences of plant histones in Section II) indicated that wheat H2A does have primary sequence variants.

The major plant histone variants have one feature that is different from the animal histone variants. For the most part, animal histone variants consist of one or more amino acid substitutions (Urban *et al.*, 1979). There are minor animal histone variants that appear to have distinct molecular weights (West and Bonner, 1980a; Bonner *et al.*, 1980; Zweidler, 1978; Newrock *et al.*, 1978a). However, these are low in relative abundance and were usually unnoticed or were considered contaminants. In plants there are major histone variants that differ not only by amino acid point substitutions but also by apparent molecular weight (Spiker, 1982). Wheat H2A has three major variants of about equal abundance. They have different electrophoretic mobilities on SDS–polyacrylamide gels, and they elute differentially in exclusion chromatography. They all have characteristic H2A amino acid compositions, yet their peptide maps are distinct. The differences in their physical properties are not due to postsynthetic modifications and they are not degradation artifacts. Based on mobilities on SDS–polyacrylamide gels, the molecular weights of the three wheat H2A variants are 19,000, 18,000, and 16,600. Wheat has two major H2B variants and four minor H2B variants. All have been characterized by amino acid content and peptide mapping. The apparent molecular weights of the two major H2B variants are 19,000 and 18,100. The minor H2B variants all have apparent molecular weights of about 15,300.

If the histone variants are hypothesized to play special roles in the control of gene expression, it is of interest to know if they are nonallelic. This has been demonstrated by Franklin and Zweidler (1977) for the mammalian histone variants. The plant histone variants that have been studied in the most detail (Spiker, 1982) were obtained from a highly inbred line of wheat. However, because the wheat is hexaploid, it must be established that the variants are nonallelic. The contention that the wheat variants are nonallelic is supported by the observation that the same histone variants are observed in diploid and tetraploid wheats, in diploid rye and barley, and in triticale (Spiker, 1976b). Inbred diploid peas also have more than two variants each of H2A and H2B, thus ruling out the possibility that they are alleles (Spiker, 1982).

If the nucleosomal histone variants have roles in the control of gene expression, one might expect them to be tissue specific and/or developmental stage specific as has been demonstrated in sea urchins (Cohen *et al.*, 1975). In plants the nucleosomal histone variants are not tissue specific and there appears to be no program of histone variant synthesis during early embryogenesis (Spiker and Krishnaswamy, 1973; Quatrano and Spiker, unpublished data). The lack of tissue specificity is easily established by the observation that the same histone variants occur in the same relative concentrations in dry embryos, germinated embryos, roots, coleoptiles, and etiolated and green mature leaves. The establishment of equal rates of synthesis during early embryogenesis is much more difficult in wheat than in sea urchins, for obvious reasons. It is essentially impossible to obtain large quantities of egg and sperm nuclei, effect *in vitro* fertilization, and follow development in wheat. The best that can be done is to study the developing seed at the earliest stage possible, about 3 days after fertilization when the embryo has about 32 cells. Histone synthesis can then be followed by exposing the developing seeds to radioactive amino acids until cell division of the embryo stops at about 16 days after fertilization. Over this time period, labeled amino acids are incorporated into all wheat histone variants and the specific activity remains constant. The experiments are complicated by the fact that because the embryos are not isolated during their development, histone synthesis in the surrounding endosperm tissue cannot be separated from that of the embryo. Some amount of histone synthesis may also be taking place in the surrounding maternal tissues. Thus, interpretation of the data is complicated. However, it is clear that in the combined embryo and endosperm tissues all the histone variants are being synthesized at the earliest measurable time. The specific activity of each variant remains constant throughout embryogenesis and upon germination (further development) of the embryo.

The lack of spatial or temporal specificity of the nucleosomal histone variants in plants does not rule out a role for them in the control of gene expression. As mentioned above, some gross changes in histone variants have been observed in animal systems. These gross changes, however, are not widespread and, in any case, mechanisms of gene control involving histone variants are likely to be much more subtle than the simple presence or absence of a specific histone variant in a specific cell type. Even meaningful quantitative changes might not realistically be expected to occur at the level of the whole cell. Specific location of histone variants within the genome is more likely to be involved.

By what mechanisms might the histone variants be involved in the construction of an active or inactive nucleosome? Newrock *et al.* (1978a) have suggested that differences in phosphorylation may be involved. Some support for this suggestion has come from the work of Halleck and Gurley (1980, 1982), who have shown that in two lines of cultured deer mice cells, the more hydrophobic H2A variant is phosphorylated to a greater extent.

Other mechanisms are possible. For example, the known locations of the amino acid substitutions that make up the histone variants have been catalogued by Isenberg (1979). It is interesting to note that the substitutions come not only in the evolutionarily variable N-terminal portion of H2A and H2B but also in the conserved C-terminal (histone–histone binding) portion. As mentioned in Section I, it has been hypothesized that one attribute of a nucleosome in active chromatin is that it must be able to disassemble transiently to allow read-through of

RNA polymerase and then to reassemble. In other words, an active nucleosome might be expected to be an unstable one. Amino acid substitutions in the evolutionarily conserved, histone–histone binding region may serve to destabilize the nucleosome. Such a destabilization based on an amino acid modification has already been observed (Spiker and Isenberg, 1977). Pea H2B has a single methionine residue in the histone–histone binding portion of the molecule. When that residue is artifactually oxidized to methionine sulfoxide, the formation of H2A–H2B histone complexes is completely precluded. The reaction is reversible. When the residue is reduced back to methionine, full complexing capacity of H2B is restored. Nucleosome destabilization by histone variant content could be investigated most easily by reconstitution experiments with purified components.

As of yet, no clear roles for the N-terminal, evolutionarily variable portions of the H2A and H2B histones have been established. It has been suggested that they may function in the specific binding of HMG proteins (Spiker, 1982). Weintraub and co-workers have established that in order to have an active chromatin structure, HMG proteins must be bound to the nucleosomes (see Section III,C). They have also demonstrated by reconstitution experiments that some factor (as yet unidentified) is necessary to direct HMG binding to the proper nucleosomes. The N-terminal sequences of the histones are candidates for such factors. The problem then would become one of how the proper histone variant becomes associated with the proper DNA sequences during development. No experimental evidence bearing on this point exists as yet.

2. Acetylation

Increased histone acetylation above the basal level has long been circumstantially associated with gene activation (Allfrey et al., 1964; Isenberg, 1979; McGhee and Felsenfeld, 1980; Mathis et al., 1980; Doenecke and Gallwitz, 1982). The discovery by Riggs et al. (1977) that hyperacetylation can be induced in animal cells by butyrate treatment led to intensive activity on the effect of hyperacetylation on gene activation and chromatin structure. Butyrate acts by inhibiting deacetylation (Hagopian et al., 1977; Sealy and Chalkley, 1978a,b; Cousens et al., 1979; Candido et al., 1978). Thus, butyrate treatment has been used to determine if hyperacetylated histones are associated with active chromatin—mainly as defined by sensitivity to DNase I, DNase II, and micrococcal nuclease. The general finding is that hyperacetylated histones are found in association with active chromatin (e.g.,

Sealy and Chalkley, 1978b; Levy *et al.*, 1979b). If these findings are valid, the question of whether hyperacetylation is a cause or an effect of gene activation must be settled. Also to be investigated is the question of developmental regulation of histone acetylation.

Histone hyperacetylation and control of gene activity have not been extensively studied in plant systems. Arfmann and Haase (1981) have presented convincing evidence that butyrate treatment of tobacco cell cultures causes increased acetylation of histone H4. However, no attempts were made to correlate the increased acetylation with gene activity or susceptibility to nucleases. The extent of butyrate-induced modification of histone H4 was less in tobacco cells than is usually recorded for animal cells in culture.

3. Phosphorylation

The postsynthetic modification of histones by phosphorylation has long been recognized as a mechanism whereby the heterogeneity (often called microheterogeneity) of the histones might be increased (DeLange and Smith, 1971). All the histones have been shown to be phosphorylated in animal systems, and increased phosphorylation has been associated with increased activities of cells—either in cell division or in RNA synthesis. In many cases the sites of phosphorylation have been studied in detail (Isenberg, 1979; Langen, 1978; Gurley *et al.*, 1978; Elgin and Weintraub, 1975).

Most of the phosphorylation and dephosphorylation is cell cycle specific and is probably related to mitotic condensation and decondensation of chromatin. H2A phosphorylation, however, appears to be different. Only a small amount is phosphorylated at any one time, and the rate of turnover does not seem to be dependent upon the replicative rate of the cells (Balhorn *et al.*, 1972; Prentice *et al.*, 1982). Some workers have suggested that H2A phosphorylation may have a role in gene control by affecting the level of condensed chromatin in the cell (Prentice *et al.*, 1982; Halleck and Gurley, 1980, 1982).

Little is known of phosphorylation in plant histones. General changes in phosphorylation levels of nuclear proteins have been noted. However, careful studies on the identity and site of phosphorylation have not been carried out. If we are to learn anything about the role of phosphorylation of histones in plant systems, it is imperative that detailed rather than superficial studies by carried out. This is especially true because of the danger that has often been cited (Murray and Key, 1978; DeLange and Smith, 1971) of mistaking phosphorylation of phospholipids, RNA, DNA, nucleotides, or other phosphorylated proteins for phosphorylation of histones. The studies mentioned below must be considered preliminary.

Murray and Key (1978) demonstrated a low incorporation of phosphate into H1 histones of auxin-treated soybeans. Any incorporation into other fractions was not detectable. *In vitro* phosphorylation of exogenous histones with chromatin-bound kinases was studied by Wielgat and Kleczkowski (1981a). They found no difference in the phosphorylation activity of the chromatin from gibberellic acid-treated and control pea embryos. No *in vivo* phosphorylation was studied. Arfmann and Willmitzer (1982) studied endogenous protein kinase activity of nuclei isolated from transformed and nontransformed tobacco cell cultures and from intact tobacco leaves. They found H1 to be the only phosphorylated histone. Transformation did not alter the endogenous kinase activity. Bannerjee *et al.* (1981) claimed decreases in the rate of histone phosphorylation with aging of rice seeds. However, the drop in histone phosphorylation correlated with drops in the ATP pools. Also, no attempt was made to determine which fractions of the histones were phosphorylated or even if histones were actually the recipients of the labeled phosphate. Their histone fraction was simply the 0.4 N H_2SO_4-soluble material from a crude nuclear pellet. Such a preparation certainly contains more than histones.

4. Poly(ADP-ribosyl)ation

Postsynthetic modification of histones by poly(ADP-ribose) was first reported by Nishizuka *et al.* (1968), who presented evidence that the polymer was covalently linked to rat liver histones H1, H2A, H2B, and H3. The discovery of such a linkage led naturally to suggestions that poly(ADP-ribosyl)ation might modify chromatin structure and thus have an effect on transcription or replication (Hayaishi and Ueda, 1977; Mandel *et al.*, 1982; Purnell *et al.*, 1980).

Studies with HeLa cell nuclei have suggested that the enzyme responsible for the synthesis of poly(ADP-ribose)—and probably its covalent linkage to the histones (Ogata *et al.*, 1980a,b)—is tightly chromatin bound and located in internucleosomal regions (Mullins *et al.*, 1977; Giri *et al.*, 1978). Poirier *et al.* (1982) have shown that poly(ADP-ribosyl)ation of polynucleosomes causes relaxation of chromatin structure by altering the interaction of histone H1 with nucleosomes. At 75 mM NaCl, control nucleosomes undergo condensation to form 250-Å fibers. Poly(ADP-ribosyl)ation prevents the formation of higher order chromatin structure and thus may bestow an active structure upon that chromatin.

Although the histones have been the most studied acceptors of poly(ADP-ribose), the HMG proteins also undergo this modification (Levy, 1981a). The trout testis HMG protein, H6, is released from chromatin by DNase I treatment, which rapidly degrades transcriptionally active sequences. Of the H6, 70% is released when only 3–4% of the DNA is degraded. The H6 that is released is highly modified by poly(ADP-ribose); the remaining 30% contains little or no poly(ADP-ribose).

There are many acceptors of poly(ADP-ribose) other than histones and HMG proteins (Mandel et al., 1982). In fact, the enzyme poly(ADP-ribose) synthetase itself is the major acceptor (Ogata et al., 1981). However, there certainly is some evidence that this interesting modification of chromosomal proteins may have a role in control of gene expression (as discussed above) or in DNA replication and repair (Creissen and Shall, 1982).

The existence of poly(ADP-ribose) has also been demonstrated in plants (Payne and Bal, 1976). Several reports have claimed that plant histones can serve as acceptors for the polymer (Willmitzer, 1979; Laroche et al., 1980; Whitby et al., 1979; Sasaki and Sugita, 1982). The general procedure used by all these workers was the same. A nuclear or purified chromatin preparation was incubated with a reaction medium that contained NAD radioactively labeled in the adenosine portion of the molecule. After the reaction was completed, the 20% trichloroacetic acid-insoluble material was collected and fractionated. Authenticity of the formation of poly(ADP-ribose) was assessed by its lability to alkali, neutral hydroxylamine, or snake venom phosphodiesterase. When poly(ADP-ribose) is digested by snake venom phosphodiesterase, a terminal AMP is generated along with the internal phosphoribosyl-AMP molecules (PR-AMP). Average chain lengths can then be determined by dividing the total AMP plus PR-AMP by the AMP. The 0.4 N H_2SO_4-soluble fraction was taken as the histone fraction and further fractionated by exclusion chromatography and/or polyacrylamide gel electrophoresis. Average chain lengths measured were in the same range as reported for animal systems—around 3 ADP units. Comigration of the histone fractions and radioactivity were taken as indications that the histones were the acceptors of the poly(ADP-ribose) chains. Follow-up investigations are needed to prove that the histones are in fact the acceptors.

In animal systems the H1 histones appear to be the major histone acceptors of poly(ADP-ribose) (Poirier and Savard, 1980). The same claim has been made for plant systems. But there are some difficulties in the interpretation of the data. An unanswered question is, does bound poly(ADP-ribose) affect histone electrophoretic mobility? Assumptions have been inconsistent in interpreting data, no determination of chain lengths on individual histone fractions has been made, and no further steps have been taken to prove that the plant histone fractions in question are in fact the receptors of the ADP polymers. For instance, Willmitzer (1979) has investigated poly(ADP-ribosyl)ation in tobacco nuclei isolated from cell culture. Low pH SDS gels were run and the stained profiles compared to autoradiograms to identify the labeled histone fractions. A heavily labeled band comigrated exactly with the stained tobacco H1 band. The interpretation was that H1 was labeled with poly(ADP-ribose) and further (unstated assumption), that the presence of the polymer did not change the electrophoretic mobility of the histone. Another heavily labeled band was interpreted to be poly(ADP-ribose) bound to tobacco H2A and H2B. However, nearly all of this broad band migrated more slowly than did stained tobacco H2A and H2B. Are we now to assume that the poly(ADP-ribose) adds sufficient size to H2A and H2B to lower their electrophoretic mobilities considerably? This contradiction must be resolved—preferably by doing the kind of experiments that would define in detail not only to which molecules the poly(ADP-ribose) is bound but also to what amino acids and by what linkages (Ogata et al., 1980a,b).

If we were to assume that poly(ADP-ribose) does not significantly affect electrophoretic mobility, then the label attributed to H2A and H2B would more likely be due to NHC proteins, such as HMG proteins, which run in the approximate area of the band visualized by autoradiography. As already mentioned, HMG proteins do act as receptors for poly(ADP-ribose) (Levy, 1981a).

Laroche et al. (1980) have made a similar study with rye histones. As in the study of Willmitzer, the work appears to be excellent except for the identification of the recipients of the poly(ADP-ribose). Laroche et al. used acetic acid–urea gels rather than low pH SDS gels. However, their data look very similar to those of Willmitzer, and the same discrepancy in the assignment of radioactivity to H1 and H2A/H2B occurs. They also separated their rye histones on a BioGel column. The profile of elution looks very unusual—especially in the separation and low quantity of histone H4. BioGel separations by other investigators have been different from the ones shown by Laroche et al. and have been consistent with each other (Nadeau et al., 1977; Spiker et al., 1976; Spiker, 1982).

Whitby et al. (1979) investigated the effect of polyamines and magnesium on in vitro poly(ADP-ribosyl)ation of histones in isolated wheat nuclei. Because a stimulation of synthesis had previously been noted in rat liver nuclei (Perella and Lea, 1979), it was of interest to see if it was a general biological mechanism. Whitby *et al.* concluded that polyamines in the presence or absence of magnesium stimulated a two- to fourfold increase in the synthesis of poly(ADP-ribose). The increase was considered to be due to new chain initiation because no change in the average chain length was noted. The distribution of poly(ADP-ribose) between histones and other acceptors did not change upon polyamine treatment. Again, the identification criteria were the same as used by Willmitzer (1979) and Laroche *et al.* (1980). On acetic acid–urea polyacrylamide gels, major peaks of radioactivity comigrated with histone H1 but ran just slower than the H2A and H2B histones.

The striking pattern of the location of incorporated radioactivity, which coincides to the electrophoretic mobility of H1 histones but is just slower than H2A and H2B histones, was continued in the observations of Sasaki and Sugita (1982). These investigators attempted to correlate the poly(ADP-ribose) synthetase activity of nuclei with their capacity for *in vitro* transcription at various times during development of wheat. They reported an inverse correlation. That is, dry mature embryos had the most poly(ADP-ribose)-synthesizing activity and the least ability to support *in vitro* transcription. As germination progressed, after 3 days the poly(ADP-ribose) synthesis activity dropped considerably whereas the transcriptional capacity of the nuclei (measured by acid-precipitable counts only) increased. Sasaki and Sugita also noted that over 40% of the radioactivity of the adenylate moiety bound to chromatin was released upon treatment with DNase I. They suggested that this may mean that the major ADP-ribosylated proteins are located in active chromatin. This aspect of the experiments carried out by Sasaki and Sugita is consistent with current theories concerning the role of ADP ribosylation and gene activity. However, their work is clearly preliminary. The receptors of poly(ADP-ribose) have not been authenticated, and the meaning of increased levels of acid-precipitable counts as an indication of changes in transcriptional activity in chromatin has never been clear (Konkel and Ingram, 1978).

5. Ubiquitinization

Animal histones have been shown to be modified by covalent attachment of a protein called "ubiquitin" (Goldknopf and Busch, 1978). We have no evidence for such a modification of plant histones. However, plants do have the protein, ubiquitin, and it seems reasonable to expect that plant histones may be ubiquitinized.

Investigation into histone modification by ubiquitinization began when a spot on two-dimensional gels of total rat liver chromosomal proteins attracted the attention of Busch and co-workers because its relative quantity decreased drastically during liver regeneration after partial hepatectomy (see Goldknopf and Busch, 1978, for review). After a long series of investigations it became apparent that the protein, called A24 because of its location on two-dimensional gels, was a branched polypeptide. One of the proteins was histone H2A. The other was an unknown protein connected to H2A by an isopeptide linkage at the lysine residue in position 119 of the histone. The sequence of the first 37 residues of that unknown protein were reported by Olson et al. (1976). Hunt and Dayhoff (1977) noticed that the first 37 residues were identical in sequence to a protein being studied by Goldstein's group (Schlesinger et al., 1975). This protein was discovered because it mimicked the action of the hormone thymopoietin. It was found to be widespread in nature, occurring in all eukaryotes (including plants) and also in prokaryotes. Thus it was called "ubiquitin."

Ubiquitinization is the most drastic and specific of the histone modifications in that a peptide of 74 amino acids is attached to the histone at a specific site. Such a drastic modification might be expected to have profound influence on chromatin structure if the molecule were indeed found in the nucleosome. Ubiquitinized H2 does indeed appear to be incorporated into nucleosomes (Goldknopf et al., 1977), and a large body of conflicting evidence has appeared that attributes a role to ubiquitinized H2A in both activating and inhibiting transcription (Kleinschmidt and Martinson, 1981; Levinger and Varshavsky, 1982; Mezquita et al., 1982; Busch and Goldknopf, 1981; Goldknopf and Busch, 1978; Andersen et al., 1981; Seale, 1981; Wu et al., 1981; Matsui et al., 1979).

H2B histones can also be ubiquitinized, although to a lesser extent (West and Bonner, 1980b).

The ubiquitin moiety itself is found bound to nucleosomes and has been termed HMG 20 (Walker et al., 1978). Free ubiquitin has been shown to have a role in ATP-dependent proteolysis (Ciechanover et al., 1981) but we have no idea yet of its role, if any, in chromatin.

D. Nonhistone Chromatin Proteins

1. HMG Proteins

The discovery and basic characterization of HMG proteins in animals and the meager evidence we have for their existence in plants were discussed in Section II,C. Some evidence that they are structural proteins of active chromatin was discussed in Section III,A, viz. the observation that when chromatin is separated into active and inactive fractions the active fraction is found to be enriched in HMG proteins. There is additional and more compelling evidence that HMG proteins are required to form an active chromatin structure. None of this evidence has come from work with plant systems. However, discussion of what we know based on animal systems is considered necessary here, because it may serve to outline important experiments that should be done with plant systems.

 a. *Evidence for a Role of HMG Proteins in Forming an Active Chromatin Structure.*   The property of DNase I of preferentially attacking active chromatin structures has been dealt with above. Also discussed was the observation of Vidali *et al.* (1977), that the HMG proteins are preferentially released from chromatin treated with DNase I. One of the important observations of Weintraub and Groudine (1976) on the DNase I sensitivity of the globin gene was that the determinant of sensitivity was at the level of the individual nucleosome. This allowed them to study a system that was less complicated than those in which an active structure was dependent upon higher order organization of chromatin. Weisbrod and Weintraub (1979) and Weisbrod *et al.* (1980) showed that when nucleosomes of chicken erythrocytes were stripped of their HMG proteins by 0.35 M NaCl the preferential sensitivity of the globin genes to DNase I digestion was lost. Furthermore, selective sensitivity could be restored to the nucleosomes by adding back either the 0.35 M NaCl extract or the purified HMG 14 or 17. Sandeen *et al.* (1980) demonstrated similar phenomena. They showed that two molecules of HMG 14 or 17 can bind tightly but reversibly to core nucleosome particles and that the affinity of the HMG proteins is greater for nucleosomes than for free DNA. Using a single-stranded probe from a cloned globin cDNA sequence, they showed preferential binding of HMG 14 or 17 to the nucleosomes that contained the globin genes. Specific binding of HMG proteins was also shown by Mardian *et al.* (1980). Thus, two things were obvious: (1) HMG proteins were necessary to confer the structure of active genes recognized by DNase I, and (2) because reconstitution of the structure recognized by DNase I could be performed, active nucleosomes must have at least one other attribute—something to allow the HMG proteins to bind to the proper nucleosomes. As of yet that additional attribute has not been discovered. Utilizing the preferential binding of HMG 14 and 17 to active nucleosomes, Weisbrod and Weintraub (1981) used an HMG-affinity column to isolate and study active nucleosomes. These nucleosomes did have some special features but none that would seem to explain clearly why they were active or why they were preferentially susceptible to digestion by DNase I. Weisbrod (1982b) found that the active nucleosomes (1) were undermethylated (see Section III,D), (2) were enriched in topoisomerase (but not as much as one molecule per nucleosome), (3) were somewhat enriched in acetylated histones, and (4) showed an altered conformation in that the sulfhydryl-containing H3 molecules were more easily dimerized and thus probably altered in their spatial relationship.

Other lines of evidence consistent with a role of HMG proteins in active chromatin have been pursued in animal systems. For example, by the use of fluorescent antibodies, proteins similar to HMG proteins have been shown to be specifically associated with transcriptionally active puffs of polytene chromosomes (Mayfield et al., 1978). In most cases the HMG proteins have not demonstrated a marked tissue specificity. Indeed, tissue specificity of the HMG proteins is not required for their action because in reconstitution studies Weisbrod et al. (1981) have shown that HMG proteins from brain or erythrocytes are equally capable of restoring a DNase I sensitivity to the erythrocyte nucleosomes containing the globin genes. Nevertheless, evidence has been presented for tissue-specific quantitative changes in HMG proteins (Gordon et al., 1980; Teng and Teng, 1981; Seyedin et al., 1981).

Even though we know there are two high-affinity binding sites for HMG proteins on nucleosomes, we do not know exactly where or how the binding occurs. Binding of HMG proteins to specific H1 subfractions has been demonstrated (Smerdon and Isenberg, 1976), but the significance of such specificity has been questioned because the binding appears to be primarily electrostatic (Cary et al., 1979). HMG proteins have been cross-linked to H1 histones (Ring and Cole, 1979), but other studies have indicated that H1 is replaced by HMG proteins (Goodwin et al., 1979). Specific binding of HMG proteins to the N-terminal regions of the nucleosomal histones has not been demonstrated but remains an intriguing possibility.

b. *Modification of the HMG Proteins.* The HMG proteins of animals undergo posttranslational modifications. They have been shown to be acetylated (Sterner et al., 1978), ADP-ribosylated (Caplan et al., 1978; Reeves et al., 1981; Levy, 1981a), phosphorylated (Saffer and Glazer, 1980, 1982; Bhorjee, 1981; Levy, 1981b), methylated (Boffa et al., 1979), and glycosylated (Reeves et al., 1981). Do these modifications serve physiological functions? As of yet we do not know. However, it is clear that modifications could bestow considerable heterogeneity upon the HMG proteins and that heterogeneity may be important in the recognition of nucleosome binding sites.

Bhorjee (1981) has demonstrated cell cycle-specific phosphorylation of HMG proteins 14 and 17 in HeLa cell cultures and has hypothesized a role in maintaining transcriptionally active chromatin structure. Levy (1981b) has shown that butyrate treatment of HeLa cells causes hyperphosphorylation of HMG 14 and 17. Because these proteins are released when the active genes are digested by DNase I, she suggested that the hyperphosphorylation of the HMG proteins may be either the cause or the consequence of the altered pattern of gene expression caused by butyrate treatment. Saffer and Glazer (1982) showed that phosphorylated HMG 14 and 17 were preferentially released from a variety of cells in culture by micrococcal nuclease. Brief digestion with DNase I, on the other hand, did not show a marked release of the phosphorylated proteins. Levy (1981a) has also shown that in trout testis chromatin poly(ADP-ribosyl)ated HMG proteins are preferentially located in DNase I-sensitive chromatin.

Postsynthetic modifications of plant HMG proteins have not been investigated.

2. Other Nonhistone Chromatin Proteins

There is an enormous literature concerning changes in NHC proteins that accompany changes in gene activity (for review, see Johns, 1982; Igo-Kemenes et al., 1982; Kleinsmith, 1978; Stein et al., 1974). Many of these investigations have involved plant systems and have been most recently reviewed by Thompson and Murray (1981). Thus, only work that has appeared since that review will be discussed here.

Hirasawa and Matsumoto (1982) and Hirasawa et al. (1978a,b, 1979a,b, 1981) studied the chromatin-bound enzyme nucleoside triphosphate dephosphatase during development in peas. The activity was found only during the early stages of germination. When chromatin was separated into putative active and inactive fractions by the method of Gottesfeld et al. (1974), most of the activity was found in the "active" fraction. Whether the method of Gottesfeld et al. was successful with the plant chromatin was not determined. What role nucleoside triphosphate dephosphatase might have in gene regulation is unknown. A speculation was made that it may act by local control of the availability of nucleoside triphosphates for transcription.

Schäfer and Kahl (1982) have continued a long-term investigation into the factors involved in activation of chromatin as a response to wounding. As in many systems, NHC proteins seem to be involved in the general activation of chromatin. Specific proteins or mechanisms of action are as yet unknown, but they have presented evidence that increased chromatin protein phosphorylation accompanies gene activation.

Some changes in phosphorylation patterns of chromosomal proteins of peas as a result of gibberellic acid treatment have also been suspected of being involved in changes in gene expression (Wielgat and Kleczkowski, 1981a). The investigators (Wielgat and Kleczkowski, 1981b) showed that NHC proteins from gibberellic acid-treated maize and pea seedlings stimulate the rate of transcription in *in vitro* systems. The results of such experiments are hard to interpret, however, because the RNA products were not characterized. Nothing is known about an effect on synthesis versus an effect of degradation. Such transcription may poorly reflect *in vivo* transcription (Konkel and Ingram, 1978), and increased rates of transcription could simply result from increased nicking of the DNA template.

Sugita *et al.* (1979) have described a series of changes that occur in NHC proteins of wheat upon germination. Some of these proteins were found to be ADP-ribosylated in an *in vitro* system (Sasaki and Sugita, 1982). An attempt was made to determine if these proteins have a role in the control of gene expression by including them in a reconstituted *in vitro* transcription system (Yoshida *et al.*, 1979). Changes in radioactivity incorporated into macromolecules were indeed found upon addition of the different fractions. The significance of these observations is at present unknown. More detailed studies would be required to determine if these proteins really do have a role in the control of gene expression *in vivo*.

E. DNA MODIFICATION

The methylated base 5-methyl cytosine (m^5C) occurs in the DNA of eukaryotic organisms, and support has been increasing in recent years for a role of this modified base in the regulation of gene expression (for reviews see Felsenfeld and McGhee, 1982; Razin and Riggs, 1980; Drahovsky and Boehm, 1980). About 90% of the methylated cytosines in vertebrates occur in the dinucleotide sequence CpG (Razin and Riggs, 1980). Patterns of methylation exist, and they are tissue specific (Waalwijk and Flavel, 1978; Sano and Sager, 1982). Furthermore, a correlation exists between gene activity and undermethylation. McGhee and Ginder (1979) showed that some methylation sites of the chicken β-globin genes are undermethylated in erythrocytes as compared to oviduct tissue. Mandel and Chambon (1979) found a similar correlation between the activity of the chicken ovalbumin gene and undermethylation. Kuo *et al.* (1979) have shown that active chromatin regions, as defined by increased susceptibility to DNase I, are also undermethylated. Weintraub *et al.* (1981) have studied the α-globin gene cluster in chickens and have shown a strict correlation between the DNase I-sensitive domain, HMG 14 and 17 content, transcription (as measured by *in vitro* runoff transcription by endogenous RNA polymerase II), and undermethylation. Naveh-Many and Cedar (1981) have presented evidence that undermethylation is a general phenomenon in all active genes by similar methods.

The idea that DNA methylation is involved in determining the site of an active chromatin structure is an attractive one because the pattern can be maintained through DNA replication by preferential methylation of a hemimethylated site (Gruenbaum *et al.*, 1982). Furthermore, the methylation is enzymically controlled and tissue-specific patterns can be formed. There is ample evidence that methylation can alter protein–DNA interaction (Razin and Riggs, 1980), but a specific alteration leading to an active chromatin structure or an indicator of active structure, e.g., HMG binding, has not been proven.

If methylation of DNA is involved in the control of gene expression, the mechanism cannot be direct and simple because expressed genes also contain methylated sequences (Sheffery *et al.*, 1982). Therefore, it is likely that a few specific sites are critical and that their effect is somehow spread through the area of the transcriptionally active chromatin—perhaps by a mechanism such as nucleosome phasing (Kornberg, 1981; Zachau and Igo-Kemenes, 1981; Weintraub, 1980).

It is probably too early to predict an involvement of Z-form DNA in changing chromatin structure. However, it is interesting to note that antibodies to Z-form DNA bind to interband regions of *Drosophila* polytene chromosomes (Nordheim *et al.*, 1981). This may indicate that Z-form DNA can exist *in vivo*. Behe and Felsenfeld (1981) have shown that when the synthetic polymer poly(GC) is methylated, the transition from B-form to Z-form DNA will occur at much lower (closer to physiological) ionic strengths than it will with the nonmethylated polymer. Does methylation favor formation of Z-form DNA *in vivo*? Could such transitions be involved in altering chromatin structure and regulating gene expression?

Higher percentages of methylated DNA have been associated with satellite and highly repetitive DNA in many vertebrate systems (Razin and Riggs, 1980). Plants in general have been noted for having high contents of repetitive DNA (Thompson and Murray, 1981) and methylated DNA (Gruenbaum *et al.*, 1981). The idea that methylation of DNA is a mechanism involved in preventing transcription of much of the repetitive DNA is consistent with these observations. Specific support for such an idea was presented by Deumling (1981), who showed that in a GC-rich satellite DNA of the monocotyledonous blue-bell (*Scilla siberica*) about 25% of the total bases were m^{5}C. The m^{5}C/C ratio was 1.5. The satellite was located in the heterochromatic regions of the chromosome by *in situ* hybridization.

In plants, m^{5}C is not restricted to the dinucleotide sequence CG. Gruenbaum *et al.* (1981) have found that a large portion of the methy-lated cytosine occurs in the trinucleotide sequence C-X-G. These inves-tigators suggested that the sequence is essential because it can provide a symmetrical arrangement of cytosines necessary to ensure inheri-tance of methylation at these sites.

As of yet there is no evidence that would correlate activation of a specific plant gene with undermethylation. There is a report by Van-yushin *et al.* (1981) that auxin sharply inhibits the replicative meth-ylation of DNA in wheat seedlings. These workers speculated that this may be important for cell differentiation in plants. Durante *et al.* (1982) have studied methylation in tobacco pith tissue that had been explanted and allowed to form callus. They showed that a satellite fraction of DNA was methylated in two waves, 24 and 72 hours after explantation. A paper by Wagner and Capesius (1981) on the deter-mination of 5-methylcytosine from plant DNA by high performance liquid chromatography should be of value to investigators interested in the biological role of DNA methylation in plants.

IV. Prospects for Chromatin Structure and Gene Regulation Studies in Higher Plants

It is obvious that we have far less information about the role of chromatin structure in gene regulation in plant systems than in ani-mal systems. What we do know about plants indicates that some of the same general mechanisms operate in both kinds of organisms, but there may be significant differences. Pursuing clues that may uncover any differences in how plants and animals regulate gene expression through chromatin structure should have a high priority. At the same time we need solid experimental evidence to determine whether mechanisms that operate in animal systems also operate in plant systems. For example, a general DNase I sensitivity of transcribed DNA has been demonstrated in a plant system (Spiker *et al.*, 1983) but, as of yet, we have no information on the sensitivity of a specific regulated gene. The existence of DNase I-hypersensitive sites has not been demonstrated in any plant system. We have some indications, but as yet no solid evidence, that the schemes for separating chromatin into active and inactive fractions based on susceptibility to micrococcal nuclease are applicable to plant systems.

We know nothing about changes in DNA methylation patterns in regulated plant genes. Our information on postsynthetic modification of plant histones is sketchy. Are hyperacetylated histones associated with active genes in plants? Are nucleosomal histones phosphorylated?

The fact that our knowledge of amino acid sequences is so limited has prevented many important physical studies with plant histones. What is the significance of molecular weight histone variants in plants? Hentschel and Birnstiel (1981) have pointed out that the histone genes of plants "have not been investigated and thus they represent a clear gap in our knowledge of histone gene structures."

We have some information concerning HMG proteins in plants but physical studies are just beginning. No sequences or even partial sequences have been published. Animal HMG proteins bind to animal nucleosomes at specific sites (Sandeen *et al.*, 1980; Mardian *et al.*, 1980). Will plant HMG proteins also have specific binding sites on plant nucleosomes? Might these binding sites be evolutionarily conserved so that plant HMG proteins would bind specifically to animal nucleosomes and vice versa? It will be of great interest to know if plant HMG-affinity columns will be as useful in isolating nucleosomes of active genes as they have been in animal systems (Weisbrod, 1982b).

It is clear that we have much to learn about the role of chromatin structure in plant gene regulation. Some of the unanswered questions mentioned in this article may provide starting points for future investigations.

ACKNOWLEDGMENTS

Research from this laboratory has been supported by Grant PCM81-05135 from the United States National Science Foundation. I thank Janet Barbour for secretarial work, Stuart Weisbrod for suggestions on the manuscript, and Roland Greimers for providing electron micrographs.

REFERENCES

Ajiro, K., Borun, T. W., and Solter, D. (1981). Quantitative changes in the expression of histone H1 and H2B subtypes and their relationship to the differentiation of mouse embryonal carcinoma cells. *Dev. Biol.* **86**, 206–211.

Alfageme, C. R., Rudkin, G. T., and Cohen, L. H. (1976). Locations of chromosomal proteins in polytene chromosomes. *Proc. Natl. Acad. Sci. U.S.A.* **73**, 2038–2042.

Allfrey, V. G., Faulkner, R., and Mirsky, A. E. (1964). Acetylation and methylation of histones and their possible role in the regulation of RNA synthesis. *Proc. Natl. Acad. Sci. U.S.A.* **51**, 786–794.

Allis, C. D., Glover, C. V. C., Bowen, J. K., and Gorovsky, M. A. (1980). Histone variants specific to the transcriptionally active, amitotically dividing macronucleus of the unicellular eukaryote, *Tetrahymena thermophila. Cell* **20**, 609–617.

Andersen, M. W., Ballal, N. R., Goldknopf, I. L., and Busch, H. (1981). Protein A24 lyase activity in nucleoli of thioacetamide-treated rat liver releases histone 2A and ubiquitin from conjugated protein A24. *Biochemistry* **20**, 1100–1104.

Arfmann, H.-A., and Haase, E. (1981). Effect of sodium butyrate on the modification of histone in cell cultures of *Nicotiana tabacum. Plant Sci. Lett.* **21**, 317–324.

Arfmann, H.-A., and Willmitzer, L. (1982). Endogenous protein kinase activity of tobacco nuclei. Comparison of transformed, non-transformed cell cultures and the intact plant of *Nicotiana tabacum. Plant Sci. Lett.* **26**, 31–38.

Balhorn, R., Oliver, D., Hohmann, R., Chalkley, R., and Granner, D. (1972). Turnover of deoxyribonucleic acid, histones, and lysine-rich histone phosphate in hepatoma tissue culture cells. *Biochemistry* **11**, 3915–3921.

Banerjee, A., Choudhuri, M. M., and Ghosh, B. (1981). Changes in nucleotide content and histone phosphorylation of aging rice seeds. *Z. Pflanzenphysiol.* **102**, 33–36.

Bassuk, J. A., and Mayfield, J. E. (1982). Major high mobility group like proteins of *Drosophila melanogaster* embryonic nuclei. *Biochemistry* **21**, 1024–1027.

Baulcombe, D. C., and Key, J. L. (1980). Polyadenylated RNA sequences which are reduced in concentration following auxin treatment of soybean hypocotyls. *J. Biol. Chem.* **255**, 8907–8913.

Bedbrook, J. R., Smith, S. M., and Ellis, R. J. (1980). Molecular cloning and sequencing of cDNA encoding the precursor to the small subunit of chloroplast ribulose-1,5-bisphosphate carboxylase. *Nature (London)* **287**, 692–697.

Behe, M., and Felsenfeld, G. (1981). Effect of methylation on a synthetic polynucleotide: The B-Z transition in poly (dG-m⁵dC)·poly(dG-m⁵dC). *Proc. Natl. Acad. Sci. U.S.A.* **78**, 1619–1623.

Bellard, M., Gannon, F., and Chambon, P. (1978). Nucleosome structure III: The structure and transcriptional activity of the chromatin containing the ovalbumin and globin genes in chick oviduct nuclei. *Cold Spring Harbor Symp. Quant. Biol.* **42**, 779–791.

Bhorjee, J. S. (1981). Differential phosphorylation of nuclear nonhistone high mobility group proteins HMG 14 and HMG 17 during the cell cycle. *Proc. Natl. Acad. Sci. U.S.A.* **78**, 6944–6948.

Blankstein, L. A., and Levy, S. B. (1976). Changes in histone f2a2 associated with proliferation of Friend leukaemic cells. *Nature (London)* **260**, 638–640.

Bloom, K. S., and Anderson, J. N. (1978). Fractionation of hen oviduct chromatin into transcriptionally active and inactive regions after selective micrococcal nuclease digestion. *Cell* **15**, 141–150.

Boffa, L. C., Sterner, R., Vidali, G., and Allfrey, V. G. (1979). Postsynthetic modifications of nuclear proteins: High mobility group proteins are methylated. *Biochem. Biophys. Res. Commun.* **89**, 1322–1327.

Bonner, J., Dahmus, M. E., Fambrough, D., Huang, R. C., Marushige, K., and Tuan, D. Y. (1968). The biology of isolated chromatin. *Science* **159**, 47–56.

Bonner, W. M., West, M. H. P., and Stedman, J. D. (1980). Two-dimensional gel analysis of histones in acid extracts of nuclei, cells and tissues. *Eur. J. Biochem.* **109**, 17–23.

Boulikas, T., Wiseman, J. M., and Garrard, W. T. (1980). Points of contact between histone H1 and the histone octamer. *Proc. Natl. Acad. Sci. U.S.A.* **77**, 127–131.

Brandt, W. F., and Von Holt, C. (1975). Isolation and characterization of the histones from cycad pollen. *FEBS Lett.* **51**, 84–87.

Brandt, W. F., Strickland, W. N., Strickland, M., Carlisle, L., Woods, D., and Von Holt, C. (1979). A histone programme during the life cycle of the sea urchin. *Eur. J. Biochem.* **94**, 1–10.

Brandt, W., Patterson, K., and Von Holt, C. (1980). The histones of yeast. *Eur. J. Biochem.* **110**, 67–76.

Brown, S. W. (1966). Heterochromatin: Heterochromatin provides a visible guide to suppression of gene action during development and evolution. *Science* **151**, 417–425.

Burkhanova, E., Staynov, D. Z., Koleva, S., and Tsanev, R. (1975). Some characteristics of chromatin from meristematic and differentiated cells of maize root. *Cell Differ.* **4**, 201–207.

Busch, H., and Goldknopf, I. L. (1981). Ubiquitin-protein conjugates. *Mol. Cell. Biochem.* **40**, 173–187.

Bustin, M., and Cole, R. D. (1968). Species and organ specificity in very lysine-rich histones. *J. Biol. Chem.* **243**, 4500–4505.

Byvoet, P., Baxter, S. C., and Sayre, D. F. (1978). Displacement and aberrant methylation in vitro of H1 histone in rat liver nuclei after half-saturation of chromatin with polycations. *Proc. Natl. Acad. Sci. U.S.A.* **75**, 5773–5777.

Candido, E. P. M., Reeves, R., and Davie, J. R. (1978). Sodium butyrate inhibits histone deacetylation in cultured cells. *Cell* **14**, 105–113.

Caplan, A., Ord, M. G., and Stocken, L. A. (1978). Chromatin structure through the cell cycle. *Biochem. J.* **174**, 475–483.

Cary, P. D., Shooter, K. V., Goodwin, G. H., Johns, E. W., Olayemi, J. Y., Hartman, P. G., and Bradbury, E. M. (1979). Does high-mobility group nonhistone protein HMG1 interact specifically with histone H1 subfractions? *Biochem. J.* **183**, 657–662.

Chambon, P. (1978). Summary: The molecular biology of the eukaryotic genome is coming of age. *Cold Spring Harbor Symp. Quant. Biol.* **42**, 1209–1234.

Cheah, K. S. E., and Osborne, D. J. (1977). Analysis of nucleosomal deoxyribonucleic acid in a higher plant. *Biochem. J.* **163**, 141–144.

Ciechanover, A., Heller, H., Katz-Etzion, R., and Hershko, A. (1981). Activation of the heat-stable polypeptide of the ATP-dependent proteolytic system. *Proc. Natl. Acad. Sci. U.S.A.* **78**, 761–765.

Cleveland, D. W., Fischer, S. G., Kirschner, M. W., and Laemmli, U. K. (1977). Peptide mapping by limited proteolysis in sodium dodecyl sulfate and analysis by gel electrophoresis. *J. Biol. Chem.* **252**, 1102–1106.

Cognetti, G., and Shaw, B. R. (1981). Structural differences in the chromatin from compartmentalized cells of the sea urchin embryo: Differential nuclease accessibility of micromere chromatin. *Nucleic Acids Res.* **9**, 5609–5621.

Cohen, L. H., Newrock, K. M., and Zweidler, A. (1975). Stage-specific switches in histone synthesis during embryogenesis of the sea urchin. *Science* **190**, 994–997.

Cousens, L. S., Gallwitz, D., and Alberts, B. M. (1979). Different accessibilities in chromatin to histone acetylase. *J. Biol. Chem.* **254**, 1716–1723.

Creissen, D., and Shall, S. (1982). Regulation of DNA ligase activity by poly(ADP-ribose). *Nature (London)* **296**, 271–272.

D'Anna, J. A., Jr., and Isenberg, I. (1974). A histone cross-complexing pattern. *Biochemistry* **13**, 4992–4997.

Darnell, J. E. (1982). Variety in the level of gene control in eukaryotic cells. *Nature (London)* **297**, 365–371.

DeLange, R. J., and Smith, E. L. (1971). Histones: Structure and function. *Annu. Rev. Biochem.* **40**, 279–314.

Deumling, B. (1981). Sequence arrangement of a highly methylated satellite DNA of a plant, *Scilla*: A tandemly repeated inverted repeat. *Proc. Natl. Acad. Sci. U.S.A.* **78**, 338–342.

Doenecke, D., and Gallwitz, D. (1982). Acetylation of histones in nucleosomes. *Mol. Cell. Biochem.* **44**, 113–128.

Drahovsky, D., and Boehm, T. L. (1980). Enzymatic DNA methylation in higher eukaryotes. *Int. J. Biochem.* **12**, 523–528.

Durante, M., Geri, C., and Ciomei, M. (1982). DNA methylation in dedifferentiating plant pith tissue. *Experientia* **38**, 451–452.

Dure, L., III, Greenway, S. C., and Galau, G. A. (1981). Developmental biochemistry of cottonseed embryogenesis and germination: Changing messenger ribonucleic acid populations as shown by in vitro and in vivo protein synthesis. *Biochemistry* **20**, 4162–4168.

Elgin, S. C. R. (1981). DNAase I-hypersensitive sites in chromatin. *Cell* **27**, 413–415.

Elgin, S. C. R., and Weintraub, H. (1975). Chromosomal proteins. *Annu. Rev. Biochem.* **44**, 725–774.

Elgin, S. C. R., Schilling, J., and Hood, L. E. (1979). Sequence of histone 2B in *Drosophila melanogaster*. *Biochemistry* **18**, 5679–5685.

Fambrough, D. M., and Bonner, J. (1966). On the similarity of plant and animal histones. *Biochemistry* **5**, 2563–2570.

Fambrough, D. M., and Bonner, J. (1968). Sequence homology and role of cysteine in plant and animal arginine-rich histones. *J. Biol. Chem.* **243**, 4434–4439.

Fambrough, D. M., and Bonner, J. (1969). Limited molecular heterogeneity of plant histones. *Biochim. Biophys. Acta* **175**, 113–122.

Fambrough, D. M., Fujimura, F., and Bonner, J. (1968). Quantitative distribution of histone components in the pea plant. *Biochemistry* **7**, 575–585.

Fazal, M., and Cole, R. D. (1977). Anomalies encountered in the classification of histones. An example using wheat germ. *J. Biol. Chem.* **252**, 4068–4072.

Felsenfeld, G. (1978). Chromatin. *Nature (London)* **271**, 115–122.

Felsenfeld, G., and McGhee, J. (1982). Methylation and gene control. *Nature (London)* **276**, 602–603.

Frado, L.-L. Y., Annunziato, A. T., and Woodcock, C. L. F. (1977). Structural repeating units in chromatin III. A comparison of chromatin subunits from vertebrate, ciliate and angiosperm species. *Biochim. Biophys. Acta* **475**, 514–520.

Franco, L., Montero, F., and Rodriguez-Molina, J. J. (1977). Purification of the histone H1 from the fruit fly *Ceratitis capitata*. Isolation of a high mobility group (HMG) non-histone protein and aggregation of H1 through a disulfide bridge. *FEBS Lett.* **78**, 317–320.

Franklin, S. G., and Zweidler, A. (1977). Non-allelic variants of histones 2a, 2b and 3 in mammals. *Nature (London)* **266**, 273–275.

Fujimura, T., Komamine, A., and Matsumoto, H. (1981). Changes in chromosomal proteins during early stages of synchronized embryogenesis in a carrot cell suspension culture. *Z. Pflanzenphysiol.* **102**, 293–298.

Galau, G. A., and Dure, L., III. (1981). Developmental biochemistry of cottonseed embryogenesis and germination: Changing messenger ribonucleic acid populations as shown by reciprocal heterologous complementary deoxyribonucleic acid-messenger ribonucleic acid hybridization. *Biochemistry* **20**, 4169–4178.

Garel, A., and Axel, R. (1976). Selective digestion of transcriptionally active ovalbumin genes from oviduct nuclei. *Proc. Natl. Acad. Sci. U.S.A.* **73**, 3966–3970.

Garel, A., Zolan, M., and Axel, R. (1977). Genes transcribed at diverse rates have a similar conformation in chromatin. *Proc. Natl. Acad. Sci. U.S.A.* **74**, 4867–4871.

Georgieva, E. I., Pashev, I. G., and Tsanev, R. G. (1981). Distribution of high mobility group and other acid-soluble proteins in fractionated chromatin. *Biochim. Biophys. Acta* **652**, 240–244.

Gigot, C., Philipps, G., Nicolaieff, A., and Hirth, L. (1976). Some properties of tobacco protoplast chromatin. *Nucleic Acids Res.* **3**, 2315–2329.

Giri, C. P., West, M. H. P., and Smulson, M. E. (1978). Nuclear protein modification and chromatin substructure 2. Internucleosomal localization of poly(adenosine diphosphate-ribose) polymerase. *Biochemistry* **17**, 3501–3504.

Goldberg, R. B., Hoschek, G., Tam, S. H., Ditta, G. S., and Breidenbach, R. W. (1981a). Abundance, diversity and regulation of mRNA sequence sets in soybean embryogenesis. *Dev. Biol.* **83**, 201–217.

Goldberg, R. B., Hoschek, G., Ditta, G. S., and Breidenbach, R. W. (1981b). Developmental regulation of cloned superabundant embryo mRNAs in soybean. *Dev. Biol.* **83**, 218–231.

Goldknopf, I. L., and Busch, H. (1978). Modification of nuclear proteins: The ubiquitin-histone 2A conjugate. *In* "The Cell Nucleus" (H. Busch, ed.), Vol. 6, Part C, pp. 149–180. Academic Press, New York.

Goldknopf, I. L., French, M. F., Musso, R., and Busch, H. (1977). Presence of protein A24 in rat liver nucleosomes. *Proc. Natl. Acad. Sci. U.S.A.* **74**, 5492–5495.

Goodwin, G. H., Sanders, C., and Johns, E. W. (1973). A new group of chromatin associated proteins with a high content of acidic and basic amino acids. *Eur. J. Biochem.* **38**, 14–19.

Goodwin, G. H., Nicolas, R. H., and Johns, E. W. (1975). An improved large scale fractionation of high mobility group non-histone chromatin proteins. *Biochim. Biophys. Acta* **405**, 280–291.

Goodwin, G. H., Woodhead, L., and Johns, E. W. (1977). The presence of high mobility group non-histone chromatin proteins in isolated nucleosomes. *FEBS Lett.* **73**, 85–88.

Goodwin, G. H., Walker, J. M., and Johns, E. W. (1978). The high mobility group (HMG) nonhistone chromosomal proteins. *In* "The Cell Nucleus" (H. Busch, ed.), Vol. 6, Part C, pp. 181–219. Academic Press, New York.

Goodwin, G. H., Mathew, C. G. P., Wright, C. A., Venkov, C. D., and Johns, E. W. (1979). Analysis of the high mobility group proteins associated with salt-soluble nucleosomes. *Nucleic Acids Res.* **7**, 1815–1835.

Gordon, J. S., Rosenfeld, B. I., Kaufman, R., and Williams, D. L. (1980). Evidence for a quantitative tissue-specific distribution of the high mobility group chromosomal proteins. *Biochemistry* **19**, 4395–4402.

Gottesfeld, J. M. (1978). Methods for fractionation of chromatin into transcriptionally active and inactive segments. *Methods Cell Biol.* **17,** 421–436.

Gottesfeld, J. M., and Butler, P. J. G. (1977). Structure of transcriptionally-active chromatin subunits. *Nucleic Acids Res.* **4,** 3155–3173.

Gottesfeld, J. M., and Partington, G. A. (1977). Distribution of messenger RNA-coding sequences in fractionated chromatin. *Cell* **12,** 953–962.

Gottesfeld, J. M., Garrard, W. T., Bagi, G., Wilson, R. F., and Bonner, J. (1974). Partial purification of the template-active fraction of chromatin: A preliminary report. *Proc. Natl. Acad. Sci. U.S.A.* **71,** 2193–2197.

Gottesfeld, J. M., Murphy, R. F., and Bonner, J. (1975). Structure of transcriptionally active chromatin. *Proc. Natl. Acad. Sci. U.S.A.* **72,** 4404–4408.

Gould, H. J., Cowling, G. J., Harborne, N. R., and Allan, J. (1980). An examination of models for chromatin transcription. *Nucleic Acids Res.* **8,** 5255–5266.

Gregor, D., Reinert, J., and Matsumoto, H. (1974). Changes in chromosomal proteins from embryo induced carrot cells. *Plant Cell Physiol.* **15,** 875–881.

Greimers, R., and Deltour, R. (1981). Organization of transcribed and nontranscribed chromatin in isolated nuclei of *Zea mays* root cells. *Eur. J. Cell Biol.* **23,** 303–311.

Grellet, F., Penon, P., and Cooke, R. (1980). Analysis of DNA associated with nucleosomes in pea chromatin. *Planta* **148,** 346–353.

Grosschedl, R., and Birnstiel, M. L. (1982). Delimitation of far upstream sequences required for maximal *in vitro* transcription of an H2A histone gene. *Proc. Natl. Acad. Sci. U.S.A.* **79,** 297–301.

Grosveld, G. C., de Boer, E., Shewmaker, C. K., and Flavell, R. A. (1982). DNA sequences necessary for transcription of the rabbit β-globin gene *in vivo. Nature (London)* **295,** 120–126.

Gruenbaum, Y., Naveh-Many, T., Cedar, H., and Razin, A. (1981). Sequence specificity of methylation in higher plant DNA. *Nature (London)* **292,** 860–862.

Gruenbaum, Y., Cedar, H., and Razin, A. (1982). Substrate and sequence specificity of a eukaryotic DNA methylase. *Nature (London)* **295,** 620–622.

Guilfoyle, T. J. (1981). DNA and RNA polymerases. *In* "The Biochemistry of Plants" (A. Marcus, ed.), Vol. 6, pp. 207–247. Academic Press, New York.

Guilfoyle, T. J., and Jendrisak, J. J. (1978). Plant DNA-dependent RNA polymerases: Subunit structure and enzymatic properties of the Class II enzyme from quiescent and proliferating tissues. *Biochemistry* **17,** 1860–1866.

Gurley, L. R., D'Anna, J. A., Barham, S. S., Deaven, L. L., and Tobey, R. A. (1978). Histone phosphorylation and chromatin structure during mitosis in Chinese hamster cells. *Eur. J. Biochem.* **84,** 1–15.

Hagopian, H. K., Riggs, M. G., Swartz, L. A., and Ingram, V. M. (1977). Effect of n-butyrate on DNA synthesis in chick fibroblasts and HeLa cells. *Cell* **12,** 855–860.

Halleck, M. S., and Gurley, L. R. (1980). Histone H2A subfractions and their phosphorylation in cultured *Peromyscus* cells. *Exp. Cell Res.* **125,** 377–388.

Halleck, M. S., and Gurley, L. R. (1982). Histone variants and histone modifications in chromatin fractions from heterochromatin rich *Peromyscus* cells. *Exp. Cell Res.* **138,** 271–285.

Hamana, K., and Iwai, K. (1979). High mobility group nonhistone chromosomal proteins also exist in *Tetrahymena. J. Biochem. (Tokyo)* **86,** 789–794.

Hayashi, H., Iwai, K., Johnson, J., and Bonner, J. (1977). Pea histones H2A and H2B. Variable and conserved regions in the sequences. *J. Biochem. (Tokyo)* **82,** 503–510.

Hayaishi, O., and Ueda, K. (1977). Poly(ADP-ribose) and ADP-ribosylation of proteins. *Annu. Rev. Biochem.* **46,** 95–116.

Hentschel, C. C., and Birnstiel, M. L. (1981). The organization and expression of histone gene families. *Cell* **25**, 301–313.

Hirasawa, E., and Matsumoto, H. (1982). Distribution and properties of chromatin-associated nucleoside triphosphate diphosphatase purified by a new method. *Plant Cell Physiol.* **23**, 417–425.

Hirasawa, E., Takahashi, E., and Matsumoto, H. (1978a). A transcription inhibitor in non-histone proteins of germinated pea cotyledon. *Plant Cell Physiol.* **19**, 599–608.

Hirasawa, E., Takahashi, E., and Matsumoto, H. (1978b). A nucleotide phosphohydrolyzing activity in non-histone proteins of pea cotyledon inhibits *in vitro* RNA synthesis. *Plant Cell Physiol.* **19**, 1095–1098.

Hirasawa, E., Matsumoto, H., and Takahashi, E. (1979a). Nucleoside triphosphate diphosphatase from germinated pea cotyledon chromatin. *Plant Cell Physiol.* **20**, 1041–1046.

Hirasawa, E., Takahashi, E., and Matsumoto, H. (1979b). Association of nucleotide-phosphohydrolyzing activity with chromatin in germinated pea cotyledon. *Plant Cell Physiol.* **20**, 219–224.

Hirasawa, E., Matsumoto, H., Ikeda, M., and Takahashi, E. (1981). Behavior of the nucleoside triphosphate diphosphatase in pea cotyledon chromatin during germination. *Plant Cell Physiol.* **22**, 283–289.

Hsu, T. C. (1962). Differential rate in RNA synthesis between euchromatin and heterochromatin. *Exp. Cell Res.* **27**, 332–334.

Hunt, L. T., and Dayhoff, M. O. (1977). Amino-terminal sequence identity of ubiquitin and the non-histone component of nuclear protein A24. *Biochem. Biophys. Res. Commun.* **74**, 650–655.

Hurley, C. K., and Stout, J. T. (1980). Maize histone H1: A partial structural characterization. *Biochemistry* **19**, 410–416.

Igo-Kemenes, T., Hörz, W., and Zachau, H. G. (1982). Chromatin. *Annu. Rev. Biochem.* **51**, 89–121.

Isenberg, I. (1979). Histones. *Annu. Rev. Biochem.* **48**, 159–191.

Jackson, J. B., Pollock, J. M., and Rill, R. L. (1979). Chromatin fractionation procedure that yields nucleosomes containing near-stoichiometric amounts of high mobility group nonhistone chromosomal proteins. *Biochemistry* **18**, 3739–3748.

Johns, E. W. (1964). Studies on histones 7. Preparative methods for histone fractions from calf thymus. *Biochem. J.* **92**, 55–59.

Johns, E. W., ed. (1982). "The HMG Chromosomal Proteins." Academic Press, New York.

Kamalay, J., and Goldberg, R. (1980). Regulation of structural gene expression in tobacco. *Cell* **19**, 935–946.

Key, J. L., Lin, C. Y., and Chen, Y. M. (1981). Heat shock proteins of higher plants. *Proc. Natl. Acad. Sci. U.S.A.* **78**, 3526–3530.

Kleinschmidt, A. M., and Martinson, H. G. (1981). Structure of nucleosome core particles containing uH2A(A24). *Nucleic Acids Res.* **9**, 2423–2431.

Kleinsmith, L. J. (1978). Phosphorylation of nonhistone proteins. *In* "The Cell Nucleus" (H. Busch, ed.), Vol. 6, Part C, pp. 222–261. Academic Press, New York.

Konkel, D. A., and Ingram, V. M. (1978). Is there specific transcription from isolated chromatin? *Nucleic Acids Res.* **5**, 1237–1252.

Kornberg, R. (1977). Structure of chromatin. *Annu. Rev. Biochem.* **46**, 931–954.

Kornberg, R. (1981). The location of nucleosomes in chromatin: Specific or statistical? *Nature (London)* **292**, 579–580.

Kornberg, R. D. (1974). Chromatin structure: A repeating unit of histones and DNA. *Science* **184**, 868–871.

Kuo, M. T., Mandel, J. L., and Chambon, P. (1979). DNA methylation: Correlation with DNase I sensitivity of chicken ovalbumin and conalbumin chromatin. *Nucleic Acids Res.* **7**, 2105–2113.

Labhart, P., and Koller, T. (1982). Structure of the active nucleolar chromatin of *Xenopus* laevis oocytes. *Cell* **28**, 279–292.

Lafontaine, J.-G. (1974). Ultrastructural organization of plant cell nuclei. *In* "The Cell Nucleus" (H. Busch, ed.), Vol. 1, pp. 149–185. Academic Press, New York.

Langen, T. A. (1978). Methods for the assessment of site-specific histone phosphorylation. *Methods Cell Biol.* **19**, 127–142.

Laroche, A., Plante, J., Beaumont, G., and Poirier, G. G. (1980). ADP-ribosylation of rye histones. *Can. J. Biochem.* **58**, 692–695.

LaRue, H., and Pallotta, D. (1976a). The selective extraction of histones from rye chromatin. *Can. J. Biochem.* **54**, 765–771.

LaRue, H., and Pallotta, D. (1976b). A study of the interaction between ethidium bromide and rye chromatin: Comparison with calf thymus chromatin. *Nucleic Acids Res.* **3**, 2193–2206.

Leber, B., and Hemleben, V. (1979a). Nucleosomal organization in active and inactive plant chromatin. *Plant Syst. Evol. Suppl.* **2**, 187–199.

Leber, B., and Hemleben, V. (1979b). Structure of plant nuclear and ribosomal DNA containing chromatin. *Nucleic Acids Res.* **7**, 1263–1281.

Levinger, L., and Varshavsky, A. (1982). Selective arrangement of ubiquinated and D1 protein-containing nucleosomes within the *Drosophila* genome. *Cell* **28**, 375–385.

Levy, B. W. (1981a). ADP-ribosylation of trout testis chromosomal proteins: Distribution of ADP-ribosylated proteins among DNase I-sensitive and -resistant chromatin domains. *Arch. Biochem. Biophys.* **208**, 528–534.

Levy, B. W. (1981b). Enhanced phosphorylation of high-mobility-group proteins in nuclease-sensitive mononucleosomes from butyrate-treated HeLa cells. *Proc. Natl. Acad. Sci. U.S.A.* **78**, 2189–2193.

Levy, B., and Dixon, G. (1977). Renaturation kinetics of cDNA complementary to cytoplasmic polyadenylated RNA from rainbow trout testis. Accessibility of transcribed genes to pancreatic DNase. *Nucleic Acids Res.* **4**, 883–898.

Levy, B., Conner, W., and Dixon, G. (1979a). A subset of trout testis nucleosomes enriched in transcribed DNA sequences contains high mobility group proteins as major structural components. *J. Biol. Chem.* **254**, 609–620.

Levy, B. W., Watson, D. C., and Dixon, G. H. (1979b). Multiacetylated forms of H4 are found in a putatively transcriptionally competent chromatin fraction from trout testis. *Nucleic Acids Res.* **6**, 259–287.

Liao, L. W., and Cole, R. D. (1981). Condensation of dinucleosomes by individual subfractions of H1 histones. *J. Biol. Chem.* **256**, 10124–10128.

Liberati-Langenbuch, J., Wilhelm, M. L., Gigot, C., and Wilhelm, F. X. (1980). Plant and animal histones are completely interchangeable in the nucleosome core. *Biochem. Biophys. Res. Commun.* **94**, 1161–1168.

Lilly, D. M. J., and Pardon, J. F. (1979). Structure and function of chromatin. *Annu. Rev. Genet.* **13**, 197–233.

Lin, P. P.-C., Wilson, R. F., and Bonner, J. (1973). Isolation and properties of nonhistone chromosomal proteins from pea chromatin. *Mol. Cell. Biochem.* **1**, 197–207.

Littau, V. C., Allfrey, V. G., Frenster, J. H., and Mirsky, A. E. (1964). Active and inactive regions of nuclear chromatin as revealed by electron microscope autoradiography. *Proc. Natl. Acad. Sci. U.S.A.* **52**, 93–100.

Lohr, D., and Hereford, L. (1979). Yeast chromatin is uniformly digested by DNase I. *Proc. Natl. Acad. Sci. U.S.A.* **76**, 4285–4288.

Lutz, C., and Nagl, W. (1980). A reliable method for preparation and electron microscopic visualization of nucleosomes in higher plants. *Planta* **149**, 408–410.

Lyon, M. F. (1961). Gene action in the X-chromosome of the mouse (*Mus musculus* L.). *Nature (London)* **190**, 372–373.

McGhee, J. D., and Engel, J. D. (1975). Subunit structure of chromatin is the same in plants and animals. *Nature (London)* **254**, 449–450.

McGhee, J. D., and Felsenfeld, G. (1980). Nucleosome structure. *Annu. Rev. Biochem.* **49**, 1115–1156.

McGhee, J. D., and Ginder, G. D. (1979). Specific DNA methylation sites in the vicinity of the chicken β-globin gene. *Nature (London)* **280**, 419–420.

Mandel, J. L., and Chambon, P. (1979). DNA methylation: Organ specific variations in the methylation pattern within and around ovalbumin and other chicken genes. *Nucleic Acids Res.* **7**, 2081–2103.

Mandel, P., Okazaki, H., and Niedergang, C. (1982). Poly(adenosine diphosphate ribose). *Prog. Nucleic Acid Res. Mol. Biol.* **27**, 1–51.

Mardian, J. K. W., and Isenberg, I. (1978). Yeast inner histones and the evolutionary conservation of histone-histone interactions. *Biochemistry* **17**, 3825–3833.

Mardian, J. K. W., Paton, A. E., Bunick, G. J., and Olins, D. E. (1980). Nucleosome cores have two specific binding sites for non-histone chromosomal proteins, HMG14 and 17. *Science* **209**, 1534–1536.

Marquez, G., Moran, F., Franco, L., and Montero, F. (1982). C1 proteins: A class of high-mobility-group non-histone chromosomal proteins from the fruit fly *Ceratitis capitata*. *Eur. J. Biochem.* **123**, 165–170.

Martinson, H. G., and True, R. J. (1979). Amino acid contacts between histones are the same for plants and mammals. Binding-site studies using ultraviolet light and tetranitromethane. *Biochemistry* **18**, 1947–1951.

Mathis, D., Oudet, P., and Chambon, P. (1980). Structure of transcribing chromatin. *Prog. Nucleic Acid Res. Mol. Biol.* **24**, 1–55.

Matsui, S., Seon, B. K., and Sandberg, A. A. (1979). Disappearance of a structural chromosomal protein A24 in mitosis: Implications for molecular basis of chromatin condensation. *Proc. Natl. Acad. Sci. U.S.A.* **76**, 6386–6390.

Mayes, E. L. V. (1982). Species and tissue specificity. *In* "The HMG Chromosomal Proteins" (E. W. Johns, ed.), pp. 7–40. Academic Press, New York.

Mayfield, J. E., Serunian, L. A., Silver, L. M., and Elgin, S. C. R. (1978). A protein released by DNAase I digestion of *Drosophila* nuclei is preferentially associated with puffs. *Cell* **14**, 539–544.

Mezquita, J., Chiva, M., Vidal, S., and Mezquita, C. (1982). Effect of high mobility group nonhistone proteins HMG-20 (ubiquitin) and HMG-17 on histone deacetylase activity assayed in vitro. *Nucleic Acids Res.* **10**, 1781–1797.

Moreno, M. L., Sogo, J. M., and de la Torre, C. (1978). Higher plant nucleosomes. A micromethod for isolating and dispersing chromatin fibers. *New Phytol.* **81**, 681–683.

Motojima, K., and Sakaguchi, K. (1981). The occurrence of N-ε-Trimethyl lysine in two histone-like proteins from wheat germ. *FEBS Lett.* **132**, 334–336.

Muller, A., Phillipps, G., and Gigot, C. (1980). Properties of condensed chromatin in barley nuclei. *Planta* **149**, 69–77.

Mullins, D. W., Giri, C. P., and Smulson, M. E. (1977). Poly(adenosine diphosphate-ribose) polymerase: The distribution of a chromosome-associated enzyme within the chromatin substructure. *Biochemistry* 16, 506–513.

Murray, M. G., and Key, J. L. (1978). 2,4-Dichlorophenoxy-acetic acid-enhanced phosphorylation of soybean nuclear proteins. *Plant Physiol.* 61, 190–198.

Nadeau, P., Pallotta, D., and Lafontaine, J.-G. (1974). Electrophoretic study of plant histones: Comparison with vertebrate histones. *Arch. Biochem. Biophys.* 161, 171–177.

Nadeau, P., Pallotta, D., and Lafontaine, J.-G. (1977). Comparative study of rye and thymus histones: Amino acid analysis and tryptic finger printing. *Can. J. Biochem.* 55, 721–727.

Nagl, W. (1976). Nuclear organization. *Annu. Rev. Plant Physiol.* 27, 39–69.

Nagl, W. (1979a). Condensed chromatin in plant and animal nuclei: Fundamental differences. *Plant Syst. Evol. Suppl.* 2, 247–260.

Nagl, W. (1979b). Nuclear ultrastructure: Condensed chromatin in plants is species-specific (karyotypical), but not tissue-specific (functional). *Protoplasma* 100, 53–71.

Naveh-Many, T., and Cedar, H. (1981). Active gene sequences are undermethylated. *Proc. Natl. Acad. Sci. U.S.A.* 78, 4246–4250.

Newrock, K. M., Cohen, L. H., Hendricks, M. B., Donnelly, R. J., and Weinberg, E. S. (1978a). Stage-specific mRNAs coding for subtypes of H2A and H2B histones in the sea urchin embryo. *Cell* 14, 327–336.

Newrock, K. M., Alfageme, C. R., Nardi, R. V., and Cohen, L. H. (1978b). Histone changes during chromatin remodeling in embryogenesis. *Cold Spring Harbor Symp. Quant. Biol.* 42, 421–431.

Nicklisch, A., Strauss, M., and Wersuhn, G. (1976). Investigations on chromatin of root tips of *Zea mays* L. *Biochem. Physiol. Pflanzen* 169, 105–119.

Nicolaieff, A., Philipps, G., Gigot, C., and Hirth, L. (1976). Ring subunit associated with plant chromatin. *J. Microsc. Biol. Cell.* 26, 1–4.

Nishizuka, Y., Ueda, K., Honjo, T., and Hayaishi, O. (1968). Enzymatic adenosine diphosphate ribosylation of histone and poly adenosine diphosphate ribose synthesis in rat liver nuclei. *J. Biol. Chem.* 243, 3765–3767.

Nordheim, A., Pardue, M. L., Lafer, E. M., Möller, A., Stollar, B. D., and Rich, A. (1981). Antibodies to left-handed Z-DNA bind to interband regions of *Drosophila* polytene chromosomes. *Nature (London)* 294, 417–422.

Ogata, N., Ueda, K., and Hayaishi, O. (1980a). ADP-ribosylation of histone H2B. *J. Biol. Chem.* 255, 7610–7615.

Ogata, N., Ueda, K., Kagamiyama, H., and Hayaishi, O. (1980b). ADP-ribosylation of histone H1. *J. Biol. Chem.* 255, 7616–7620.

Ogata, N., Ueda, K., Kawaichi, M., and Hayaishi, O. (1981). Poly(ADP-ribose) synthetase, a main acceptor of poly(ADP-ribose) in isolated nuclei. *J. Biol. Chem.* 256, 4135–4137.

Olins, A. L., and Olins, D. E. (1974). Spheroid chromatin units (ν bodies). *Science* 183, 330–332.

Oliver, D., Sommer, K. R., Panyim, S., Spiker, S., and Chalkley, R. (1972). A modified procedure for fractionating histones. *Biochem. J.* 129, 349–353.

Olson, M. O. J., Goldknopf, I. L., Guetzow, K. A., James, G. T., Hawkins, T. C., Mays-Rothberg, C. J., and Busch, H. (1976). The NH₂- and COOH-terminal amino acid sequence of nuclear protein A24. *J. Biol. Chem.* 251, 5901–5903.

Oudet, P., Gross-Bellard, M., and Chambon, P. (1975). Electron microscopic and bio-chemical evidence that chromatin structure is a repeating unit. *Cell* **4**, 281–300.

Palter, K. B., Foe, V. E., and Alberts, B. M. (1979). Evidence for the formation of nucleosome-like histone complexes on single-stranded DNA. *Cell* **18**, 451–467.

Panyim, S., and Chalkley, R. (1969a). The heterogeneity of histones. I. A quantitative analysis of calf thymus histones in very long polyacrylamide gels. *Biochemistry* **8**, 3972–3979.

Panyim, S., and Chalkley, R. (1969b). A new histone found only in mammalian tissues with little cell division. *Biochem. Biophys. Res. Commun.* **37**, 1042–1049.

Panyim, S., Chalkley, R., Spiker, S., and Oliver, D. (1970). Constant electrophoretic mobility of the cysteine-containing histone in plants and animals. *Biochim. Biophys. Acta* **214**, 216–221.

Panyim, S., Bilek, D., and Chalkley, R. (1971). An electrophoretic comparison of verte-brate histones. *J. Biol. Chem.* **246**, 4206–4215.

Patthy, L., Smith, E. L., and Johnson, J. (1973). The complete amino acid sequence of pea embryo histone III. *J. Biol. Chem.* **248**, 6834–6840.

Payne, J. F., and Bal, A. K. (1976). Cytoplasmic detection of poly(ADP-ribose) poly-merase. *Exp. Cell Res.* **99**, 428–432.

Perella, F. W., and Lea, M. A. (1979). Spermine-induced variations in the adenosine 5′-diphosphate ribosylation patterns of the nuclear proteins from rat liver and hepatoma. *Cancer Res.* **39**, 1382–1389.

Philipps, G., and Gigot, C. (1977). DNA associated with nucleosomes in plants. *Nucleic Acids Res.* **4**, 3617–3626.

Pipkin, J. L., and Larson, D. A. (1972). Characterization of the very lysine-rich histones of active and quiescent anther tissues of *Hippeastrum belladonna*. *Exp. Cell Res.* **71**, 249–260.

Pipkin, J. L., and Larson, D. A. (1973). Changing patterns of nucleic acids, basic and acidic proteins in generative and vegetative nuclei during pollen germination and pollen tube growth in *Hippeastrum belladonna*. *Exp. Cell Res.* **79**, 28–42.

Pitel, J. A., and Durzan, D. J. (1978). Chromosomal proteins of conifers. 1. Comparison of histones and nonhistone chromosomal proteins from dry seeds of conifers. *Can. J. Biochem.* **56**, 1915–1927.

Poirier, G. G., and Savard, P. (1980). ADP-ribosylation of pancreatic histone H1 and of other histones. *Can. J. Biochem.* **58**, 509–515.

Poirier, G. G., DeMurcia, G., Jongstra-Bilen, J., Niedergang, C., and Mandel, P. (1982). Poly(ADP-ribosyl)ation of polynucleosomes causes relaxation of chromatin struc-ture. *Proc. Natl. Acad. Sci. U.S.A.* **79**, 3423–3427.

Prentice, D. A., Loechel, S. C., and Kitos, P. A. (1982). Histone H2A phosphorylation in animal cells: Functional considerations. *Biochemistry* **21**, 2412–2420.

Purnell, M. R., Stone, P. R., and Whish, W. J. D. (1980). ADP-ribosylation of nuclear proteins. *Biochem. Soc. Trans.* **8**, 215–227.

Razin, A., and Riggs, A. D. (1980). DNA methylation and gene function. *Science* **210**, 604–610.

Reeves, R., Chang, D., and Chung, S.-C. (1981). Carbohydrate modifications of the high mobility group proteins. *Proc. Natl. Acad. Sci. U.S.A.* **78**, 6704–6708.

Riggs, M. G., Whittaker, R. G., Neuman, J. R., and Ingram, V. M. (1977). n-Butyrate causes histone modification in HeLa and Friend erythroleukaemia cells. *Nature (London)* **268**, 462–464.

Ring, D., and Cole, R. D. (1979). Chemical cross-linking of H1 histone to the nucleosomal histones. *J. Biol. Chem.* **254**, 11688–11695.

Rodrigues, J. D. A., Brandt, W. F., and Von Holt, C. (1979). Plant histone 2 from wheat germ, a family of histone H2A variants. Partial amino acid sequences. *Biochim. Biophys. Acta* **578**, 196–206.

Russanova, V., Venkov, C., and Tsanev, R. (1980). A comparison of histone variants in different rat tissues. *Cell Differ.* **9**, 339–350.

Rykowski, M. C., Wallis, J. W., Choe, J., and Grunstein, M. (1981). Histone H2B subtypes are dispensable during the yeast cell cycle. *Cell* **25**, 477–487.

Saffer, J. D., and Glazer, R. I. (1980). The phosphorylation of high mobility group proteins 14 and 17 from Ehrlich ascites and L1210 *in vitro*. *Biochem. Biophys. Res. Commun.* **93**, 1280–1285.

Saffer, J. D., and Glazer, R. I. (1982). The phosphorylation of high mobility group proteins 14 and 17 and their distribution in chromatin. *J. Biol. Chem.* **257**, 4655–4660.

Sandeen, G., Wood, W. I., and Felsenfeld, G. (1980). The interaction of high mobility proteins HMG14 and 17 with nucleosomes. *Nucleic Acids Res.* **8**, 3757–3778.

Sanders, M. M. (1978). Fractionation of nucleosomes by salt elution from micrococcal nuclease-digested nuclei. *J. Cell Biol.* **79**, 97–109.

Sano, H., and Sager, R. (1982). Tissue specificity and clustering of methylated cytosines in bovine satellite I DNA. *Proc. Natl. Acad. Sci. U.S.A.* **79**, 3584–3588.

Sasaki, K., and Sugita, M. (1982). Relationship between poly(adenosine 5′-diphosphate-ribose) synthesis and transcription activity in wheat embryo chromatin. *Plant Physiol.* **69**, 543–545.

Savic, A., Richman, P., Williamson, P., and Poccia, D. (1981). Alterations in chromatin structure during early sea urchin embryogenesis. *Proc. Natl. Acad. Sci. U.S.A.* **78**, 3706–3710.

Schäfer, W., and Kahl, G. (1982). Phosphorylation of chromosomal proteins in resting and wounded potato tuber tissues. *Plant Cell Physiol.* **23**, 137–146.

Schlesinger, D. H., Goldstein, G., and Niall, H. D. (1975). The complete amino acid sequence of ubiquitin, an adenylate cyclase stimulating polypeptide probably universal in living cells. *Biochemistry* **14**, 2214–2218.

Seale, R. L. (1981). Rapid turnover of the histone-ubiquitin conjugate, protein A24. *Nucleic Acids Res.* **9**, 3151–3158.

Sealy, L., and Chalkley, R. (1978a). The effect of sodium butyrate on histone modification. *Cell* **14**, 115–121.

Sealy, L., and Chalkley, R. (1978b). DNA associated with hyperacetylated histones is preferentially digested by DNase I. *Nucleic Acids Res.* **5**, 1863–1876.

Seyedin, S. M., Pehrson, J. R., and Cole, R. D. (1981). Loss of chromosomal high mobility group proteins HMG1 and HMG2 when mouse neuroblastoma and Friend erythroleukemia cells become committed to differentiation. *Proc. Natl. Acad. Sci. U.S.A.* **78**, 5988–5992.

Shaw, B. R., Cognetti, G., Sholes, W. M., and Richards, R. G. (1981). Shift in nucleosome populations during embryogenesis: Microheterogeneity in nucleosomes during development of the sea urchin embryo. *Biochemistry* **20**, 4971–4978.

Sheffery, M., Rifkind, R. A., and Marks, P. A. (1982). Murine erythroleukemia cell differentiation: DNase I hypersensitivity and DNA methylation near the globin genes. *Proc. Natl. Acad. Sci. U.S.A.* **79**, 1180–1184.

Sidorova, V. V., and Konarev, V. G. (1981). Isolation and purification of plant histones. *Biokhimiya* **46**, 1047–1054.

Simon, J. H., and Becker, W. M. (1976). A polyethylene glycol/dextran procedure for the isolation of chromatin proteins (histones and nonhistones) from wheat germ. *Biochim. Biophys. Acta* **454,** 154–171.

Simpson, R. (1981). Modulation of nucleosome structure by histone subtypes in sea urchin embryos. *Proc. Natl. Acad. Sci. U.S.A.* **78,** 6803–6807.

Sledziewski, A., and Young, E. T. (1982). Chromatin conformational changes accompany transcriptional activation of a glucose-repressed gene in *Saccharomyces cerevisiae. Proc. Natl. Acad. Sci. U.S.A.* **79,** 253–256.

Smerdon, M. J., and Isenberg, I. (1976). Interactions between the subfractions of calf thymus H1 and nonhistone chromosomal proteins HMG1 and HMG2. *Biochemistry* **15,** 4242–4247.

Smith, B. J., Walker, J. M., and Johns, E. W. (1980). Structural homology between a mammalian H1^0 subfraction and avian erythrocyte-specific histone H5. *FEBS Lett.* **112,** 42–44.

Sommer, A. (1978). Yeast chromatin: Search for histone H1. *Mol. Gen. Genet.* **161,** 323–331.

Sommer, K. R., and Chalkley, R. (1974). A new method for fractionating histones for physical and chemical studies. *Biochemistry* **13,** 1022–1027.

Spiker, S. (1975). An evolutionary comparison of plant histones. *Biochim. Biophys. Acta* **400,** 461–467.

Spiker, S. (1976a). Identification of histones F2b and F1 in stained gels. *J. Chromatog.* **128,** 244–248.

Spiker, S. (1976b). Expression of parental histone genes in the intergeneric hybrid *Triticale hexaploide. Nature (London)* **259,** 418–420.

Spiker, S. (1982). Histone variants in plants: Evidence for primary structure variants differing in molecular weight. *J. Biol. Chem.* **257,** 14250–14255.

Spiker, S., and Chalkley, R. (1971). Histones of gibberellic acid-treated peas. *Plant Physiol.* **47,** 342–345.

Spiker, S., and Isenberg, I. (1977). Cross-complexing pattern of plant histones. *Biochemistry* **16,** 1819–1826.

Spiker, S., and Isenberg, I. (1978). Evolutionary conservation of histone-histone binding sites: Evidence from interkingdom complex formation. *Cold Spring Harbor Symp. Quant. Biol.* **42,** 157–163.

Spiker, S., and Krishnaswamy, L. (1973). Constancy of wheat histones during development. *Planta* **110,** 71–76.

Spiker, S., Key, J. L., and Wakim, B. (1976). Identification and fractionation of plant histones. *Arch. Biochem. Biophys.* **176,** 510–518.

Spiker, S., Mardian, J. K. W., and Isenberg, I. (1978). Chromosomal HMG proteins occur in three eukaryotic kingdoms. *Biochem. Biophys. Res. Commun.* **82,** 129–135.

Spiker, S., Murray, M., and Thompson, W. F. (1983). DNAase I sensitivity of transcriptionally active genes in intact nuclei and isolated chromatin of plants. *Proc. Natl. Acad. Sci. U.S.A.* **80,** 815–819.

Stalder, J., Larsen, A., Engel, J. D., Dolan, M., Groudine, M., and Weintraub, H. (1980). Tissue-specific DNA cleavages in the globin chromatin domain introduced by DNAase I. *Cell* **20,** 451–460.

Stedman, E., and Stedman, E. (1950). Cell specificity of histones. *Nature (London)* **166,** 780–781.

Stein, G. S., Spelsberg, T. C., and Kleinsmith, L. J. (1974). Nonhistone chromosomal proteins and gene regulation. *Science* **183,** 817–824.

Sterner, R., Boffa, L. C., and Vidali, G. (1978). Comparative structural analysis of high mobility group proteins from a variety of sources. *J. Biol. Chem.* **253,** 3830–3836.

Stout, J. T., and Phillips, R. L. (1973). Two independently inherited electrophoretic variants of the lysine-rich histones of maize (*Zea mays*). *Proc. Natl. Acad. Sci. U.S.A.* **70,** 3043–3047.

Strickland, M., Strickland, W. N., and VonHolt, C. (1981). The occurrence of sperm isohistones H2B in single sea urchins. *FEBS Lett.* **135,** 86–88.

Sugita, M., and Sasaki, K. (1982). Transcriptional activation and structural alteration of wheat chromatin during germination and seedling growth. *Physiol. Plant* **54,** 41–46.

Sugita, M., Yoshida, K., and Sasaki, K. (1979). Germination-induced changes in chromosomal proteins of spring and winter wheat embryos. *Plant Physiol.* **64,** 780–785.

Teng, C. T., and Teng, C. S. (1981). Changes in quantities of high-mobility-group protein 1 in oviduct cellular fraction after oestrogen stimulation. *Biochem. J.* **198,** 85–90.

Tessier, A., Roland, B., Gauthier, C., Anderson, W. A., and Pallotta, D. (1980). Yeast, rye and calf histones. Similarities and differences detected by electrophoretic and immunological methods. *Can. J. Biochem.* **58,** 405–409.

Thomas, J. O. (1978). Chromatin structure. *In* "Biochemistry of Nucleic Acids" (B. F. C. Clark, ed.), Vol. 17, pp. 181–232. Univ. Park Press, Baltimore, Maryland.

Thompson, W. F., and Murray, M. G. (1981). The nuclear genome: Structure and function. *In* "The Biochemistry of Plants" (A. Marcus, ed.), Vol. 6, pp. 1–81. Academic Press, New York.

Towill, L. E., and Nooden, L. D. (1973). An electrophoretic analysis of the acid-soluble chromosomal proteins from different organs of maize seedlings. *Plant Cell Physiol.* **14,** 851–863.

Towill, L. E., and Nooden, L. D. (1975). An electrophoretic analysis of the acid-insoluble chromosomal proteins from different organs of maize seedlings. *Plant Cell Physiol.* **16,** 1073–1084.

Urban, M. K., Franklin, S. G., and Zweidler, A. (1979). Isolation and characterization of the histone variants in chicken erythrocytes. *Biochemistry* **18,** 3952–3960.

Vanyushin, B. F., Bashkite, E. A., Fridrich, A., and Chvoika, L. A. (1981). Methylation of DNA in wheat seedlings and the influence of phytohormones. *Biokhimiya* **46,** 35–40.

Vidali, G., Boffa, L. C., and Allfrey, V. G. (1977). Selective release of chromosomal proteins during limited DNAase I digestion of avian erythrocyte chromatin. *Cell* **12,** 409–415.

Von Holt, C., Strickland, W. N., Brandt, W. F., and Strickland, M. S. (1979). More histone structures. *FEBS Lett.* **100,** 201–218.

Waalwijk, C., and Flavell, R. A. (1978). DNA methylation at a CCGG sequence in the large intron of the rabbit β-globin gene: Tissue-specific variations. *Nucleic Acids Res.* **5,** 4631–4641.

Wagner, I., and Capesius, I. (1981). Determination of 5-methylcytosine from plant DNA by high-performance liquid chromatography. *Biochim. Biophys. Acta* **654,** 52–56.

Walker, J. M., Goodwin, G. H., and Johns, E. W. (1978). The isolation and identification of ubiquitin from the high mobility group (HMG) non-histone protein fraction. *FEBS Lett.* **90,** 327–330.

Watson, D. C., Wong, N. C. W., and Dixon, G. H. (1979). The complete amino-acid sequence of a trout-testis non-histone protein, H6, localized in a subset of nucleosomes and its similarity to calf thymus non-histone proteins HMG-14 and HMG-17. *Eur. J. Biochem.* **95,** 193–202.

Weber, S., and Isenberg, I. (1980). High mobility group proteins of *Saccharomyces cerevisiae*. *Biochemistry* **19**, 2236–2240.

Weintraub, H. (1980). Recognition of specific DNA sequences in eukaryotic chromosomes. *Nucleic Acids Res.* **8**, 4745–4753.

Weintraub, H., and Groudine, M. (1976). Chromosomal subunits in active genes have an altered conformation. *Science* **193**, 848–856.

Weintraub, H., Worcel, A., and Alberts, B. (1976). A model for chromatin based upon two symmetrically paired half-nucleosomes. *Cell* **9**, 409–417.

Weintraub, H., Flint, S. J., Leffak, I. M., Groudine, M., and Grainger, R. M. (1978). The generation and propagation of variegated chromosome structures. *Cold Spring Harbor Symp. Quant. Biol.* **42**, 401–407.

Weintraub, H., Larsen, A., and Groudine, M. (1981). α-Globin-gene switching during the development of chicken embryos: Expression and chromatin structure. *Cell* **24**, 333–344.

Weisbrod, S. (1982a). Active chromatin. *Nature (London)* **297**, 289–295.

Weisbrod, S. T. (1982b). Properties of active nucleosomes as revealed by HMG 14 and 17 chromatography. *Nucleic Acids Res.* **10**, 2017–2042.

Weisbrod, S., and Weintraub, H. (1979). Isolation of a subclass of nuclear proteins responsible for conferring a DNase I-sensitive structure on globin chromatin. *Proc. Natl. Acad. Sci. U.S.A.* **76**, 630–634.

Weisbrod, S., and Weintraub, H. (1981). Isolation of actively transcribed nucleosomes using immobilized HMG 14 and 17 and an analysis of α-globin chromatin. *Cell* **23**, 391–400.

Weisbrod, S., Groudine, M., and Weintraub, H. (1980). Interaction of HMG 14 and 17 with actively transcribed genes. *Cell* **19**, 289–301.

West, M. H. P., and Bonner, W. M. (1980a). Histone 2A, a heteromorphous family of eight protein species. *Biochemistry* **19**, 3238–3245.

West, M. H. P., and Bonner, W. M. (1980b). Histone 2B can be modified by the attachment of ubiquitin. *Nucleic Acids Res.* **8**, 4671–4680.

Whitby, A. J., Stone, P. R., and Whish, W. J. D. (1979). Effect of polyamines and Mg^{++} on poly(ADP-ribose) synthesis and ADP-ribosylation of histones in wheat. *Biochem. Biophys. Res. Commun.* **90**, 1295–1304.

Wielgat, B., and Kleczlowski, K. (1981a). Gibberellic acid-enhanced phosphorylation of pea chromatin proteins. *Plant Sci. Lett.* **21**, 381–388.

Wielgat, B., and Kleczkowski, K. (1981b). Nonhistone chromosomal proteins from gibberellic acid treated maize and pea plants, and their effect on transcription *in vitro*. *Int. J. Biochem.* **13**, 1201–1203.

Wilhelm, M. L., Langenbuch, J., Wilhelm, F. X., and Gigot, C. (1979). Nucleosome core particles can be reconstituted using mixtures of histones from two eukaryotic kingdoms. *FEBS Lett.* **103**, 126–132.

Willmitzer, L. (1979). Demonstration of *in vitro* covalent modification of chromosomal proteins by poly(ADP) ribosylation in plant nuclei. *FEBS Lett.* **108**, 13–16.

Willmitzer, L., and Wagner, K. G. (1981). The isolation of nuclei from tissue-cultured plant cells. *Exp. Cell Res.* **135**, 69–77.

Wu, R. S., Kohn, K. W., and Bonner, W. M. (1981). Metabolism of ubiquitinated histones. *J. Biol. Chem.* **256**, 5916–5920.

Yakura, K., Fukuei, K., and Tanifuji, S. (1978). Chromatin subunit structure in different tissues of higher plants. *Plant Cell Physiol.* **19**, 1381–1390.

Yoshida, K., and Sasaki, K. (1977). Changes of template activity and proteins of chromatin during wheat germination. *Plant Physiol.* **59**, 497–501.

Yoshida, K., Sugita, M., and Sasaki, K. (1979). Involvement of nonhistone chromosomal proteins in transcriptional activity of chromatin during wheat germination. *Plant Physiol.* **63**, 1016–1021.

Zachau, H. G., and Igo-Kemenes, T. (1981). Face to phase with nucleosomes. *Cell* **24**, 597–598.

Zalenskaya, I. A., Zalensky, A. O., Zalenskaya, E. O., and Vorob'ev, V. I. (1981). Heterogeneity of nucleosomes in genetically inactive cells. *FEBS Lett.* **128**, 40–42.

Zurfluh, L. L., and Guilfoyle, T. J. (1982). Auxin-induced changes in the population of translatable messenger RNA in elongating sections of soybean hypocotyl. *Plant Physiol.* **69**, 332–337.

Zweidler, A. (1978). Resolution of histones by polyacrylamide gel electrophoresis in the presence of nonionic detergents. *Methods Cell Biol.* **17**, 223–233.

THE MOLECULAR GENETICS OF CROWN GALL TUMORIGENESIS

P. J. J. Hooykaas and R. A. Schilperoort

Laboratory of Biochemistry, University of Leiden, Leiden, The Netherlands

ADVANCES IN GENETICS, Vol. 22

I. Introduction

The phytopathogenic bacteria *Agrobacterium tumefaciens* (Smith and Townsend, 1907) and *A. rhizogenes* (Riker, 1930) are the causative agents of the widespread plant diseases "crown gall" and "hairy root," respectively. It is now well established that virulent strains of these bacterial species transfer a piece of bacterial DNA into plant cells, thereby transforming these into tumor cells (Fig. 1). In research much attention has been paid to the agrobacteria for several reasons. First is the desire to develop a system for the genetic engineering of plant cells based on the natural system for gene transfer between *Agrobacterium* species and plant cells. Second, there is a striking resemblance between the etiology of animal cancers and the plant cancer crown gall that was recognized as early as in 1927 (Stapp). This led to basic studies on the process of plant tumor induction and on the recovery of plant cells from the tumorous state. A third important interest lies in crown gall as a disease that is the cause of economically important losses in agriculture and horticulture in Europe (Kerr and Panagopoulos, 1977; Süle, 1978), North America (Dhavantari, 1978; Kennedy and Alcorn, 1980), and Australia (New and Kerr, 1972). Research has been aimed at finding means to prevent crown gall and to cure plants of this disease (reviewed by Kerr, 1980; El-Fiki and Giles, 1981; Moore and Cooksey, 1981).

II. Bacterial Taxonomy

Bacteria from the genus *Agrobacterium* together with those from the genus *Rhizobium* belong to the bacterial family of Rhizobiaceae (see Bergey's Manual; Allen and Holding, 1974). Bacteria in this family are gram-negative rods that occur abundantly in soil, but agrobacteria have also recently been encountered in hospitals (Riley and Weaver, 1977; Plotkin, 1980). The most important characteristic that immedi-

Fig. 1. Tumors induced by *Agrobacterium tumefaciens* on *Kalanchöe daigremontiana*.

ately distinguishes agrobacteria from most other bacteria is their ability to induce tumor formation in plants (Smith and Townsend, 1907; Riker, 1930), although avirulent *Agrobacterium* strains have also been isolated from nature. For the isolation of agrobacteria from soil or from tumors a number of selective media have been developed (Schroth *et al.,* 1965; New and Kerr, 1971; Kerr and Panagopoulos, 1977). Furthermore, minimal media for growth of agrobacteria have been designed. Most strains can grow on a minimal salts medium with added carbon, nitrogen, and energy sources; but some require the addition of biotin, nicotinic acid, and pantothenate (Starr, 1946; Lippincott and Lippincott, 1969). The Nile Blue test has been recommended as an easy method for distinguishing between agrobacteria and rhizobia (Skinner, 1977). A further difference between agrobacteria and rhizobia is that the former produce H_2O_2 from amino acids, whereas rhizobia usually are negative in peroxide formation, in accordance with their symbiotic status (Clark, 1972). A unique characteristic of many (biotype 1) *Agrobacterium* strains is that they are able to reduce sugars such as lactose to ketoderivatives (Bernaerts and De Ley, 1963; Chern *et al.,* 1976).

The *Agrobacterium* strains that are able to cause crown gall have been classified as the species *A. tumefaciens,* those inducing hairy root as the species *A. rhizogenes,* those inducing "cane gall"—identical to crown gall, but in nature mainly induced on aerial parts of plants—as *A. rubi,* and nonpathogenic ones as *A. radiobacter* (see Bergey's Manual; Allen and Holding, 1974). However, genetic experiments have shown that the phytopathogenic character of agrobacteria is determined largely by plasmids that can freely move among the different "species" (see Section VII). Therefore, in fact, the phytopathogenicity trait cannot be used for taxonomic purposes. Numerical taxonomic studies have shown that in agrobacteria three main clusters or biotypes can be distinguished, but strains of intermediate or aberrant biotype have also been isolated (Keane *et al.,* 1970; White, 1972; Kersters *et al.,* 1973; Panagopoulos and Psallidas, 1973; Kerr and Panagopoulos, 1977). Some *Agrobacterium* strains seem more closely related to *Rhizobium meliloti* or to *R. leguminosarum* than to certain other agrobacteria (Graham, 1964; White, 1972). In fact, at least some *Rhizobium* strains become phytopathogenic after conjugation with a virulent *Agrobacterium* strain (Hooykaas *et al.,* 1977; Hooykaas, 1979). Bacteriophages have been isolated for many *Agrobacterium* and *Rhizobium* strains, but lysis of agrobacteria by *Rhizobium* phages has not been observed (Conn *et al.,* 1945; Bruch and Allen, 1957; Parker and Allen, 1957; Hooykaas, 1979). With one exception, lysis of rhizobia by *Agrobacterium* phages has also not been seen (Conn *et al.,* 1945; Roslycky *et al.,* 1962; Boyd *et al.,* 1970; Hooykaas, 1979). One exceptional phage isolated by Roslycky *et al.* (1962) was reported as being able to lyse many *Agrobacterium* strains as well as some *R. meliloti*

TABLE 1
Agrobacterium Biotypes[a]

	Biotype 1	Biotype 2	Biotype 3
Ketolactose production	+	−	−
Maximum growth temperature	37	29	35
Growth on erythritol as a carbon source	−	+	−
Growth in the presence of 2% NaCl	+	−	+
Sensitivity to phages LPB81, LPB82, LPB83	+ or −	−	−
Nile Blue test	+	+	+

[a]Data from Kerr and Panagopoulos (1977), Skinner (1977), and Hooykaas (1979).

and *R. leguminosarum* strains. Bacteriophages have been isolated for biotype 1 *Agrobacterium* strains only, and these do not lyse strains of other biotypes (Hooykaas, 1979). A number of characteristic differences between *Agrobacterium* strains of different biotypes can be used for classifying a new strain. Production of ketolactose from lactose, growth at 29, 35, or 37°C, sensitivity to NaCl, and utilization of sugars such as erythritol are the best characteristics by which biotyping can be done (Table 1).

III. Plant Tumor Induction

In nature, crown gall is mostly induced at sites on plants where wounds frequently originate, such as the root crown. The formation of the plant tumor may lead to severe stunting or even to death of the plant due to the interference with the flow of water and nutrients. Hairy root is characterized—on some plants—by abundant root formation from wound sites. It is not certain whether root formation interferes with plant development or not (Munnecke *et al.*, 1963; Jaynes and Strobel, 1981). In general, all parts of a plant can become susceptible to crown gall and hairy root after wounding. In the laboratory the stems of fast-growing plants, such as tomato and sunflower, are usually used for oncogenicity tests, but pea stem segments (Kurkdjian *et al.*, 1974), pinto bean leaves (Lippincott and Heberlein, 1965), and carrot or potato disks (Klein and Tenenbaum, 1955; Anand and Heberlein, 1977) are also frequently used. In order to study tumor morphology, *Kalanchöe* and tobacco stems (Braun, 1953; Bopp and Resende, 1966; Hooykaas *et al.*, 1977, 1982a; Gresshoff *et al.*, 1979; Ooms *et al.*, 1981) have been used. In fact, crown gall can be induced by *A. tumefaciens* on a wide range of dicotyledonous plants (Tamm, 1954; De Cleene and De Ley, 1977), although some host specificity of strains has been observed (Stapp, 1938; Panagopoulos and Psallidas, 1973; Anderson and Moore, 1979; Hooykaas *et al.*, 1979a; Loper and Kado, 1979). Hairy root can also be induced on many different plant species (Munnecke *et al.*, 1963; Jaynes and Strobel, 1981); numerous plant species, however, do not react with root formation, but rather with tumor formation (e.g., *Nicotiana rustica* and *N. glauca*) or with a combination of tumor and root formation (e.g., *Pisum sativum* and *Catharanthus roseus*) (Mitter and Beiderbeck, 1980; White *et al.*, 1982; Costantino and Hooykaas, unpublished). Monocotyledonous plants are not sensitive to *A. tumefaciens,* possibly because the bacterium cannot attach to their cell walls (see below).

The status of the plant is crucial to the question of whether agrobacteria will be able to induce tumors. Bopp and Resende (1966) have reported that on certain varieties of *Kalanchöe* tumors can be induced on flowering plants, but not on nonflowering plants. Furthermore, the presence of plant viruses may interfere with tumor induction. It has been found that *A. tumefaciens* is not capable of inducing tumors on *Vigna sinensis* plants that are systemically infected with RNA viruses (Saedi *et al.*, 1979). Double infection on plants with *A. tumefaciens* and citrus exocortis viroid is possible. This results in tumor formation, whereby viroid RNA accumulates in the tumor cells (Semancik *et al.*, 1978).

Wounding of plants is essential for tumor induction to be possible, first of all because agrobacteria are not invasive, and second for "conditioning" of the healthy plant cells. After wounding, 30 hours at 25°C is necessary before plant cells become sensitive to tumor induction by *A. tumefaciens*, and thereafter they remain susceptible only during a limited period of time (Braun, 1947; Lipetz, 1966). Only bacteria metabolically active in the wound sap of the plant can induce tumors; dead bacteria or metabolically inactive bacteria, such as some multiple auxotrophs, are unable to do so (Langley and Kado, 1972). The bacteria can induce tumors only after they have been present in the wound for at least 16 hours (Lipetz, 1966). During this period of time (the "bacterial adjustment" phase) functions possibly necessary for plant transformation have to be induced in the bacteria. Agrobacteria do not invade healthy plant cells (Riker, 1923), but rather attach to their cell walls (Lippincott and Lippincott, 1969; Schilperoort, 1969; Glogowski and Galsky, 1978; Lippincott and Lippincott, 1980).

During attachment, virulent and certain avirulent agrobacteria compete for a limited number of attachment sites. When such competing avirulent agrobacteria are inoculated into wound sites prior to the virulent bacteria, no tumors are formed, but if they are added after the virulent bacteria normal tumor formation occurs. Preincubation of wound sites with dead agrobacteria or with lipopolysaccharides (LPS) isolated from agrobacterial cell walls leads to inhibition of tumor formation by subsequently inoculated virulent agrobacteria, indicating that attachment occurs via bacterial LPS (Whatley *et al.*, 1976). Recent data suggest that N-acetyl-D-galactosamine and β-D-galactose present in the O-antigenic region of agrobacterial LPS are involved in the attachment to plant cell wall components (Banerjee *et al.*, 1981). Isolated plant cell walls have also been fractionated. It seems that the polygalacturonic acid fraction is responsible for binding of *A. tumefaciens*, because preincubation of agrobacteria with this cell wall fraction prior to inoculation into fresh plant wound sites prevents tumor formation (Rao *et al.*, 1982). Similar experiments with the polygalacturonic acid fraction from the cell walls of monocotyledonous plants or of crown gall cells did not result in inhibition of tumor formation, showing that this does not contain attachment sites (Lippincott and Lippincott, 1978a; Rao *et al.*, 1982). However, the cell walls of monocots and crown gall cells become inhibitory for tumor induction after treatment with the enzyme pectinesterase; this can be partially reversed by subsequent treatment with the enzyme pectin methyl transferase (Rao *et al.*, 1982). This suggests that the degree of methylation of the polygalacturonic acid fraction of the cell wall is critical to adherence of agrobacteria to plant cells. In agreement, after treatment of the cell walls of dicots with pectin methyl transferase these become less inhibitory in tumor induction, and this is reversible by treatment with pectinesterase.

After attachment to plant cells, agrobacteria form cellulose fibrils that surround them and anchor them to the plant cell surface (Matthysse *et al.*, 1981). Additional bacteria are entrapped in this network of fibrils, and this eventually results in the formation of large clusters of bacteria at or in the vicinity of the plant cell wall. Agrobacteria then transform "conditioned" plant cells into tumor cells by introducing into them a piece of genetic information which in established plant tumor cells can be recovered as T-DNA integrated into the nuclear plant DNA. Until now the molecular mechanism of plant tumor induction has been largely unknown. It is known, however, that compounds that prevent tumor induction by leukemia viruses in mice have a profound inhibitory effect on tumor induction by *A. tumefaciens* (Galsky *et al.*, 1980). Plant tumor induction requires no more than 18 hours (Lipetz, 1966) and contains at least one step that is thermosensitive in certain plants (Riker, 1924). For instance, in tomato no tumor induction occurs at a temperature higher than 30°C, although agrobacteria and plants are not inhibited in growth at this temperature. After tumor induction has occurred the bacteria can be killed, but tumor formation continues even if the plants are incubated at more than 30°C (Braun, 1943).

The resulting plant tumor cells, crown gall cells, have been studied intensively, and some of their characteristics are now known. The most important characteristic, and probably the reason for their tumorous character, is that in tissue culture they can grow in the absence of phytohormones, which are essential for the growth (control) of normal plant cells (Braun, 1958). Grafting of tumor tissue onto a healthy plant is followed by renewed tumor formation. Crown gall tumor cells possess some specific antigenic sites on their cell walls (Galsky *et al.*, 1981), and the pectic portion of their cell walls is more severely methylated than that of normal cells (Rao *et al.*, 1982). Moreover, it has been demonstrated that tumor cells have an enhanced permeability for certain inorganic ions as compared to normal cells (Wood and Braun, 1965). Tumor cells characteristically contain a set of unusual amino acid derivatives (opines) that have never been found in normal plant cells (recently reviewed by Tempé and Goldmann, 1982). Crown gall cells carry a piece of DNA that is homologous to a part of *A. tumefaciens* DNA integrated into their nuclear DNA (e.g., Chilton *et al.*, 1977; Thomashow *et al.*, 1980a; Lemmers *et al.*, 1980; Ooms *et al.*, 1982b). The T-DNA is responsible for the phytohormone-independent growth of tumor cells (Ooms *et al.*, 1981) and encodes enzymes that synthesize opines in the tumor cells (Schröder *et al.*, 1981a,b; Murai and Kemp, 1982b; Schröder and Schröder, 1982). Opine-synthesizing activity can be detected as soon as 29–36 hours after inoculation of the plants with the bacteria (Douglas *et al.*, 1979; Otten, 1982); it is not clear whether or not integration of T-DNA into host DNA has already occurred at that time. Hairy root is less well characterized than crown gall (Fig. 2). However, it seems so far that the mechanism of hairy root induction is very similar to crown gall induction. Both processes require attachment of the inducing bacteria to plant cell walls (Lippincott and Lippincott, 1978b), and both have a thermosensitive step in certain plants (Beiderbeck, 1973a). A mixed infection of plants with an *A. tumefaciens* and an *A. rhizogenes* strain results in the formation of crown galls as well as hairy roots (Beiderbeck, 1973b). The addition of kinetin

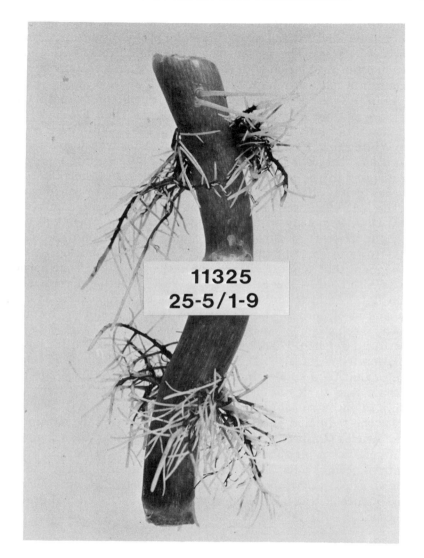

FIG. 2. Hairy roots induced by *Agrobacterium rhizogenes* on a *Kalanchöe* stem segment.

(a cytokinin) during hairy root induction by *A. rhizogenes* suppresses root formation, and instead more tumorlike proliferations are formed (Beiderbeck, 1973a). The reason why some plants react with tumor formation after inoculation with *A. rhizogenes,* therefore, may be that they contain a naturally high cytokinin level, at least at the wound sites. As in the case of crown gall induction, during hairy root induction a piece of DNA homologous to *A. rhizogenes* DNA becomes integrated into the plant DNA (Chilton *et al.,* 1982; Willmitzer *et al.,* 1982a; White *et al.,* 1982; Spano *et al.,* 1982). After transformation by *A. rhizogenes,* the cells of some plant species differentiate into roots that can grow in the absence of phytohormones. In other plants *A. rhizogenes* incites an amorphous callus that is phytohormone independent. The T-DNA present in hairy root cells bears no (or only very limited) DNA homology with the T-DNA present in crown gall cells (White and Nester, 1980b; Risuleo *et al.,* 1982; Willmitzer *et al.,* 1982a). However, part of the hairy root T-DNA is homologous to a DNA segment that is naturally present in the genome of some normal plants (White *et al.,* 1982; Spano *et al.,* 1982). It is therefore possible that the T-DNAs introduced into plant cells by agrobacteria at least partly consist of normal plant sequences that in evolution, by some unknown mechanism of gene transfer, have become linked to the agrobacterial genome. If such normal plant genes carried important information for regulation of the growth cycle of the plant, it can be envisaged that their sudden introduction into plant cells via agrobacteria could lead to deviations from the normal growth pattern and therefore to uncontrolled growth. A similar mechanism has been proposed for the induction of animal and human cancers by certain oncogenic viruses (Dulbecco, 1979), some of which have been found to carry a gene for a phosphorylating enzyme that is present in small amounts in normal cells. Transformation of an animal cell by the oncogenic virus possibly leads to a higher production of the enzyme, and in consequence of this possibility deviations from the normal growth cycle occur.

IV. Tumor-Inducing and Root-Inducing Plasmids

A. DETECTION AND ISOLATION OF PLASMIDS IN *Agrobacterium* SPECIES

Besides one large replicon, the chromosome, a number of smaller, independently replicating DNA molecules, plasmids, or extrachromosomal elements may be present in bacteria. Physical and genetic techniques have been developed for the detection of plasmids. Analytical procedures are now available that allow for the rapid screening of large numbers of bacterial strains for small (Telford *et al.*, 1977; Barnes, 1977), large (Hansen and Olsen, 1978; Casse *et al.*, 1979; Kado and Liu, 1981), and extremely large plasmids (Eckhardt, 1978; Rosenberg *et al.*, 1982). Also, preparative methods have been developed for the isolation of small and large plasmids (Clewell and Helinski, 1969; Currier and Nester, 1976a; Ledeboer *et al.*, 1976; Schwinghamer, 1980; Prakash *et al.*, 1981).

Many *Agrobacterium* and *Rhizobium* strains have been analyzed for their plasmid content. Zaenen *et al.* (1974) and Nuti *et al.* (1977) were the first to detect large plasmids in agrobacteria and rhizobia, respectively. Initially, large plasmids were demonstrated only in tumorigenic *A. tumefaciens* strains and not in avirulent strains, but later research has shown that nontumorigenic strains also contain one or more large plasmids (Lippincott *et al.*, 1977; Merlo and Nester, 1977; Sciaky *et al.*, 1978; Sheikholeslam *et al.*, 1979; Casse *et al.*, 1979). Plasmids of a size smaller than 100 Md are uncommon in agrobacteria, and small, high-copy-number plasmids have never been found. As in *A. tumefaciens* strains, large plasmids have been detected in *A. rhizogenes* strains (Currier and Nester, 1976b; White and Nester, 1980a; Costantino *et al.*, 1981).

B. ROLE OF PLASMIDS IN TUMOR AND ROOT INDUCTION

In some *Agrobacterium* strains the virulence trait is an unstable characteristic, oncogenicity being lost during growth at temperatures higher than 29°C, the temperature at which agrobacteria are usually cultured in the laboratory (Hamilton and Fall, 1971). When plants were inoculated by a mixture of such an avirulent strain and a virulent one, and agrobacteria were recovered from the resulting tumor, it was found that some of the avirulent agrobacteria had been converted into pathogenic strains (Kerr, 1969, 1971). In the same manner, the ability to induce hairy root was transferred from *A. rhizogenes* to avirulent *A. tumefaciens* strains (Albinger and Beiderbeck, 1977) and the ability to induce crown gall was transferred from *A. tumefaciens* to an avirulent *A. rhizogenes* strain (Cölak and Beiderbeck, 1979). The above data strongly suggested a predominant role of plasmid genes in crown gall and hairy root induction, but definite proof that this is the case only came when the plasmid profiles of virulent strains and their avirulent derivatives were compared. In summary, the results showed that in *A.*

TABLE 2

Properties of Strains Carrying or Lacking an Octopine Ti-Plasmid

	Agrobacterium tumefaciens		Rhizobium trifolii	
	LBA 202	LBA 661	LPR 5002	LPR 5011
Presence of Ti-plasmid	−	+	−	+
Oncogenicity	−	+	−	+
LpDH activity in tumors		+		+
Utilization of octopine	−	+	−	+
Sensitivity to				
Phages LPB 1, LPB 51	−	−	+	+
Phage LPB 84	+	+	−	−
Ketolactose production	+	+	−	−
Growth in the presence of				
2% NaCl	+	+	−	−
Growth on erythritol as a				
carbon source	−	−	+	+
Nodulation of clovers	−	−	+	+
Presence of Sym-plasmid	−	−	+	+

tumefaciens strains a plasmid of 120–150 Md determines the on-cogenicity trait (Van Larebeke *et al.*, 1974, 1975; Watson *et al.*, 1975) whereas in *A. rhizogenes* a plasmid of a similar size is responsible for hairy root induction (Moore *et al.*, 1979; White and Nester, 1980a; Chilton *et al.*, 1982; Hooykaas *et al.*, 1982b) (Table 2). These plasmids have therefore been called tumor-inducing (Ti-) plasmids and root-inducing (Ri-) plasmids, respectively. In order to unequivocally demon-strate the role of Ti- and Ri-plasmids in tumor and root induction, these plasmids have been transferred into related bacterial species. It was found that *Rhizobium trifolii* strains became tumorigenic upon introduction of a Ti-plasmid, whereas after receiving a Ri-plasmid they became rhizogenic (Hooykaas *et al.*, 1977; Hooykaas, 1979, un-published). However, introduction of Ti-plasmids into *R. meliloti* strains did not convert these into oncogenic bacteria, indicating either that the plasmid is not properly expressed in these strains, or that some virulence functions are chromosomally determined in *A. tumefa-ciens* and that these are present in *R. trifolii* but absent in *R. meliloti* (Hooykaas, 1979, unpublished). In *Escherichia coli* no expression of Ti-plasmid-determined functions was observed (Holsters *et al.*, 1978a). Recently, a large number of transposon insertion mutants of a virulent

A. *tumefaciens* strain have been obtained. Although most of the insertions that convert the strain into an avirulent one were localized in the Ti-plasmid, some were found to be located in the chromosome, underscoring the role of chromosomally determined functions in tumor induction (Garfinkel and Nester, 1980).

C. FUNCTIONS DETERMINED BY TI- AND RI-PLASMIDS

Because plasmid genes play a role in crown gall and hairy root induction by agrobacteria, it is plausible to ask whether these plasmids determine the ability of agrobacteria to attach to plant cell walls. A number of reports have been published stating that Ti-plasmids determine the ability of their host *Agrobacterium* strain to attach to plant cell walls (Matthysse *et al.*, 1978; Whatley *et al.*, 1978), but what role chromosomal genes play has not become clear. In this respect it is interesting to note that *R. meliloti* strains do not attach to plant cell walls, even if they carry a Ti-plasmid (F. Krens, personal communication).

Agrobacteria and rhizobia produce plant hormones viz. indoleacetic acid (IAA) and some cytokinins (Clark, 1974; Kaiss-Chapman and Morris, 1977). It is not yet known whether the production of these compounds plays a role in the plant–microorganism interaction. Recently it was found that besides chromosomal genes, Ti-plasmid genes play a role in the production of the above-mentioned phytohormones. One specific Ti-plasmid (pTiC58) was found to contain a locus for the production of *trans*-zeatin (Regier and Morris, 1982) as well as a locus for the production of IAA (Liu and Kado, 1979; Liu *et al.*, 1982). Mutants in which the locus for IAA production was abolished turned out to be avirulent, indicating a role of the production of IAA in tumor induction (Liu *et al.*, 1982).

Ti- and Ri-plasmids determine certain of the characteristics of the induced tumors, such as tumor morphology or the opine content of the tumors. During studies on the nitrogen metabolism of tumor cells as compared to that of normal cells, the group headed by Morel in France found certain tumor-specific compounds—"opines" (Fig. 3). The first opine to be identified was lysopine [N^2-(D-1-carboxyethyl)-L-lysine] (Biemann *et al.*, 1960). In a similar tumor tissue, octopine [N^2-(D-1-carboxyethyl)-L-arginine], octopinic acid or ornopine [N^2-(D-1-carboxyethyl)-L-ornithine], and histopine [N^2-(D-1-carboxyethyl)-L-histidine] were subsequently discovered (Ménagé and Morel, 1964, 1965; Kemp, 1977). All these compounds are synthesized by a specific enzyme,

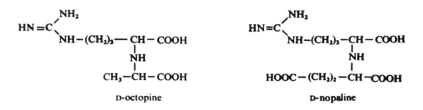

FIG. 3. Structural formulas of octopine and nopaline.

which has been called lysopine dehydrogenase (LpDH) or octopine synthase (Goldmann, 1977; Hack and Kemp, 1977; Otten *et al.*, 1977). The substrate range of LpDH is wide, making it likely that other, so far undetected, opines are present in tumor cells (Otten *et al.*, 1977). More recently two opines were detected that structurally differ greatly from the above-mentioned opines (Firmin and Fenwick, 1978; Tempé *et al.*, 1980; Tempé and Goldmann, 1982). The first of these, agropine, is presumed to be formed by lactonization of the second compound, called mannopine [N^2(1-D-mannityl)-L-glutamine]. In tumors induced by other bacterial strains none of the above-mentioned compounds is present, but other compounds, such as nopaline [N^2-(1,3-dicarboxypropyl)-L-arginine] and nopalinic acid or ornaline [N^2-(1,3-dicarboxypropyl)-L-ornithine] have been detected (Goldmann *et al.*, 1969; Firmin and Fenwick, 1977). These compounds are synthesized by a tumor-specific enzyme called nopaline dehydrogenase (NpDH) (Goldmann, 1977; Sutton *et al.*, 1978; Otten, 1979). Recently, two more tumor-specific compounds were found in nopaline tissues, viz. agrocinopines A and B. The chemical formulas of these compounds are unknown, but they are phosphorylated sugar derivatives (Ellis and Murphy, 1981). Besides "octopine" and "nopaline" tumors, a third class of tumors is known. Initially these were called "null-type" tumors because neither octopine nor nopaline were detected in their cells. Now we know that they contain agropine (Guyon *et al.*, 1980) and two phosphorylated sugar derivatives called agrocinopines C and D (Ellis and Murphy, 1981). In roots induced by an *A. rhizogenes* strain, agropine (Tepfer and Tempé, 1981) and agrocinopines (Tempé, cited in Willmitzer *et al.*, 1982a) have been detected.

Virulent *Agrobacterium* strains are able to degrade opines and use them as sources of nitrogen, carbon, or energy. However, strains that induce octopine tumors can only catabolize the opines found in the tumors they induce, viz. octopine, octopinic acid, lysopine, histopine, mannopine, and agropine (Petit *et al.*, 1970; Kemp, 1977; Firmin and Fenwick, 1978). Likewise, nopaline strains can only feed on opines found in nopaline tumors, null-type strains only on opines found in null-type tumors, and *A. rhizogenes* strains only on opines found in the roots they induce. The catabolic functions are encoded by the Ti- and Ri-plasmids. Strains cured of their Ti-plasmid no longer are able to consume opines, and (re)introduction of a Ti- or Ri-plasmid not only restores the capacity to induce tumors, but also the ability to consume opines (Bomhoff *et al.*, 1976; Kerr and Roberts, 1976; Hooykaas *et al.*, 1977; Montoya *et al.*, 1977; Guyon *et al.*, 1980). If a Ti-plasmid from an octopine strain is introduced into a derivative of a Ti-plasmid-cured nopaline strain, the resulting transconjugants are octopine strains able to induce octopine tumors and catabolize the opines present in octopine tumors. Vice versa, if a Ti-plasmid from a nopaline strain is introduced into a Ti-plasmid-free octopine strain, the transconjugants are nopaline strains able to induce nopaline tumors and catabolize the opines present in nopaline tumors. Ti- and Ri-plasmids can therefore be considered as special types of catabolic plasmids, namely, ones that transform plant cells and direct these to produce compounds that their bacterial host can catabolize via plasmid-determined enzymes. The role of chromosomal genes in octopine utilization is not clear. Spontaneous mutants derived from Ti-plasmid-free strains can degrade octopine (Montoya *et al.*, 1978). Besides virulent agrobacteria, some avirulent agrobacteria have been described that are able to consume nopaline or octopine (Merlo and Nester, 1977) and an octopine- and nopaline-catabolizing nonfluorescent *Pseudomonas* strain has also been isolated from soil (Hooykaas, unpublished) (Table 3).

Bacteriocins are produced by some *Agrobacterium* strains (Stonier, 1960; Kerr and Htay, 1974; Kerr and Panogopoulos, 1977; Cooksey and Moore, 1980a). The best known of these is a bacteriocin called agrocin 84, which is produced by the avirulent nopaline strain K84 and by the virulent null-type strains 396 and 542 (Kerr, 1972; Vervliet *et al.*, 1975; Slota and Farrand, 1982). The bacteriocin is produced via genes located on a 25–30-Md plasmid present in the above three strains (Ellis *et al.*, 1979; Cooksey and Moore, 1980b; Slota and Farrand, 1982). Strains lacking a Ti-plasmid and strains with an octopine Ti-plasmid are resistant to agrocin 84, but strains with a nopaline Ti-plasmid are sensitive to it (Engler *et al.*, 1975; Watson *et al.*, 1975; Kerr and Roberts, 1976). Agrocin-resistant mutants of nopaline strains either had lost the entire Ti-plasmid, or had large deletions or

TABLE 3

Properties Determined by Various Types of *Agrobacterium* Plasmids

Plasmids originating from strain	Characters conferred upon hosts by the plasmids					
	Utilization of				AP1 exclusion	Agrocin 84 sensitivity
	Octopine	Nopaline	Agropine	Oncogenicity		
4, 15955, B6, Ach5	+	−	+	+	+	−
AG60, AG67	+	−	ND[a]	−	−	−
K108, 1651, C58	−	+	−	+	+	+
T37, EIII9.6.1	−	+	−	+	−	+
K14	−	+	ND	−	−	+

[a]ND, Not determined.

small rearrangements in the Ti-plasmid (Engler *et al.,* 1975; Liu and Kado, 1977; Süle and Kado, 1980; Cooksey and Moore, 1982). The reason for the sensitivity of nopaline strains to agrocin 84 is that on the nopaline Ti-plasmid a gene is present that determines an uptake system for agrocin 84 (Murphy and Roberts, 1979; Murphy *et al.,* 1981). The uptake system is meant for the transport of agrocinopines A and B into the cell, and agrocin 84 probably enters because it has some structural analogy with these agrocinopines (Tate *et al.,* 1979; Ellis and Murphy, 1981; Murphy *et al.,* 1981). The toxicity of agrocin 84 is probably due to the fact that it acts as a fraudulent nucleotide, and interferes with DNA synthesis (Murphy *et al.,* 1981). Agrocin 84 and strain K84 have been used extensively for the biological control of crown gall in many parts of the world (recently reviewed by Kerr, 1980; Moore and Cooksey, 1981).

Agrobacterium strains with a Ri-plasmid induce rooting on *Kalanchöe* and tobacco, and strains with a Ti-plasmid give tumor formation on these plants. The appearance of the tumors that are induced may vary, and is dependent on the bacterium that is used for inoculation (Braun, 1948, 1951; Hooykaas *et al.,* 1977; Gresshoff *et al.,* 1979; Hooykaas, 1979). Braun (1948, 1951) reported that octopine strain B2 induces unorganized tumors on *Kalanchöe* and tobacco, but that nopaline strain T37 forms teratomas, tumors from which leaf structures and shoots appear. It was later noted that strains with an octopine or a null-type Ti-plasmid induce unorganized tumors with a "rough" surface on *Kalanchöe,* whereas strains with a nopaline Ti-plasmid induce tumors with a "smooth" surface from which leaflike structures may appear, depending on the site of inoculation and the specific nopaline strain used (Hooykaas *et al.*, 1977; Hooykaas, 1979). The tumors induced by octopine and nopaline strains are surrounded by adventitious roots, whereas those of null-type strains almost lack such surrounding roots. Gresshoff *et al.* (1979) reported that teratoma formation on *Nicotiana* species is controlled by the plant species or variety used and by the plasmid content of the bacterial strain used for tumor induction. On tomato, different *Agrobacterium* strains induce tumors of a similar appearance and size; however, those induced by null-type strains turn black after some time (Hooykaas, unpublished). Some *Agrobacterium* strains have a restricted host range, inducing tumors only on grape and a few other plant species (Panagopoulos and Psallidas, 1973). It has been found that host range is determined by the Ti-plasmid in these strains (Loper and Kado, 1979; Thomashow *et al.*, 1980c).

A number of different bacteriophages have been isolated that are infectious for *Agrobacterium* strains. One particular phage, called AP1, was isolated by the group headed by Schell and Van Montagu (Ghent, Belgium). This phage has a wide host range among biotype 1 *Agrobacterium* strains (Hooykaas, 1979). Phage AP1 was found to form plaques on Ti-plasmid-free strains and strains with a null-type Ti-plasmid or Ri-plasmid, but not on strains with an octopine Ti-plasmid or (with some exceptions) a nopaline Ti-plasmid (Hooykaas *et al.*, 1977; Van Larebeke *et al.*, 1977; Hooykaas, 1979). The octopine and nopaline Ti-plasmids probably interfere with phage multiplication, and this phenomenon has been called phage exclusion. More recently it has been demonstrated that octopine Ti-plasmids inhibit the intracellular development of another phage, called ψ, that has a restricted host range among *Agrobacterium* strains (Expert and Tourneur, 1982; Expert *et al.*, 1982).

V. Genetic Systems for A. *tumefaciens*

A. Introduction

Genetic techniques developed mainly for *E. coli* have been modified to be applicable for *Agrobacterium*. At the moment, methods are available for the mutagenesis of *A. tumefaciens,* the enrichment and isolation of mutants, and the exchange of genetic material between different agrobacterial strains and between *A. tumefaciens* and other bacteria, such as those belonging to the bacterial families of Entero-bacteriaceae (*E. coli, Klebsiella pneumoniae,* and *Salmonella typhimurium*), Rhizobiaceae (*R. trifolii, R. meliloti, R. leguminosarum, R. phaseoli,* and *R. japonicum*), and Pseudomonadaceae (*Pseudomonas aeruginosa*).

B. MUTAGENESIS OF *A. tumefaciens* AND ISOLATION OF MUTANTS

Procedures have been described for the mutagenesis of *A. tumefaciens* with the chemical mutagens ethyl methane sulfonate (EMS) (Klapwijk *et al.,* 1976) or *N*-methyl-*N'*-nitro-*N*-nitrosoguanidine (NTG) (Langley and Kado, 1972; Klapwijk *et al.,* 1975), with ultraviolet irradiation (Hooykaas *et al.,* 1979b), or with transposable elements (see Section V,C). After such mutagenic treatments, auxotrophic, antibiotic-resistant, sugar fermentation-deficient, recombination-deficient, opine utilization-negative, agrocin-resistant, and nontumorigenic mutants have been isolated. It has recently been found that *Agrobacterium* phage ψ induces mutations during lysogenization. About 1–3% of the ψ lysogens have (leaky) auxotrophic mutations (Expert and Tourneur, 1982). A technique has been developed for the enrichment of specific *Agrobacterium* mutants (Klapwijk *et al.,* 1975). The following auxotrophs have been obtained from *A. tumefaciens* strain C58: *cys, glu, his, ilv, leu, met, pur, pyr, ser, thr, trp,* and *tyr* (Hooykaas *et al.,* 1982c, unpublished; Bryan *et al.,* 1982). Among the mutants unable to utilize certain sugars as a carbon source were strains unable to utilize mannitol (*mtl*), galactose (*gal*), and arabinose (*ara*) (Hooykaas and Klapwijk, unpublished). Recombination-deficient derivatives of strains C58 (Molbas, unpublished) and Ach5 (Klapwijk *et al.,* 1979) have been isolated after NTG treatment. They were discovered among isolates selected for their enhanced UV and methyl methane sulfonate (MMS) sensitivity.

C. TRANSFORMATION, TRANSDUCTION, AND CONJUGATION IN *A. tumefaciens*

In 1972 transfection of a competent *A. tumefaciens* strain with a temperate phage was reported (Milani and Heberlein, 1972), and later the freeze–thaw technique initially developed for the transfection of *E. coli* was modified for the transfection of *A. tumefaciens* with a temperate phage and for transformation with plasmid DNA (Holsters *et al.,* 1978c). Unfortunately, the method is inefficient and gives only about

1000 transformants per μg DNA, making it unsuitable for most cloning experiments. Moreover, some strains, such as Ach5, have so far remained refractory to transformation (B. P. Koekman, personal communication).

Lysogenic phages have been described for *A. tumefaciens* (Beardsley, 1960; Zimmerer *et al.,* 1966; Vervliet *et al.,* 1975), but until now no transduction system with these phages has been described. The *E. coli* phage Mu has been introduced into *A. tumefaciens* via conjugation in cointegrate structures with R plasmids (Boucher *et al.,* 1977; Van Vliet *et al.,* 1978; Hooykaas, 1979; Murooka *et al.,* 1981). Phage Mu production can be induced with low efficiency in the new host, but this system has not yet been used for the transfer of genetic material between agrobacteria. *A. tumefaciens* is resistant to the *E. coli* general transducing phage P1, but recently *A. tumefaciens* mutants sensitive to P1 have been isolated (Murooka and Harada, 1979). With such strains P1 might be used in transduction experiments.

Many of the large plasmids that are naturally present in *A. tumefaciens* are self-transmissible. Although Ti-plasmids are conjugative only under specific circumstances resembling those that occur in plant tumors (see Section VI,B), the very large 300-Md cryptic plasmids in *A. tumefaciens* (e.g., plasmid pAtC58) have also been found to be self-transmissible *in vitro* (Hooykaas and Den Dulk-Ras, unpublished). Thus, conjugation between different *A. tumefaciens* strains can be brought about by plasmids that naturally occur in *A. tumefaciens*.

D. Introduction of Plasmids into *A. tumefaciens* from Other Bacterial Species and Their Expression

In order to develop efficient systems for gene transfer in *A. tumefaciens,* different types of plasmids have been examined for their transfer to and expression in *A. tumefaciens*. Bacterial plasmids with a limited host range, such as those of *inc* group F, could not be introduced into and be maintained in *A. tumefaciens*. Certain other plasmids, such as those of the *inc* groups N and Iα, could be transferred to agrobacteria by conjugation, but could not be stably maintained in the new host (Leemans *et al.,* 1981; J. Leemans, personal communication). Wide host range R plasmids, however, could be introduced into and be stably maintained in *A. tumefaciens*. Among these were the *inc* C plasmid R55 (Hernalsteens *et al.,* 1977), the *inc* P plasmids RP4, R702, R772, R751:Tn5, Rp1.pMG1, and pUZ8 (Datta and Hedges, 1972; Hooykaas, 1979; Hooykaas *et al.,* 1980a, unpublished), the nonconjugative *inc* Q plasmids pKT212 and pKT214 (Hille and Schilperoort, 1981a), and the *inc* W plasmid Sa (Hernalsteens *et al.*, 1977; Farrand *et al.*, 1981). These R plasmids can be introduced into *A. tumefaciens* from various hosts; the conjugative plasmids among them express their transfer genes in *A. tumefaciens* and, therefore, can bring about their own transfer to other agrobacterial hosts and to other bacterial species. *A. tumefaciens* strains carrying an *inc* P group R plasmid are sensitive to phage U5, which also lyses *E. coli* and *P. aeruginosa* strains with an *inc* P plasmid (Van Larebeke *et al.*, 1977; Hooykaas, 1979). This phage uses the sex pili determined by the *inc* P plasmid transfer genes as the attachment site. It can therefore be concluded that similar pili are formed in *A. tumefaciens* and other bacterial species, and that these provide a phage attachment site in all these bacteria. Plasmids of incompatibility group P have been studied thoroughly, and the regions that are necessary for replication in *E. coli* have been identified (Thomas, 1981). These plasmids carry regions that are necessary for replication in certain hosts, but not in others (Thomas *et al.*, 1982). A deletion of a specific segment of the plasmid makes it unstable in *Methylophilus methylotrophus* and *A. tumefaciens,* but not in *E. coli* (Windass *et al.*, 1980; Hooykaas *et al.*, 1982e). The antibiotic resistance markers determined by the R plasmids are expressed in *A. tumefaciens,* albeit the resistance level of R^+ agrobacteria may differ from that of R^+ *E. coli* strains. Expression in *A. tumefaciens* has been obtained for Cb^r (not in some strains), Cm^r, Gm^r, Hg^r, Km^r, Sm^r, Sp^r, and Tc^r determinants. Trimethoprim resistance (Tp^r) determined by *inc* P plasmid R751 could not be tested, because the strain investigated (C58) appeared to be naturally resistant to high levels of trimethoprim (Hooykaas, unpublished).

On many of the R plasmids introduced into *A. tumefaciens* transposable elements were present. In this way the insertion sequences IS8, IS21, and IS70, and the transposons Tn1 (Cb^r), Tn3 (Cb^r), Tn5 (Km^r), Tn7 ($Sm^rSp^rTp^r$), Tn9 (Cm^r), Tn10 (Tc^r), Tn402 (Tp^r), Tn732 (Gm^r), Tn904 (Sm^r), Tn1771 (Tc^r), and Tn1831 ($Sm^rSp^rSu^rHg^r$) have been introduced into *A. tumefaciens*. Transposition of IS8 (De Picker *et al.*, 1980a), IS70 (Hooykaas *et al.*, 1980a; Hille *et al.*, 1983), Tn1 (Hooykaas *et al.*, 1979b; Holsters *et al.*, 1980), Tn5 (Hooykaas, 1979; Garfinkel and Nester, 1980), Tn7 (Holsters *et al.*, 1980), Tn904 (Klapwijk *et al.*, 1980), and Tn1831 (Hooykaas *et al.*, 1980a) has been demonstrated to occur in *A. tumefaciens,* indicating the expression of the transposition functions encoded by these elements in *A. tumefaciens*.

Some plasmids present in *Rhizobium* species have been found to be transferable to *A. tumefaciens*. Besides an R plasmid of unknown *inc* group from *R. japonicum* (Cole and Elkan, 1973), plasmids determining the nitrogen-fixing symbiotic association between rhizobia and leguminous plants (Sym-plasmids) have been introduced into *Agrobacterium* strains (Hooykaas *et al.*, 1981, 1982d; Hooykaas and Den Dulk-Ras, unpublished) (Table 4). The Sym-plasmids were somewhat unstable in *A. tumefaciens*, but they were expressed in the new host. The Sym-plasmid from *R. trifolii* gave *A. tumefaciens* and *A. rhizogenes* the ability to induce root nodules on clover species. Whereas sparse nodulation occurred on red clover (*Trifolium pratense*), abundant nodulation occurred on one variety of subterranean clover (*Trifolium subterraneum*). *R. leguminosarum* Sym-plasmids enabled *A. tumefaciens* to nodulate some species of the pea vetch cross-inoculation group. The common vetch (*Vicia sativa*) was a better host than other species of that cross-inoculation group (Hooykaas *et al.*, 1982d; Van Brussel *et al.*, 1982). The *R. phaseoli* Sym-plasmid provided *A. tumefaciens* with the capacity to form root nodules on beans (*Phaseolus vulgaris*) (Hooykaas, Van Brussel, and Den Dulk-Ras, unpublished). No nitrogen was fixed in the root nodules induced by agrobacteria with Sym-plasmids (Hooykaas *et al.*, 1981, 1982d; Van Brussel *et al.*, 1982), although the structural *nif*-genes were present on the introduced Sym-plasmids (Prakash *et al.*, 1981; Hooykaas, unpublished). This shows either that

TABLE 4
Properties of Strains Carrying or Lacking Sym-Plasmid pSym5

	Rhizobium trifolii		*Agrobacterium tumefaciens*	
	LPR 5039	LPR 5035	LBA 953	LBA 2703
Presence of Sym-plasmid	−	+	−	+
Nodulation of clovers	−	+	−	±
N fixation in nodules		+		−
Sensitivity to				
Phages LPB1, LPB51	+	+	−	−
Phage LPB84	−	−	+	+
Growth in the presence of				
2% NaCl	−	−	+	+
Growth on erythritol as a				
carbon source	+	+	−	−
Presence of Ti-plasmid	−	−	+	+
Oncogenicity	−	−	+	+
Nopaline utilization	−	−	+	+

a function necessary for nitrogen-fixation is encoded by the *Rhizobium* chromosome and is absent in the *A. tumefaciens* chromosome, or that the nitrogen-fixation genes on the Sym-plasmids are not properly expressed in agrobacteria. A cryptic plasmid from *R. phaseoli* turned out to be extremely unstable in *A. tumefaciens,* whereas it was stable in the original *R. phaseoli* host. This suggests that plasmid replicators suitable for *Rhizobium* species are not necessarily appropriate for *Agrobacterium* species.

The R plasmids and Sym-plasmids introduced into *A. tumefaciens* so far have all been found to be compatible with Ti- and Ri-plasmids. With the exception of R plasmid Sa, plasmids introduced into *A. tumefaciens* did not interfere with tumor induction. Plasmid Sa inhibits tumor induction by agrobacteria (Farrand *et al.*, 1981), and the responsible plasmid segment has been accurately mapped (Tait *et al.*, 1982).

E. A Chromosomal Linkage Map of *A. tumefaciens*

In an early report, recombination between R^- auxotrophs of *A. tumefaciens* strain 37400 was mentioned (Bezděk *et al.*, 1976). For other

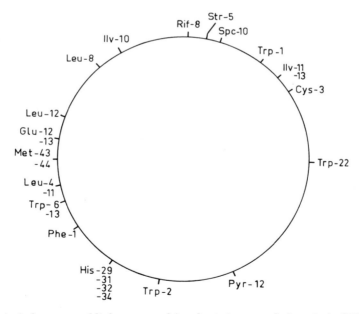

Fig. 4. A chromosomal linkage map of *Agrobacterium tumefaciens* strain C58. (Taken from Hooykaas *et al.*, 1982c.)

strains, such as 15955, C58, and B6, no recombination was found in the absence of an R plasmid. Different R plasmids were introduced into them to see whether they could mediate chromosome mobilization. R plasmid R68.45 was found to be the most efficient at mobilizing the chromosomes of strains 15955 (Hamada *et al.*, 1979) and C58 (Fig. 4) (Hooykaas, 1979; Bryan *et al.*, 1982; Hooykaas *et al.*, 1982c). The circular linkage map shown in Fig. 2 was obtained by performing 3- and 4-factor crosses for *A. tumefaciens* strain C58 (Hooykaas *et al.*, 1982c). Preliminary comparative studies indicate that the *A. tumefaciens* map is similar to the maps of *R. meliloti* and *R. leguminosarum*. R primes with parts of the *R. meliloti* chromosome have been found to be capable of complementing some *A. tumefaciens* auxotrophs in a Rec$^+$ as well as in a Rec$^-$ background (Hooykaas *et al.*, 1982c, unpublished). Thus, *R. meliloti* chromosomal genes can be functional in *A. tumefaciens*.

VI. The Genetic Map of Ti-Plasmids

A. OCTOPINE AND NOPALINE UTILIZATION

As the octopine and nopaline utilization traits were the first markers for the octopine and nopaline Ti-plasmids, respectively, research has been done to obtain insight into these functions and to develop them into selectable markers for these Ti-plasmids. For selection of Ti-containing strains, octopine or nopaline can be added to the medium as a nitrogen source or as a carbon and energy source. If solid medium is required, the agar added must be washed extensively because agar contains impurities that can be utilized by agrobacteria as a nitrogen source. Alternatively, an indicator medium specifically developed to distinguish octopine (nopaline)-utilizing strains from nonutilizing strains may be used (Hooykaas *et al.*, 1979b). Two enzymes are involved in the catabolism of octopine and nopaline, a specific permease and a membrane-bound oxidase (for recent reviews see Klapwijk and Schilperoort, 1982; Tempé and Petit, 1982). The genes for these two proteins are encoded by the Ti-plasmid (Klapwijk *et al.*, 1977, 1978), and are coordinately controlled by a repressor (Klapwijk *et al.*, 1978; Petit and Tempé, 1978). Assays for permease activity (Klapwijk *et al.*, 1977; Petit and Tempé, 1978) and for oxidase activity (Klapwijk *et al.*, 1978) have been developed. The octopine system can be induced by octopine, lysopine, ornopine, and *meso*-alanopine [N^2-(D-l-carboxyethyl)-L-alanine], but not by histopine, nopaline, noroctopine (N^2-1-carboxymethyl-L-arginine), strombine (N^2-1-carboxymethyl-L-alanine), or arginine (Klapwijk et al., 1977, 1978; Petit and Tempé, 1978; Tempé and Petit, 1982). After induction, octopine strains can degrade histopine, noroctopine, and strombine, but not nopaline. Nopaline is not a substrate for the octopine oxidase (Hooykaas et al., 1979b). Mutants able to degrade histopine or noroctopine without induction have been isolated. Such mutants were either constitutive for octopine catabolism, or had an altered repressor that was also sensitive to histopine and/or noroctopine (Klapwijk et al., 1978; Petit and Tempé, 1978). The nopaline system can be induced by nopaline, but not by octopine, lysopine, ornopine, noroctopine, or arginine. However, after induction by nopaline, nopaline strains can degrade octopine (Klapwijk et al., 1977; Petit and Tempé, 1978). Spontaneous mutants capable of catabolizing octopine in the absence of an inducer (nopaline) have been isolated from nopaline strains (Petit and Tempé, 1975; Hooykaas et al., 1977). These mutants were constitutive, or had a repressor with altered specificity (Petit and Tempé, 1975, 1978; Klapwijk et al., 1977). Besides octopine, the constitutive mutants can at least also utilize ornopine and lysopine (Hooykaas et al., 1979b). Competition experiments have shown that Ti-plasmids determine one transport system and one oxidase system, and that these are used for the catabolism of all the above-mentioned opines as long as they are substrates for the Ti-carrying bacterium (Klapwijk et al., 1977, 1978; Petit and Tempé, 1978). Recently it has been established that octopine and nopaline Ti-plasmids also encode a system for the degradation of arginine (Ellis et al., 1979). This system is coordinated by the same regulator that also controls the opine utilization genes.

Mutants have been isolated that have point mutations or small deletions in the octopine utilization (occ) genes (Klapwijk et al., 1976; Montoya et al., 1977). Such mutants were still tumorigenic, and they induced tumors in which octopine was synthesized, showing that the octopine utilization trait (occ) and the octopine synthesis trait (ocs) are governed by different genes.

B. Conjugation Mediated by Ti-Plasmids

Ti-plasmids and Ri-plasmids behave as conjugative plasmids in plant tumors (Kerr and Roberts, 1976; Albinger and Beiderbeck, 1977). In vitro, however, no transmissibility was initially observed, and therefore systems were developed for the mobilization of these plasmids by R plasmids (Chilton et al., 1976; Hooykaas et al., 1977, 1980a, 1982b). Then it was found that Ti-plasmids become conjugative if their host bacteria are incubated on specific media. For octopine Ti-plasmids no transfer was observed in incubations on rich media, but transfer was found if incubation took place on minimal medium plus octopine (Genetello *et al.,* 1977; Kerr *et al.,* 1977). Furthermore, preincubation of octopine bacteria with octopine, but not with pyruvate plus arginine, resulted in Ti-plasmid transfer (Petit *et al.,* 1978). An extensive investigation of the conditions under which octopine Ti-plasmids become transferable showed that no transfer occurred on minimal medium or rich medium, but conjugative ability was induced if octopine, lysopine, or ornopine had been included in the minimal medium (Hooykaas *et al.,* 1979b). Addition of noroctopine or nopaline did not result in plasmid transfer, and sulfur-containing amino acids such as methionine and cysteine inhibited transfer, even if octopine was also present in the medium (Hooykaas *et al.,* 1979b). The transfer system of octopine Ti-plasmids is inactivated above 32°C (Tempé *et al.,* 1977; Hooykaas *et al.,* 1979b).

As octopine, lysopine, and ornopine are inducers both of the Ti-plasmid *occ*-genes and of the *tra*-genes, these functions could be coordinately regulated. To investigate this possibility, mutant Ti-plasmids have been isolated that show transfer even in the absence of an inducer (Petit *et al.,* 1978; Hooykaas *et al.,* 1979b). These derepressed mutants appeared to be either constitutive or inducible for octopine catabolism (Klapwijk *et al.,* 1978; Hooykaas *et al.,* 1979b). On the other hand, some but not all mutants constitutive for octopine catabolism turned out to be derepressed for transfer (Klapwijk *et al.,* 1978; Petit *et al.,* 1978). Together, these data indicate that one regulator molecule controls two different operons, one with the *occ*-genes, the other with the *tra*-genes. Strains diploid for the *occ*- and *tra*-genes have been constructed in order to find out whether there is negative or positive control. The combination of a set of inducible genes with a set of constitutive genes in one cell gave an inducible phenotype, showing that the *occ*- and *tra*-genes are negatively controlled by a repressor (Klapwijk and Schilperoort, 1979). For many nopaline Ti-plasmids, and for one null-type plasmid, the situation has been found to be analogous to that for octopine Ti-plasmids. At least some nopaline Ti-plasmids are induced for transfer by agrocinopines A and B (Ellis *et al.,* 1982a); one transfer-constitutive mutant of a nopaline Ti-plasmid was isolated, and found to be constitutive also for agrocinopine catabolism (Ellis *et al.,* 1982b), suggesting a coordinated regulation of the *tra*-genes and the agrocinopine catabolism genes. For null-type Ti-plasmid pTi542, transfer is induced by agrocinopines C and D (Ellis *et al.,* 1982a).

To study whether nopaline Ti-plasmids carry a repressor gene that can substitute for the gene present in octopine Ti-plasmids, a strain was constructed carrying a cointegrate of a wild-type nopaline Ti-plasmid and an Occ^c Tra^c octopine plasmid. This resulted in a plasmid with an *Occ^c Tra^c* phenotype, showing that the regulator molecules determined by the nopaline Ti-plasmid are not able to repress the *occ-* and *tra*-genes of the octopine Ti-plasmid (Hooykaas *et al.*, 1980b).

C. Isolation of Ti-Plasmid Transposon Insertion and Deletion Mutants

For the localization of genes, transposon insertion mutants and deletion mutants have been very useful, because their site of insertion can be determined accurately if physical maps are available of the mutated DNA segment and of the transposon.

A method has been developed for the isolation of random transposon insertions in *R. leguminosarum* that is also applicable for the isolation of transposon inserts in *A. tumefaciens* and some other organisms (Beringer and Beynon, 1978). The procedure consists of introducing into *A. tumefaciens* a "suicide" plasmid, which is a plasmid that cannot establish itself in *A. tumefaciens,* but from which transposons may survive in the new host by jumping from the suicide plasmid into a site in the genome before the plasmid is lost. Initially, *inc* P R plasmids with an insertion of *E. coli* bacteriophage Mu were used as suicide plasmids, but more recently it was found that *inc* N R plasmids can also be used (Leemans *et al.*, 1981). *inc* N plasmids do not establish themselves at all in *A. tumefaciens*. Intact R::Mu plasmids are but rarely maintained. More often stable or unstable derivatives with large deletions are obtained. With the aid of the suicide plasmid pJB4JI many transposon Tn5 insertion mutants of *A. tumefaciens* strains C58 (Hooykaas, 1979) and A6 (Garfinkel and Nester, 1980) have been isolated.

A number of methods have been developed specifically for the transposon mutagenesis of Ti-plasmids or of parts of them. The first procedure consists of transferring a cointegrate of Ti- and a wide host range R plasmid to an *E. coli* strain with a chromosomally located transposon insertion, and subsequently crossing back the R::Ti cointegrate into *A. tumefaciens* with selection for transfer of the transposon originally present in the *E. coli* chromosome (Hernalsteens *et al.*, 1978). Disadvantages of this method are that R::Ti cointegrates are sometimes unstable in *E. coli* (Holsters *et al.*, 1978a), and insertion of the R plasmid into the Ti-plasmid may affect virulence even if integration has taken place at a site that is not essential for tumor induction (Ooms and Hooykaas, unpublished). The second method is similar to the first one, but now transposon mutagenesis is performed in *A. tumefaciens*. A (transfer-constitutive) Ti-plasmid is introduced into an *A. tumefaciens* strain with a transposon located either in the chromosome or in a transfer-deficient R plasmid. The Ti-plasmid is transferred from this strain to another *A. tumefaciens* strain with selection for co-transfer of the transposon originally present in the chromosome or the Tra$^-$ plasmid in the *A. tumefaciens* donor strain (Holsters *et al.*, 1980; Klapwijk *et al.*, 1980; De Greve *et al.*, 1981). A disadvantage of this procedure is that for each transconjugant it has to be proved that transfer of the transposon was not due to cotransfer with the *A. tumefaciens* 300-Md cryptic plasmid, or to mobilization of the entire Tra$^-$ plasmid by the Ti-plasmid. The third method is based on the findings that per se Ti-plasmid transfer does not occur if the host cells are incubated on rich medium (Hooykaas *et al.*, 1979b), and that Ti-plasmid transfer can be accomplished under such conditions via mobilization by an R plasmid (Chilton *et al.*, 1976; Hooykaas *et al.*, 1977, 1980a; Van Larebeke *et al.*, 1977) (Fig. 5). Mobilization is due to replicon fusion between the Ti-plasmid and the R plasmid. These (transient) cointegrates are transposition intermediates in which R and Ti-plasmid are fused via two copies of a transposon or an insertion sequence that originally was present in the R plasmid (Hooykaas, 1979; Hooykaas *et al.*, 1980a). Resolution of the transient cointegrate in a Rec$^+$ or Rec$^-$ recipient results in the formation of an unaltered R plasmid and a Ti-plasmid now carrying a copy of a transposable element that originally was present in the R plasmid only. Transposons Tn904 (Sm^r) and Tn1831 (Hg^{r}Sm^{r}Sp^{r}Su^r) have appeared to be very efficient at jumping into octopine Ti-plasmids from R plasmids, thereby causing Ti-plasmid mobilization at the reasonably high frequency of 10^{-3} per transferred R plasmid (Hooykaas *et al.*, 1980a). By mechanisms similar to those described above, relatively stable and completely stable cointegrates between R and Ti-plasmids have been obtained (Holsters *et al.*, 1978b; Klapwijk *et al.*, 1978). In cointegrates, linkage between R and Ti-plasmid occurs through insertion sequences such as IS8 (De Picker *et al.*, 1980a) and IS70 (Hille *et al.*, 1983). In completely stable cointegrates one of the two copies of the linking insertion sequences is partly or completely deleted (Hille *et al.*, 1983).

Methods have been developed for the site- and region-directed mutagenesis of Ti-plasmids. These procedures make use of clones consisting

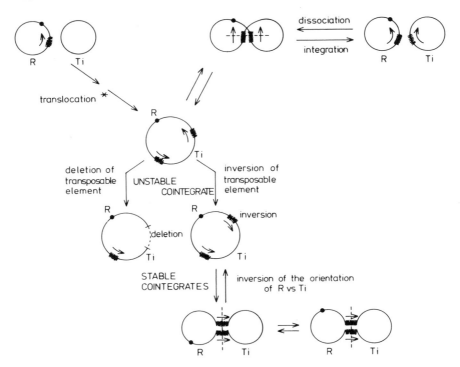

FIG. 5. Molecular mechanism of Ti-plasmid mobilization by R plasmids and the formation of stable R::Ti cointegrates. (Taken from Hooykaas *et al.,* 1980a.)

of an *E. coli* vector and the segment of the Ti-plasmid to be mutated. A
transposon insertion or an *in vitro* mutation is introduced into the Ti-
plasmid segment of the clone. This segment is then recloned in a wide
host range vector and is subsequently introduced into an *Agrobacterium* strain with the wild-type Ti-plasmid. In the final step recombinants are selected for in which the mutation has been transferred to
the wild-type Ti-plasmid (Leemans *et al.,* 1981, 1982; Matzke and
Chilton, 1981). Recently we found that recloning in a wide host range
vector is not necessary. Mutated Ti-clones derived from vectors such as
pBR322 can be introduced efficiently in Ti-carrying *Agrobacterium*
strains via mobilization by an R plasmid if plasmid ColK is present in
the *E. coli* donor strain. Plasmid ColK contains *mob*-genes, which recognize the *bom*-sequence of ColE1 derivatives (Young and Poulis,
1978). As plasmid ColE1 derivatives cannot replicate in *A. tumefa-*

ciens, they can survive in this new host only after recombination with a resident replicon. If a Ti-clone is used this survives by homologous recombination with the resident Ti-plasmid. A second recombination event can result in a mutation being introduced into the Ti-plasmid. Alternatively, a cointegrate of Ti with a wide host range R plasmid is introduced into an *E. coli* strain with the above-mentioned mutated clone, and directly in *E. coli* recombinants are sought in which the mutation in the Ti-plasmid segment of the clone is introduced into the R::Ti cointegrate (Hille *et al.,* 1983).

Besides transposon insertion mutants, deletion mutants of octopine and nopaline Ti-plasmids have also been isolated. From nopaline Ti-plasmids deletion mutants have been obtained after plating nopaline strains on plates with agrocin 84. Among the agrocin 84-resistant mutants that were isolated some turned out to have large deletions in their nopaline Ti-plasmid (Lin and Kado, 1977; Süle and Kado, 1980; Cooksey and Moore, 1982).

Homooctopine [N^2-(D-1-carboxyethyl)-L-homoarginine] is a toxic analog of octopine; its breakdown results in the production of L-homoarginine, which is toxic for cells (Petit and Tempé, 1976). As homooctopine is not an inducer of the *occ*-genes, strains with a wild-type octopine Ti-plasmid are insensitive to it. In contrast, mutants with constitutive *occ*-genes take up homooctopine and break it down into homoarginine and pyruvate, which results in growth inhibition (Petit and Tempé, 1976, 1978; Klapwijk *et al.,* 1977, 1978). Among the homooctopine-resistant mutants that were isolated from Occ^c octopine strains, some appeared to have large deletions in the Ti-plasmid (Koekman *et al.,* 1979; Ooms *et al.,* 1982a). Insertion of a transposable element into the Ti-plasmid prior to the isolation of agrocin 84-resistant or homooctopine-resistant mutants resulted in the isolation of deletion mutants, in which most frequently the deletion started at the transposable element (Koekman *et al.,* 1979; Ooms *et al.,* 1982c).

Recently, a technique has been described for the isolation of well-defined deletions in Ti-plasmids (Hille and Schilperoort, 1981b). The method consists of introducing into a cell already carrying a Ti-plasmid with a transposon insertion a second Ti-plasmid with an insertion of the same transposon at another site in the Ti-plasmid. Recombination between the two Ti-plasmids results in an intermediate in which two copies of the same transposon are present. If the transposons are in a direct repeat (orientation) recombination between them will produce a deletion.

D. Physical and Genetic Maps of Ti-Plasmids

Treatment of Ti- or Ri-plasmids with restriction enzymes followed by agarose gel electrophoresis of fragments gives complex patterns of 20 or more fragments of different sizes. For some of these plasmids the order of the fragments in the plasmid has been established, resulting in physical maps of these plasmids. For the wide host range octopine Ti-plasmids pTiAch5, pTiA6, and pTiB6, which are similar or identical, physical maps have been constructed for the restriction enzymes SmaI, HpaI (Chilton et al., 1978a), KpnI, XbaI (Ooms et al., 1980), BamHI, EcoRI, HindIII (De Vos et al., 1981), XhoI, SstI, and SalI (Knauf and Nester, 1982). For nopaline Ti-plasmid pTiC58 a map has been made for the enzymes BamHI, EcoRI, HindIII, HpaI, KpnI, SmaI, and XbaI (De Picker et al., 1980b). For the Ri-plasmids pRi8196 and pRi1855 the construction of physical maps is in progress (J. Tempé and M. D. Chilton, personal communication; P. Costantino, personal communication).

The physical maps have been used for the construction of genetic maps. To this end the positions of deletions and transposon insertions in the Ti-plasmids from mutants were determined by restriction enzyme analysis of plasmid DNAs. Subsequently, the positions of the mutations were correlated with the phenotype of the mutants. Both octopine and nopaline Ti-plasmids were found to contain two regions with genes that are involved in tumor induction, viz. the T-region, which is the region that is introduced into plant cells during tumor formation, and the virulence (Vir) region, which contains genes that are expressed in the bacterium, and which are possibly involved in transfer of the T-region into plant cells (see Section VII). The T-region and the Vir-region are located next to each other, and occupy about one-third of the Ti-plasmid. In the remaining part of the Ti-plasmid genes are located that are involved in the catabolism of opines and arginine, in phage AP1 exclusion, in agrocin 84 sensitivity, and in conjugative transfer (Koekman et al., 1979; Holsters et al., 1980; Ooms et al., 1980, 1981, 1982a; De Greve et al., 1981) (Fig. 6). None of the above-mentioned functions is necessary for tumor induction, because their deletion does not result in avirulence. Recently, the genes for octopine and arginine catabolism were more accurately mapped by the isolation of clones with parts of the Ti-plasmid that gave Ti-plasmid-free Agrobacterium strains the ability to grow on octopine or arginine as a sole carbon source (Knauf and Nester, 1982). The replicator region of the octopine Ti-plasmid has been mapped accurately by the isolation in vitro of small, replicating derivatives of this Ti-plasmid (Koekman et

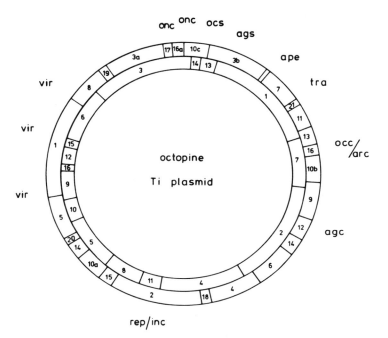

FIG. 6. Genetic map of an octopine Ti–plasmid. Restriction sites for *Hpa*I (inner circle) and *Sma*I (outer circle) are shown.

al., 1980, 1982). A sequence in the Ti replicator region responsible for stability and copy number control of the Ti-plasmid was mapped (Koekman *et al.*, 1982). Some transposon insertions in the replicator region of the octopine Ti-plasmid affect virulence (Koekman *et al.*, 1982), but the entire replicator region can be deleted from an R::Ti cointegrate without any effect on tumorigenicity (Hille *et al.*, 1982). This shows that transposon insertions in the Ti-plasmid may affect its oncogenic functions indirectly; the same has been observed when R plasmids were integrated into the Ti-plasmid (Ooms and Hooykaas, unpublished).

VII. Relationship between Ti-, Ri-, and Other *Agrobacterium* Plasmids

A. INCOMPATIBILITY STUDIES

Incompatibility is the inability of two (related) plasmids to be maintained by one cell as independent replicons in the absence of selection

TABLE 5

Incompatibility between Different *Agrobacterium* and *Rhizobium* Plasmids

Name	Plasmid type	pTiB6	pTiC58	pAtAg67	pRi1855	pSym5	*Inc* group
pTiB6	A[a]	+	+	−	−	−	Rh-1
pTiAg57	B	+	+	−	−	−	Rh-1
pTiC58	C	+	+	−	−	−	Rh-1
pTi542	D	−	−	+	−	−	Rh-2
pAtAg67	E	−	−	+	−	−	Rh-2
pRi1855	F	−	−	−	+	−	Rh-3
pSym5	G	−	−	−	−	+	Rh-4

[a]A, Octopine pTi (wide host range); B, octopine pTi (limited host range); C, nopaline pTi; D, null-type pTi; E, octopine non-Ti; F, agropine pRi; and G, *R. trifolii* pSym.

pressure. The exact molecular mechanism(s) of the incompatibility reaction between plasmids have still to be solved, but functions determined by the replicator region of the plasmids are involved. Because the replication units of plasmids are conservative in evolution, incompatibility is probably a conservative trait. Usually, incompatible plasmids share large regions of DNA homology (Grindley *et al.*, 1973), but exceptions to this rule have been found. The latter may be due to plasmid rearrangements by large transposable elements (Table 5).

In the case of Ti-plasmids it has been found that octopine Ti-plasmids, including those from limited host range strains, belong to the same incompatibility group (*inc* Rh-1) as nopaline Ti-plasmids (Hooykaas, 1979; Hooykaas *et al.*, 1980b, unpublished; Knauf and Nester, 1982). Null-type Ti-plasmids form an independent group (*inc* Rh-2) to which the non-Ti octopine catabolic plasmid pAtAG60 also belongs (Hooykaas, 1979; Hooykaas, unpublished). Ri-plasmids are compatible with plasmids of incompatibility groups *inc* Rh-1 and *inc* Rh-2 and form a third group (*inc* Rh-3) (Hooykaas, 1979; Costantino *et al.*, 1980; White and Nester, 1980a). It has not yet been determined whether all Ri-plasmids belong to the same *inc* group. All the above-mentioned plasmids are compatible with a number of cryptic plasmids that are present in different *A. tumefaciens* strains, so plasmids of more than three different types do naturally occur in *A. tumefaciens*. The Sym-plasmids from *R. leguminosarum* (pSym1), *R. trifolii* (pSym5), and *R. phaseoli* (pSym9), which are of different incompatibility groups, were found to be compatible with the plasmids of *inc* groups *inc* Rh-1, *inc* Rh-2, and *inc* Rh-3 (Hooykaas *et al.*, 1981, 1980d; Hooykaas, unpublished).

The incompatibility reactions between different octopine Ti-plasmids and those between different nopaline Ti-plasmids were as follows. Usually the resident Ti-plasmid was expelled after establishment of the incoming Ti-plasmid. In Rec$^+$ hosts in up to 10% of the events, recombination occurred between the resident and the incoming Ti-plasmid (Hooykaas *et al.,* 1980b; Klapwijk *et al.,* 1979). However, the incompatibility reactions between an octopine and a nopaline Ti-plasmid were different. Only in exceptional cases was loss of the resident octopine (nopaline) Ti-plasmid observed upon introduction of a nopaline (octopine) Ti-plasmid. Establishment of a nopaline (octopine) Ti-plasmid in an octopine (nopaline) Ti-plasmid-carrying strain was rare, and usually was accompanied by cointegrate formation between the resident octopine (nopaline) Ti-plasmid and the incoming nopaline (octopine) Ti-plasmid (Hooykaas *et al.,* 1980b). Such cointegrates were stable and could be transferred to other hosts as one entity. The cointegration event usually occurred somewhere in the *T*-region of both plasmids, which is the region of the strongest DNA homology between different Ti-plasmids (see Section VII,B). The incompatibility reaction between a null-type Ti-plasmid and the non-Ti catabolic octopine plasmid pAtAG60 resulted in loss of either plasmid, or in cointegrate formation between the two plasmids. It remains to be determined in which region of the plasmids cointegration occurred in this case.

The low level of establishment of octopine (nopaline) Ti-plasmids in nopaline (octopine) Ti-plasmid-carrying strains is not specific for conjugation experiments. The same was observed when the plasmids were introduced by transformation. When nopaline Ti-plasmid DNA was used for the transformation of octopine Ti-plasmid harboring strains, no transformants were obtained (Montoya *et al.,* 1978). Similarly, octopine Ti-plasmid DNA could be used for the transformation of strains harboring a Ri-plasmid, a null-type Ti-plasmid, or a non-Ti catabolic plasmid, but not for the transformation of strains with a nopaline Ti-plasmid (Hooykaas, unpublished). We have proposed to call the responsible function the *ein*- (establishment inhibition) function (Koekman *et al.,* 1982). The *ein*-functions of different octopine Ti-plasmids are identical, but different nopaline Ti-plasmids may have different *ein*-functions. For instance, establishment of nopaline Ti-plasmid pTiC58 in strains that already carry either of the nopaline plasmids pTiT37 or pTiEIII9.6.1 occurs only rarely, showing that the *ein*-functions of these plasmids are different (Hooykaas, 1979). *Ein*$^-$ deletion derivatives of an octopine Ti-plasmid have been obtained. Such plasmids do not prevent the establishment of a nopaline Ti-plasmid in the cell, and they are expelled after the introduction of a nopaline Ti-plasmid. The *ein*⁻ plasmids are unable to expel nopaline and octopine Ti-plasmids from the cell, and they therefore cannot establish themselves in nopaline strains, and only rarely in octopine strains, namely via cointegrate formation at the replicator regions (Koekman *et al.*, 1982). Remarkably, after cointegrate formation between a wild-type Ti-plasmid and a small *ein*⁻ copy number mutant of the same plasmid, the integrated copy number mutant plasmid retains its susceptibility to selective amplification, resulting in an increase in size of the cointegrate under conditions of amplification (Koekman *et al.*, 1982).

B. DNA Homology Studies

Hybridization experiments have shown that Ti-plasmids share varying amounts of DNA homology. Different wide host range octopine Ti-plasmids share at least 74% sequence homology, whereas this is 27–95% for different nopaline Ti-plasmids (Currier and Nester, 1976b). Although limited host range octopine Ti-plasmids share much homology (more than 64%) among themselves, they have only 6–15% sequence homology with wide host range octopine and nopaline Ti-plasmids (Thomashow *et al.*, 1981). Octopine and nopaline Ti-plasmids share only limited DNA homology (6–36%) (Currier and Nester, 1976b; Drummond and Chilton, 1978). These data are in agreement with restriction enzyme analyses of Ti-plasmids, which show that the restriction enzyme profiles of different wide host range octopine Ti-plasmids are (almost) identical, but different from those of nopaline Ti-plasmids, and limited host range octopine Ti-plasmids (Sciaky *et al.*, 1978; Thomashow *et al.*, 1981). The sequence homology existing between null-type Ti-plasmids and octopine Ti-plasmids is not clear; in the literature values of 11–27% (Currier and Nester, 1976b) and of 62% sequence homology (Drummond and Chilton, 1978) have appeared. Ri-plasmids from different *Agrobacterium* strains give restriction profiles that show varying degrees of similarity. Southern blotting experiments revealed that they all share large regions of strong DNA homology (Costantino *et al.*, 1981). Ri-plasmids have only limited DNA homology with wide host range octopine and nopaline plasmids (Currier and Nester, 1976b; White and Nester, 1980b; Risuleo *et al.*, 1982) (Fig. 7).

The regions of sequence homology that different Ti- and Ri-plasmids share have been mapped on the physical map of a wide host range octopine Ti-plasmid and on that of a wide host range nopaline Ti-plasmid. The first results showed that all wide host range Ti-plasmids

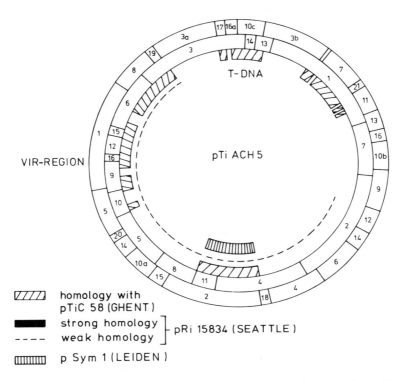

FIG. 7. Regions of DNA homology between an octopine Ti-plasmid and nopaline Ti-plasmid pTiC58, Ri-plasmid pRi15834, and the *Rhizobium leguminosarum* plasmid pSym1. [Taken from Engler *et al.* (1981), White and Nester (1980b), and Prakash and Schilperoort (1982).]

share one region of strong DNA homology, which is part of the *T*-region (Chilton *et al.*, 1978b; De Picker *et al.*, 1978; Hepburn and Hindley, 1979; Perry and Kado, 1982). In contrast, Ri-plasmids and limited host range Ti-plasmids lack strong DNA homology with the *T*-region of wide host range Ti-plasmids (White and Nester, 1980b; Thomashow *et al.*, 1981; Risuleo *et al.*, 1982). Aside from the *T*-region, octopine and nopaline wide host range Ti-plasmids have three other regions in common (Drummond and Chilton, 1978; Hepburn and Hindley, 1979; Engler *et al.*, 1981). These three regions embrace genes involved in tumor induction (virulence region), in replication, in conjugative transfer, and in phage AP1 exclusion. Limited host range Ti-plasmids have homology with the wide host range Ti-plasmids in the same three regions (Thomashow *et al.*, 1981). For Ri-plasmids, strong DNA homology is restricted to the virulence region (White and Nester, 1980b; Risuleo *et al.*, 1982).

The possible presence of DNA homology between the rhizobial Sym-plasmids and the wide host range Ti-plasmids has also been investigated (Prakash and Schilperoort, 1982). Some scattered areas of sequence homology were found. These correspond to a piece of the Ti-plasmid virulence region, to a piece of the replicator region, and to the region with the genes for conjugative transfer and AP1 exclusion. No sequence homology was detected with the *T*-region of the Ti-plasmid.

All the above data show that Ti-plasmids have been subject to severe alterations during evolution, resulting in a variety of types. Moreover, one can speculate that the genes involved in tumor induction have become linked to different types of (catabolic) plasmids by (illegitimate?) recombination. Only this can explain the finding that the null-type Ti-plasmids belong to the same incompatibility class as the catabolic (non-Ti) octopine plasmid pAtAG60, and that they belong to a different class than the wide host range octopine and nopaline Ti-plasmids. More research is required with different Ri-plasmids before any speculations can be made about their evolutionary origin.

C. Genetic Analysis and Complementation Studies of Virulence Functions

The virulence regions of octopine Ti-plasmid pTiAch5 and of nopaline Ti-plasmid pTiC58 have been compared precisely by heteroduplex analysis (Engler *et al.*, 1981). This study revealed that the virulence regions of these plasmids have extensive DNA homology, but nevertheless also contain small areas of nonhomology. Transposon insertions in the virulence region are known to cause avirulence (Garfinkel and Nester, 1980; Holsters *et al.*, 1980; Ooms *et al.*, 1980; De Greve *et al.*, 1981; Hooykaas, unpublished). However, more recently mutants have also been isolated that retain their virulence in spite of a transposon insertion in this region (Ooms *et al.*, 1981; Hooykaas, unpublished). This suggests that a number of different genes are present in this area and that there are sequences between these genes that are not involved in tumor induction.

The entire *Vir*-region of the octopine Ti-plasmid pTiB6, as well as parts of it, have been cloned *in vivo* onto the wide host range R plasmid R772 (Hille *et al.*, 1982), and parts of the octopine Ti-plasmid pTiA6 have been cloned *in vitro* onto the wide host range cosmid vector pHK17 (Klee *et al.*, 1982). These R primes and cosmid clones have been introduced into avirulent strains with transposon insertions in the *Vir*-region. In contrast with some of the results obtained in complementation experiments with cosmid clones (Klee *et al.*, 1982), all transposon insertion mutants studied were restored to virulence upon the introduction of an R prime with the wild-type *Vir*-region segment (Hille *et al.*, 1982; J. Hille, personal communication).

As the virulence region of the Ti-plasmid has never been detected in crown gall cells, and all of the genes in this region can be complemented *in trans* in *Agrobacterium* strains, it is plausible to presume that the *Vir*-genes are expressed in the bacterium. Genes in this region could be involved in the processing of the *T*-region during tumor induction, and/or in the transfer of the *T*-region to the plant cells. Data are accumulating that genes in the *Vir*-region might also be involved in production or secretion of the plant hormones IAA (Liu *et al.*, 1982) and *trans*-zeatin (Regier and Morris, 1982).

Wild-type nopaline Ti-plasmids were introduced into strains with octopine Ti-plasmids that had become avirulent by transposon insertions (Hooykaas, unpublished). The virulence of transconjugants was examined in which cointegrates had been formed between the mutated octopine Ti-plasmid and the wild-type nopaline Ti-plasmid. Such cointegrates were virulent and formed tumors in which both octopine and nopaline were present. Wild-type Ri-plasmids were transferred to strains with avirulent octopine Ti-plasmids. It was found that genes present in the *Vir*-region of the Ri-plasmid could complement many, but not all, mutations in the octopine Ti-plasmid *Vir*-region (Hooykaas, unpublished). This is in agreement with the partial sequence homology between the *Vir*-regions of the octopine Ti-plasmids and the Ri-plasmids (Table 6).

TABLE 6
Complementation Experiments with *Vir⁻* Transposon Insertion Mutants

Insertion mutant (LBA number)	Site of insertion (*Hpa*I site)	Tumor induction after introduction of		
		pAL1818	pTiC58	pRi1855
1582	9	+	+	+
1512	12	+	+	+
1551	6	+	+	+
1586	6	+	+	−
1514	3	+	+	+
1517	3	+	+	+

Three avirulent transposon insertion mutants with mutations in the nopaline Ti-plasmid *Vir*-region have been used in complementation experiments (Hooykaas, unpublished). All three were restored in virulence by an R prime with the octopine Ti-plasmid *Vir*-region. Tumors were formed in which nopaline was synthesized. In the same three mutants null-type Ti-plasmid pTi542 and Ri-plasmid pRi1855 were also introduced. The presence of plasmid pTi542 restored virulence of the mutant nopaline Ti-plasmids, because in tumors formed by these strains nopaline was present. The Ri-plasmid was able to restore tumor induction only in two of the three mutants. In the cases that tumors were formed these contained nopaline. The three avirulent nopaline Ti-plasmids have also been used for the construction of cointegrates with avirulent octopine Ti-plasmids. In most cases strains harboring such cointegrates were virulent and formed tumors synthesizing octopine and nopaline. However, no tumors were formed when in both plasmids the same function had been affected. In this way homologous functions were mapped in the *Vir*-regions of the octopine Ti-plasmid pTiB6 and the nopaline Ti-plasmid pTiC58.

All the data obtained so far indicate that the virulence regions of octopine, nopaline, and null-type Ti-plasmids determine similar functions, some of which are also present in the virulence regions of Ri-plasmids. It remains to be seen whether all the functions encoded by the virulence region are common for all of the Ti-plasmids.

VIII. T-DNA Structure and Expression

A. Structure of T-DNA

Much research has been devoted to the search for the presence of bacterial DNA in crown gall cells. Unsuccessful and false positive experiments were followed by the demonstration of T-DNA in a crown gall cell line by Chilton *et al.* (1977). In these first successful experiments octopine Ti-plasmid restriction enzyme fragments were used separately as probes in reassociation kinetic experiments with the total DNA from an octopine crown gall tissue. These tests revealed the presence of part of *Sma*I fragment 3b in about 18 copies per diploid plant cell. This DNA was called T-DNA. Whether the T-DNA caused the transformation of normal cells into tumor cells was questioned, because it was found that octopine Ti-plasmid deletion mutants that lacked fragment 3b entirely were still tumorigenic (Koekman *et al.*,

1979; Ooms *et al.*, 1982a). This ambiguity disappeared when the results from Southern blotting experiments became available. In such tests it was shown that in all octopine tumor lines another segment of the Ti-plasmid is present in one or only a few copies. This region is adjacent to and partly overlaps fragment 3b in the octopine Ti-plasmid (Thomashow *et al.*, 1980b; De Beuckeleer *et al.*, 1981; Ooms *et al.*, 1981, 1982b). Although the exact boundaries may differ somewhat, the core T-DNA (or T_L-DNA) of octopine lines comprises a piece of DNA of 8 Md corresponding to the region of the octopine Ti-plasmid between *Sma*I fragment 17 and *Eco*RI fragment 19a. In some octopine crown gall lines the region between *Eco*RI fragment 19a and *Eco*RI fragment 20 is also present, albeit that this segment (the T_R-DNA) is usually not linked to the T_L-DNA. As already stated above, deletion of the T_R-region from the octopine Ti-plasmid does not lead to avirulence (Koekman *et al.*, 1979; Ooms *et al.*, 1982b). However, deletion of the T_L-region from the Ti-plasmid does result in the loss of tumorigenicity. All these data indicate that the T_L-DNA is responsible for the tumorous behavior of crown gall cells. Part of the T_L-region of the wide host range octopine Ti-plasmid has strong sequence homology with wide host range nopaline and null-type Ti-plasmid DNA (Drummond and Chilton, 1978). It is therefore not surprising that the 15-Md T-DNA present in nopaline tumor lines encompasses the sequence common to all wide host range Ti-plasmids (Lemmers *et al.*, 1980; Yadav *et al.*, 1980). Limited host range Ti-plasmids do not have strong sequence homology with the *T*-region of wide host range Ti-plasmids (Thomashow *et al.*, 1981). This indicates that limited host range Ti-plasmids have a completely different *T*-region, and suggests the possibility that the *T*-region of Ti-plasmids determines host range (see also Section IX). Recently, sterile root cultures were started from hairy roots induced by *A. rhizogenes*. Such root lines were found to contain a piece of DNA originating from the *A. rhizogenes* Ri-plasmid (Chilton *et al.*, 1982; Spano *et al.*, 1982; Willmitzer *et al.*, 1982a). Tumors induced by *A. rhizogenes* on *N. glauca* also contained T-DNA (White *et al.*, 1982). The *T*-region of the Ri-plasmid was found to have only very limited DNA homology with the *T*-region of wide host range Ti-plasmids (Willmitzer *et al.*, 1982a)

The question of where the T-DNA is located in the crown gall cells was addressed in Southern blotting experiments. The *T*-region of the Ti-plasmid was used as a probe on DNA preparations from various plant cell organelles, viz. nuclei, mitochondria, and chloroplasts. A positive signal was obtained only in hybridization experiments with nuclear DNA. This shows that the T-DNA is present in the nucleus of plant tumor cells (Chilton *et al.*, 1980; Willmitzer *et al.*, 1980).

Visualizing the T-DNA in tumor lines with T-region probes in Southern blotting experiments results in the appearance of fragments that are identical or not identical in size to certain T-region fragments. Fragments identical in size to T-region fragments are "internal" fragments of the T-DNA; the others are "border" fragments. Some border fragments were found to share DNA homology with left-end T-region probes as well as with right-end T-region probes. It was therefore postulated that such borders are "fusion fragments" in which tandemly arranged T-DNA copies are linked together (Lemmers *et al.*, 1980; Zambryski *et al.*, 1980; Ooms *et al.*, 1982b). Alternatively, such fusion fragments could have been formed during the hypothetical formation of circular T-region copies either in the bacterium or in the transformed plant cell during tumor induction. However, such circular intermediates consisting of T-DNA have so far never been detected in agrobacteria or transformed plant cells (Lemmers *et al.*, 1980). Recently, such T-region circles were found when Ti-plasmid DNA was used for the transformation of human and *Tupaia* fibroblasts (Harth *et al.*, 1982). Unlike the fusion fragments discussed above, true border fragments may have DNA homology only with a right-end T-region probe or a left-end T-region probe. Such fragments were cloned and subsequently used as probes in Southern blotting experiments with DNA from untransformed plant cells. Either unique bands or multiple bands appeared in these tests. This shows that these border fragments consist in part of Ti-plasmid DNA and in part of plant DNA. This proves that the T-DNA is integrated into plant DNA, and indicates that integration had not occurred at a fixed site in the plant genome (Thomashow *et al.*, 1980a; Yadav *et al.*, 1980; Zambryski *et al.*, 1980, 1982).

The finding that the T-DNA found in different tumor tissues is of a fixed size poses the question of whether T-DNA integrates into plant DNA via a mechanism similar to that by which transposable elements integrate into foreign DNA. The answer to this question is not known. However, it is known that the ends of the T-region are important for tumor induction. Deletion of the T_R-region of the Ti-plasmid does not affect virulence, but if the deletion also removes the right border region of the T_L-region virulence is strongly attenuated (Koekman *et al.*, 1979; Ooms *et al.*, 1982a). However, if the right border region of the T_L-region is deleted but the T_R-region is left intact, virulence is not affected (Leemans *et al.*, 1981). In tumor lines induced by such a mutant,

T_L-DNA and T_R-DNA are probably integrated as one piece of DNA (J. Leemans, personal communication). This shows that a T_L-region right border sequence can be substituted by a T_R-region right border sequence. Recently, left and right border regions from a nopaline Ti-plasmid were sequenced (Yadav *et al.*, 1982; Zambryski *et al.*, 1982), as well as the left border region from an octopine Ti-plasmid (Simpson *et al.*, 1982). It was found that a 25-bp imperfect direct repeat is present at the ends of the *T*-region of the nopaline Ti-plasmid. Strikingly, a similar 25-bp sequence was found at the left end of the octopine Ti-plasmid *T*-region. Border fragments in which Ti-plasmid DNA is linked to plant DNA have been isolated and sequenced. Comparison of the sequences present in these border fragments with the Ti-plasmid sequences revealed at which nucleotide Ti-plasmid DNA had been linked to plant DNA in the border fragments. Integration turned out to have occurred within a variation of a single base pair for various right borders. For left borders the integration site varied over at least 70 bp (Zambryski *et al.*, 1982). The above-mentioned 25-bp repeats were either completely or partly absent from the T-DNA. It has to be kept in mind, however, that the experiments have been done with established tumor lines, and therefore rearrangements at the ends of the T-DNA may have occurred after integration of T-DNA into plant DNA.

Petunia and tobacco cells can be transformed *in vitro* with octopine Ti-plasmid DNA, and tumor lines have been obtained from such experiments (Draper *et al.*, 1982; Krens *et al.*, 1982). The T-DNA present in such tumor lines was found to be different from that present in tumor lines established *in vivo*. Occasionally, much larger pieces of DNA than the 8-Md core T-DNA had been integrated into plant DNA (F. A. Krens, personal communication), and consequently other Ti-plasmid sequences had been used for integration. This suggests that the 25-bp border signals in the Ti-plasmid play a role in directing the "processing" of the Ti-plasmid in the bacterium rather than in directing T-DNA integration into plant DNA. "Processing" of the Ti-plasmid might lead to the creation of circular *T*-region copies, which are subsequently transferred from bacterium to plant cell. Such circular T-DNA might then multiply in plant cells before integration takes place. Possibly multimerization can also occur, resulting in the formation of large multimers that can be found as tandem repeats after integration. *In situ* hybridization experiments of newly transformed plant cells with probes consisting of different Ti-plasmid segments could unambiguously show whether the entire Ti-plasmid or only the *T*-region is transferred to plant cells during tumor induction. Such experiments have already been carried out successfully with the entire Ti-plasmid as a probe (Nuti *et al.*, 1980).

B. Expression of T-DNA

Animal oncogenic viruses are of two types. Those of the first type only transform normal cells into tumor cells if integration of the virus occurs at a specific site close to a specific genomic gene. Integration of the virus causes enhanced expression of this specific genomic gene, and this in turn is the cause of the tumorous character of the cells (Dulbecco, 1979). Oncogenic viruses of the second type transform normal cells into tumor cells regardless of the site of integration. These viruses can be considered as special transducing viruses carrying a gene in their genome that probably originated from one of their hosts. The increased expression of the gene present in the virus compared to the low expression of the host gene is the cause of the tumorous character of the cells.

The finding that T-DNA is integrated at different sites in the plant DNA in different tumor lines suggests that T-DNA is comparable with type 2 oncogenic viruses. If this were true the T-DNA must contain genes, the expression of which causes the tumorous character of the crown gall cells. Furthermore, one might expect to find sequence homology between the Ti-plasmid *onc*-genes and genes present in normal plant cells.

Southern blotting experiments of normal plant DNA with *T*-region probes have been carried out. Very faint hybridizing bands were observed if the octopine Ti-plasmid *T*-region was used as a probe on tobacco DNA (Thomashow *et al.*, 1980b). Similar experiments with the *T*-region from the Ri-plasmid revealed strongly hybridizing bands in total DNA from *N. glauca* (White *et al.*, 1982), and in that from pea and carrot (Spano *et al.*, 1982; P. Costantino, personal communication).

In order to find out whether T-DNA is expressed as mRNA in tumor cells, ^{32}P-labeled RNA was isolated from tumor and normal tobacco callus tissues, and used as a probe in Southern blotting experiments on restriction fragments of octopine Ti-plasmid DNA (Drummond *et al.*, 1977; Ledeboer, 1978). Bands were only observed with labeled tumor RNA as a probe. Initially only RNA homologous to the T_R-region was found (with a tumor line containing 18 copies per diploid genome of T_R-DNA). After hybridization, fragments *Sma*I 3b, *Hin*dIII 1 and 5, and *Eco*RI 12 and 20 were visible (Drummond *et al.*, 1977). Ledeboer (1978) was the first to detect transcription also of T_L-DNA, and his results were confirmed and extended by others (Gurley *et al.*, 1979; Gelvin *et al.*, 1981; Willmitzer *et al.*, 1981a). Yang *et al.* (1980a) measured transcription in a nopaline tumor line. The results obtained show that almost all of the T-DNA is transcribed, although not all segments are transcribed with an equal intensity. Similar data were obtained whether poly(A)$^+$ or poly(A)$^-$ RNA was used as a probe. Using purified active nuclei from crown gall cells, it could be demonstrated that transcription of T-DNA was inhibited by low concentrations of α-amanitin, indicating that T-DNA is transcribed by plant RNA polymerase II (Willmitzer *et al.*, 1981b). The number, size, and location of transcripts from the T-DNA were determined in Northern blotting experiments (Gelvin *et al.*, 1982; Murai and Kemp, 1982a; Willmitzer *et al.*, 1982b). In such experiments the poly(A)$^+$ RNAs isolated from tumor cells are separated on an agarose gel, blotted onto nitrocellulose, and hybridized with T-region probes. In total, seven transcripts are encoded by the T_L-region of the T-DNA. Their direction of transcription has been determined by using strand-separated probes (Willmitzer *et al.*, 1982b). From the T_R-region of the T-DNA at least one 1600-base RNA was transcribed (Murai and Kemp, 1982a). Evidence has been presented that initiation and termination of transcription of T-DNA sequences occurs within the T-DNA, indicating that the prokaryotic T-region carries all the necessary signals for transcription by plant RNA polymerase II (Gelvin *et al.*, 1982; Leemans *et al.*, 1982; Willmitzer *et al.*, 1982b). Some of the T-DNA transcripts present in octopine tumors have homologous counterparts in nopaline tumors, as would be expected on the basis of the homology observed between the nopaline and octopine Ti-plasmid T-regions (Willmitzer *et al.*, 1982b). However, some transcripts present in octopine tumor lines had no homologous counterparts in nopaline tumor lines, and vice versa. Among the latter are probably the mRNAs for the enzymes involved in octopine, nopaline, agrocinopine, and agropine synthesis.

Recently, some work was reported on the transcription of T-DNA in an axenic tobacco hairy root tissue induced by *A. rhizogenes* strain 15834 (Willmitzer *et al.*, 1982a). Some of the regions transcribed appeared to have weakly homologous counterparts in octopine and nopaline tumors. The function of most of the transcripts observed in the different tumor lines has not been established yet, although some clues toward their possible activity have been obtained from the analysis of mutants (see Section IX). One transcript present in octopine tumor lines has been investigated in more detail. This transcript of about 1400 bases was translated *in vitro* in the wheat germ and rabbit reticulocyte systems into a protein of about M_r 40,000 that could be immunoprecipitated with antibody against lysopine dehydrogenase (Schröder et al., 1981a; Murai and Kemp, 1982b; Schröder and Schröder, 1982). Translation was inhibited by the cap analog pm⁷G (Schröder and Schröder, 1982). In agreement with the genetic evidence that has been obtained (see Section IX) these data show that the octopine Ti-plasmid T-DNA carries the structural gene for the enzyme lysopine dehydrogenase. Recently, the gene for lysopine dehydrogenase was fused to the chloramphenicol acetyltransferase gene present in the cloning vector pACYC 184. In E. coli minicells a protein of M_r 18,000 was produced. This turned out to be a fusion product of chloramphenicol acetyltransferase and lysopine dehydrogenase (Schröder et al., 1981b). Besides the mRNA for lysopine dehydrogenase, some other T-DNA-specific RNAs have been translated into proteins in a wheat germ system (McPherson et al., 1980; Schröder and Schröder, 1982). The function of these proteins remains to be elucidated.

Bacterial extracts have been prepared to find out whether the T-region is also expressed when it is present in A. tumefaciens. It was found that all sections of the T-region are weakly transcribed by bacterial RNA polymerase (Gelvin et al., 1981). T-region fragments cloned in E. coli vectors directed the production of some proteins in E. coli minicells (Schweitzer et al., 1980) as well as in E. coli and A. tumefaciens cell extracts (Kartasova et al., 1981). The significance of the above results is still unknown.

IX. Suppression of the Tumorous Character of Crown Gall Cells: Effect of T-DNA Mutations

A. ISOLATION OF T-REGION MUTANTS

A few years ago we discovered that certain transposon insertion mutants of an octopine strain of A. tumefaciens formed tumors that were morphologically aberrant (Hooykaas, 1979; Klapwijk, 1979). The transposon insertions in three of these mutants were localized, and their tumorigenic behavior was carefully examined on a variety of plants (Ooms et al., 1980, 1981). All three mutants had an insertion in the T-region. They induced weak tumors on tomato, but tumors of a "wild-type" size on N. rustica and N. debneyi. This shows that host range is at least partially controlled by genes in the T-region of the Ti-plasmids. Strains LBA 1501 and LBA 4060, which have a transposon insertion in the left part of the T-region, formed small, unorganized tumors on *Kalanchöe* stems. *Kalanchöe* tumors induced by LBA 4060 caused the loss of apical dominance, resulting in the development of shoots from the sites of the axillary buds above the tumors (Fig. 8). On tobacco, both LBA 1501 and LBA 4060 induced tumors from which shoots appeared. Strain LBA 4210, which has a transposon insertion in the right part of the T_L-region, formed tumors from which roots developed on tobacco. Tumors induced by LBA 4210 on *Kalanchöe* were surrounded by massive amounts of adventitious roots. Inoculation of plants with a mixture of a *shooter* mutant plus a *rooter* mutant (LBA 4060 + LBA 4210) resulted in the formation of tumors that were identical in all respects to those induced by the wild type (Ooms *et al.*, 1981). Tobacco tumor tissue induced by LBA 4060 + LBA 4210 was brought into tissue culture, and after some growth had occurred subclones were separated after protoplast isolation (Ooms *et al.*, 1982c). Six subclones that were able to grow in the absence of phytohormones and produce octopine were analyzed for their T-DNA content. All six clones appeared to contain T-DNA that originated from each mutant, suggesting that they were derived from doubly infected cells. Genetic complementation in the tumor cells, therefore, probably forms at least part of the basis for the observed enhancement of tumor induction by the two types of weakly virulent mutants. It was observed that certain mixtures of tumorigenic and nontumorigenic wild-type strains had an enhanced tumor initiation capacity. This may have had the same molecular basis (Lippincott and Lippincott, 1970, 1978b; Lippincott *et al.*, 1977).

Recently, Garfinkel *et al.* (1981) isolated a large set of transposon insertion mutants in the T-region of the octopine Ti-plasmid, and Leemans *et al.* (1982) isolated a set of site-specific deletions and substitutions in the T-region of the same Ti-plasmid. All these mutants were mapped and examined for tumorigenicity. With new *shooter* and *rooter* mutants the auxin and cytokinin loci were positioned more accurately on the transcription map of the T-DNA of the octopine Ti-plasmid (see Section VIII). In *shooter* mutants the transposon turned out to abolish either transcript 1 or transcript 2, or both of these transcripts, whereas in *rooter* mutants transcript 4 was inactivated (Leemans *et al.*, 1982). These transcripts are probably involved in establishing a tumorous phenotype in transformed plant cells, because inactivation of the areas for transcript 1 or 2 and transcript 4 at the same time leads to the loss of oncogenicity of the Ti-plasmid (Leemans *et al.*, 1982; Hille *et al.*, 1983). In nopaline Ti-plasmids the regions for transcripts 1, 2, and 4

FIG. 8. Tumors induced on *Kalanchöe* by *rooter* mutant LBA 4210 (left); wild-type LBA 4001; a mixture of LBA 4060 and LBA 4210; and *shooter* mutant LBA 4060 (right). (Taken from Ooms *et al.*, 1981.)

have been conserved (Willmitzer *et al.*, 1982a; Willmitzer *et al.*, 1983). Mutants with a mutation in the region for transcript 1 were complemented in tumorigenicity by mutants with a mutation in the region for transcript 2 (Van Slogteren, 1983).

Plant hormones (e.g., auxins, cytokinins, and gibberellins) regulate the growth and differentiation of plant cells. For the growth of dicotyledonous plant cells *in vitro* a cytokinin as well as an auxin usually has to be added to the culture medium. If the level of cytokinins in the medium is high compared to the level of auxins, tobacco callus will develop shoots, whereas in the reverse case roots will appear. In the absence of cytokinin or auxin dominance an amorphous callus will be formed (Skoog and Miller, 1957). We have seen above that tumors induced on tobacco by *A. tumefaciens* may develop roots or shoots, or remain unorganized depending on the mutant strain that incited the tumor. It was therefore postulated by Ooms *et al.* (1981) that T-DNA influences the auxin and cytokinin activities in transformed plant cells. From the analysis of transcripts from octopine T-DNA it is evident that transcripts 1 and 2 together have an auxinlike effect in plant cells, whereas transcript 4 has a cytokininlike effect in plant cells (Willmitzer *et al.*, 1982b). Inactivation of the region for transcript 1 or that for transcript 2 (e.g., in LBA 4060) would then result in a mutated T-DNA with a cytokininlike effect in plant cells. This leads to shoot formation. In contrast, inactivation of transcript 4 (e.g., in LBA 4210) would result in a mutated T-DNA with an auxinlike effect in plant cells. This gives root formation (Figs. 8 and 9).

In accordance with the theory, the presence of an auxin during tumor induction stimulated tumor formation by *shooter* mutants on tomato (Ooms *et al.*, 1981) and on carrot slices (Leemans *et al.*, 1982). Likewise, tumor formation on tomato by *rooter* mutants was strongly enhanced by the presence of a cytokinin (Ooms *et al.*, 1981). Cytokinins (ribosyl-*trans*-zeatin) and auxin (IAA) levels were determined in tumors induced on tobacco by a wild-type octopine strain and by *shooter* and *rooter* mutants (Akiyoshi *et al.*, 1983). The IAA levels were not significantly different in the different tumors, but there were drastic differences in the ribosyl-*trans*-zeatin content of the tumors. This cytokinin is the major cytokinin present in crown gall tissue (Miller, 1974). In primary *Kalanchöe* tumors induced by *rooter* mutants the ribosyl-*trans*-zeatin level was much lower than in tumors induced by the wild type, whereas in tumors induced by *shooter* mutants this level was much higher than in tumors induced by the wild type (Akiyoshi *et al.*, 1983).

The T-DNA of the Ri-plasmid has an auxinlike effect on plant cells. This suggests the possibility that the T-DNA of Ri is a naturally occurring derivative of the Ti-plasmid T-DNA with a mutation in the cytokinin locus. This turned out not to be true, because no strong DNA homology has been found between the T-DNAs of Ti- and Ri-plasmids

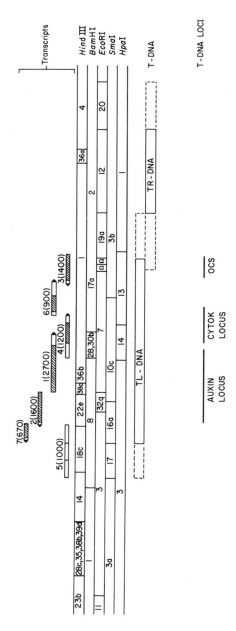

FIG. 9. A map of the *T*-region of the octopine Ti-plasmid.

(White and Nester, 1980b; Risuleo *et al.*, 1982). However, weak sequence homology has been detected between Ri-plasmid T-DNA transcripts and the auxin locus of the Ti-plasmid T-DNA (Willmitzer *et al.*, 1982a). On tomato, both Ti-plasmid *shooter* mutants and *A. rhizogenes* strains form only small swellings. However, strains carrying a Ri-plasmid plus a Ti-plasmid with a *shooter* mutation form tumors of a size comparable to that of those induced by wild-type *A. tumefaciens* strains (Van Slogteren, 1983). This could be due to genetic complementation, as described above for the combination of a *shooter* and a *rooter* mutant. On *Kalanchöe* leaves *A. rhizogenes* induces hairy roots. The addition of a cytokinin, however, leads to tumor rather than to root formation (Beiderbeck, 1973a).

A new locus affecting tumor formation by Ti-plasmids was found by Garfinkel *et al.* (1981). Insertion of a transposon into this locus, which covers the region of T-DNA transcripts 6a or 6b, led to the formation of tumors of a much larger than normal size on *Kalanchöe* stems. On other plants, such as tobacco, this effect was not seen. From experiments with Ti-plasmid deletion mutants (Koekman *et al.*, 1979; Ooms *et al.*, 1982a) as well as with transposon insertion mutants (Garfinkel *et al.*, 1981), and site-specific insertion and deletion mutants (Leemans *et al.*, 1981, 1982) it has become clear that the structural gene for lysopine dehydrogenase is located at the right extremity of the T_L-region. Transposon insertions mapping in the T_R-region led to the formation of tumors in which neither agropine nor mannopine was synthesized (Dessaux and Hooykaas, unpublished). This suggests that genes coding for enzymes involved in the synthesis of agropine and mannopine are located in T_R-DNA. Deletions leading to the removal of the entire T_L-region from the Ti-plasmid result in the loss of oncogenicity (Koekman *et al.*, 1979; Leemans *et al.*, 1982). When mutants lacking the Ti-plasmid T_L-region (but still having the T_R-region) were used for infection of plants, no tumors were formed. However, after some time agropine synthesis was detected in the wound area (Leemans *et al.*, 1982). This indicates that the T_R-region was transferred to plant cells and expressed there. Moreover, these data provide evidence that genes involved in agropine and mannopine synthesis are not located in T_L-DNA, but rather in T_R-DNA. The data also show that the virulence functions of the Ti-plasmid and the T_R-region functions can operate even in the absence of the core T-region. These results do not prove that T_R-DNA has the capacity to integrate into plant DNA because T-DNA expression in plant cells could occur before T-DNA integration.

B. The Fate of T-DNA in Plants Regenerated from Tumor Cells

In contrast to animals, complete plants can be regenerated from single cells. Regeneration experiments have been carried out with crown gall and hairy root cells in order to find out whether their tumorous phenotype could be suppressed. If Ti- and Ri-plasmids are to be used for the genetic manipulation of plant cells, it is essential (at least for some purposes) that (fertile) plants can be regenerated from the transformed cells.

Uncloned nopaline tumors generally generate shoots, but uncloned octopine tumors only rarely form shoots. However, tumors that had been induced by octopine *shooter* mutants retained the capacity to produce shoots in tissue culture for a shorter or longer period of time (Leemans *et al.*, 1982; Ooms *et al.*, 1982b; Van Slogteren, 1983). An interesting question in this respect is whether or not tumors consist solely of tumor cells. With respect to the regenerated shoots, one can ask whether they originated from tumor cells, from normal cells present along with the tumor cells in the tumor, or from a mixture of both cell types.

Most shoots obtained from uncloned tumor tissues do not contain opines, are sensitive to infection by agrobacteria, and form roots. This indicates that predominantly normal cells present in the tumor tissue are stimulated to regenerate. However, in a few exceptional cases opine-positive shoots were obtained from uncloned tumor tissues (Scott, 1979; Leemans *et al.*, 1982; Ooms *et al.*, 1982; Owens, 1982; Van Slogteren, 1983; G. J. Wullems, personal communication). These shoots might consist entirely of a mixture of differentiated normal cells and nondifferentiated opine-producing tumor cells. It has been shown that such chimera formation can indeed occur (Turgeon *et al.*, 1976; Binns *et al.*, 1981). To avoid the latter ambiguity, tumor tissues were cloned and then analyzed.

Cloning was first used to see whether all vegetatively growing cells in tumor tissues are tumorous or not. Single cells were isolated from an axenic null-type tumor, and these were grown until they had formed calli. Only 3 of the 253 clones obtained turned out to be fully phytohormone independent (Sacristan and Melchers, 1977). Similarly, clones obtained from a number of different axenic octopine tumors were studied. More than 90% of the isolated clones did not produce octopine and most of these were phytohormone dependent (Van Slogteren, 1983). These results show that the vast majority of cells in tumors and uncloned tumor tissues can be nontumorous cells. During growth on phytohormone-free media the nontumorous cells present in uncloned tumor tissues were probably provided with phytohormones by the tumor cells that were present in their vicinity. Consistent with this, normal tobacco calli can grow on phytohormone-free media if crown gall tissue is present in the same petri dish (e.g., Ooms, 1981).

Braun (1959) was the first to obtain shoots from a cloned nopaline tumor tissue. Almost all of these shoots synthesized nopaline. Moreover, leaf tissue from the opine-positive shoots was found to consist at least partially of cells capable of growing in the absence of phytohormones (Braun and Wood, 1976; Wood et al., 1978). The nopaline-positive shoots were somewhat different in appearance from normal shoots. The stems were usually swollen, the leaves small, and apical dominance was incomplete (Turgeon, 1981; Wullems et al., 1981b). However, stems and leaves found on teratoma shoots were indistinguishable in histological and functional aspects from those found on normal tobacco shoots (Braun and Wood, 1976). Because such teratoma shoots did not form roots, they were grafted onto healthy tobacco plants (Turgeon et al., 1976; Wood et al., 1978). This led to the development of teratomatous outgrowths, or to the formation of normal shoots that ultimately flowered and set seed (Turgeon et al., 1976). Tumors frequently erupted at the graft junctions (Turgeon, 1981). The flowers formed were normal except that the stigmas extended 2–4 mm beyond the anthers (Binns et al., 1981; Wullems et al., 1981b). Therefore, these flowers had to be pollinated by hand in order to obtain seed. In the case of tobacco line SR1, male sterility was observed (Wullems et al., 1981b). In spite of the fact that the above-mentioned nopaline-positive shoots originated from a cloned nopaline tumor tissue, they might still be chimeras partially consisting of normal cells that could have arisen through loss of part or all of their T-DNA at a later stage of culture. In order to check this possibility, Binns et al. (1981) isolated leaf protoplasts from nopaline-positive shoots, and characterized the calli derived from these protoplasts. From 560 calli obtained, 3 turned out to have lost the ability to grow on media devoid of phytohormones, and these 3 calli did not produce nopaline. Because the vast majority of cells in the nopaline-positive shoot were tumorous, this indicates that the tumorous character of the T-DNA in plant cells can be suppressed. Furthermore, after suppression of their tumorous character the cells can differentiate into all the different types of specialized cells in stems and leaves (e.g., Huff and Turgeon, 1981). However, differentiation into root cells seems to be impossible (see below). The results of Binns *et al.* (1981) also indicate that a (permanent) recovery from the tumorous state occurs constantly during growth in the vegetative state at a low frequency. In agreement with this, some nopaline-negative shoots were obtained after transfer of a cloned nopaline tumor tissue to a medium containing a high level of kinetin (1 mg/liter). Such shoots could be induced to root (Yang and Simpson, 1981).

Cloned octopine tumors are usually refractory to regeneration, but infrequently octopine-positive shoots were obtained from such lines (Ooms *et al.,* 1981). However, among octopine lines that were obtained in *in vitro* transformation (cocultivation) experiments with *A. tumefaciens* a high percentage did show regeneration (Marton *et al.,* 1979; Wullems *et al.,* 1981a). Such octopine-positive shoots were morphologically similar to the nopaline-positive shoots described above (Wullems *et al.,* 1981b). These shoots did not form roots, and they were resistant to infection by *A. tumefaciens.* After grafting onto normal tobacco plants the shoots (derived from tobacco lines) flowered, but they set seed only after pollination with pollen from normal tobacco plants. This indicates that the flowers, which showed heterostyly, had male sterility (Wullems *et al.,* 1981b). These data indicate that in the case of octopine tumor cells a suppression of the tumorous state is also possible.

An important question is, of course, whether the suppression of or the recovery from the tumorous condition is due to an alteration in or a loss of the T-DNA. Therefore, some opine-positive and opine-negative shoots regenerated from cloned tumor tissues were studied for their T-DNA content. Nopaline-positive shoots were found to have a T-DNA that was almost identical to the T-DNA present in the callus from which they were regenerated (Lemmers *et al.,* 1980; Yang *et al.,* 1980b). In the case that nopaline-positive shoots were derived from a callus in which the left part of the T-DNA was already missing, there was no change in the T-DNA content after regeneration into plants (Memelink *et al.,* 1982). In nopaline-negative shoots that were obtained after kinetin treatment of a nopaline-positive callus, no T-DNA was found except for some very small parts derived from the ends of normal T-DNA (Yang and Simpson, 1981). From *in vitro* transformation experiments, octopine-positive calli were obtained that displayed either shooting or nonshooting behavior. The calli that generated shoots contained T-DNA from which the auxin region was absent. The octopine-positive shoots that were analyzed had the same T-DNA as the callus from which they were derived (Wullems *et al.,* 1981b). These results suggest that the absence of an intact auxin locus effectuates the suppression of the tumorous condition in octopine crown gall cells. Some nopaline-positive shoots were found to have an intact auxin locus (Lemmers *et al.*, 1980). This shows that deletion of the auxin locus is not the only mechanism by which the tumorous state can be suppressed in crown gall cells. Octopine tumors induced by *A. tumefaciens shooter* mutants were cloned and examined for their growth characteristics and regeneration potential. In spite of the fact that in these lines the auxin locus had been inactivated by a transposon insertion, they were able to grow on media devoid of phytohormones and continued to generate shoots (Van Slogteren, 1983). As shoot apices are centers for the production of auxins, the formation of shoots may be essential for these calli to survive in auxin-free media. Out of 15 cloned lines, 14 gave rise only to octopine-positive shoots. The remaining line (TS038) generated octopine-positive as well as octopine-negative shoots. None of the octopine-positive or octopine-negative shoots formed roots, which suggests that in all the shoots T-DNA was present (G. M. S. Van Slogteren, personal communication). Protoplast isolation experiments showed that the octopine-positive shoots from lines other than TS038 consisted entirely of octopine-positive, tumorous cells. Tissue TS038 was subcloned via a second protoplast isolation step. The calli obtained could be divided into phytohormone autotrophic octopine-positive and octopine-negative calli. The octopine-negative calli generated octopine-negative and a few octopine-positive shoots. The octopine-negative shoots became octopine positive after grafting onto normal tobacco understems. Regulation of the octopine synthesis trait in TS038 appeared to be exerted at the level of transcription (Van Slogteren, 1983). Instability of the T-DNA in an uncloned tumor tissue induced by a *shooter* mutant has been observed by Otten *et al.* (1981). They recovered an octopine-positive shoot upon which root formation could be induced. Moreover, tumors could be formed on this shoot by *A. tumefaciens*. The shoot flowered and set seed after cross-pollination as well as after selfing. These data indicate that the cells in this shoot have lost their tumorous state entirely. In conclusion, it can be said that in transformed plant cells the tumorous character can largely be suppressed if the auxin locus in the T-DNA is inactivated or deleted. Deletion of almost the entire core T-DNA, including the auxin and cytokinin loci, seems necessary for a complete recovery from the tumorous state. However, it is still possible that the partial or complete recovery from the tumorous condition can be accomplished via other mechanisms than deletion of T-DNA.

An important question is whether T-DNA can go through meiosis and be transmitted to the progeny. In order to answer this question, seeds obtained from nopaline-positive shoots were germinated and F_1 plants were raised. None of the F_1 plants produced nopaline, and they lacked T-DNA (Yang et al., 1980b). This shows either that T-DNA was lost during meiosis, or that only cells which had lost T-DNA could participate in meiosis. However, nopaline-positive plants that had only small border fragments of T-DNA in their genome were found to be able to transmit these sequences to their progeny (Yang and Simpson, 1981). This shows that foreign DNA introduced into plant cells via A. tumefaciens may be stable during meiosis. Otten et al. (1981) found that an octopine-positive plant that appeared to be completely normal was able to transmit the octopine synthesis trait to its progeny after selfing as well as after cross-pollination. The data obtained showed that the octopine synthesis trait behaves as a single dominant mendelian marker. Similar data were obtained by Wullems et al. (1981b) for octopine T-DNA characteristics as well as for nopaline T-DNA traits. They used shoots grafted onto normal tobacco plants, because root induction was not possible on these shoots. F_1 plants obtained after sowing in soil were all opine negative. However, after sowing on agar plates opine-positive and -negative F_1 plants were acquired in a ratio of about 1:1. For the opine-positive F_1 plants the development of the root system stopped 2 weeks after germination, and the main root degenerated into a callus. This is probably the reason why after sowing in soil no opine-positive shoots could be recovered. All of the F_1 plants examined, including the opine-negative ones, produced heterostylous flowers with sterile pollen. Therefore, these latter traits may at least partially be an inherent property of the tobacco line (SR1) that was used. The nopaline-positive F_1 plants were analyzed for their T-DNA content. They were found to contain a T-DNA identical in structure to that present in the callus from which the shoot set to flower was derived (Memelink et al., 1983). The auxin locus was absent in this T-DNA. This might be the reason why this T-DNA could be transmitted to an F_1 progeny in contrast to the T-DNA present in the nopaline-positive shoots analyzed by Yang et al. (1980b). In summary, the data known today show that a large part of the T-DNA (including the gene for octopine synthase and the cytokinin locus) can go through meiosis intact.

Hairy roots induced by A. rhizogenes have also been brought into tissue culture. Such roots kept growing as roots in phytohormone-free media (Tepfer and Tempé, 1981). It is possible that the T-DNA of the Ri-plasmid lacks a region that is functionally equivalent to the cytokinin locus of the Ti-plasmid. Because apices of roots are production centers for cytokinins, the root structure of the Ri-transformed cells may be essential to sustain growth on media devoid of cytokinins. In media with a cytokinin hairy roots tend to form a callus (Spano *et al.*, 1981). Plants have been regenerated from hairy roots and hairy root callus (Ackermann, 1977; Chilton *et al.*, 1982; Spano and Costantino, 1982). In such plants mannopine is present (Chilton *et al.*, 1982). The flowers of Ri-transformed tobacco plants (line SR1) showed heterostyly and for the most part their pollen was sterile (Spano and Costantino, 1982). Recently it was shown that the complete T-DNA of the Ri-plasmid can go through plant regeneration and meiosis (Tepfer, 1983.

X. Prospects for the Genetic Engineering of Plants with Ti- and Ri-Plasmids

Recent interest in *A. tumefaciens* and *A. rhizogenes* stems primarily from the finding that these bacteria have systems for the horizontal gene transfer from a prokaryote to a eukaryote. In fact, agrobacteria have been found to genetically engineer plant cells to their own advantage. There are good prospects that Ti- and Ri-plasmids can be exploited in the near future as the first vectors for the genetic manipulation of plants by man. The development of other vehicles for plants (recently reviewed by Howell, 1982) probably will require more time. The introduction of foreign genes into plants would not only be useful for the improvement of certain crops, but also would facilitate fundamental studies of gene expression in plants. Furthermore, the way in which the *onc*-genes in the T-DNA direct differentiation in plants naturally deserves further study.

In fact, the Ti-plasmid has already been used for the transfer of foreign DNA to plant cells. Transposons and other DNA fragments have been introduced into the *T*-region of the Ti-plasmid, and this foreign DNA is integrated into the nuclear DNA of the plant along with the *T*-region (Hernalsteens *et al.*, 1980; Garfinkel *et al.*, 1981; Ooms *et al.*, 1981). In this way, besides the *T*-region, at least 10 Md of foreign DNA (the size of transposon Tn1831) was introduced into plant cells. Foreign DNA integrated along with the T-DNA into plant DNA was recovered from total plant DNA by molecular cloning. Biochemical and genetic analysis showed that this recovered DNA did not carry mutations (Holsters *et al.*, 1982).

Obviously, genes that are to be introduced into plant cells have to be inserted into the *T*-region of a Ti- or Ri-plasmid first. Because direct cloning is impossible, alternative methods have been developed for the transfer of DNA segments to the *T*-region using a combination of *in vitro* and *in vivo* methods (see Section VIII). The Ti-plasmid with the *T*-region containing the desired gene is introduced into *A. tumefaciens,* which then transfers the *T*-region, including the desired gene, to the plant. Alternatively, the manipulated Ti-plasmid DNA can be used directly for the transformation of plant protoplasts (Draper *et al.,* 1982; Krens *et al.,* 1982).

Plant genetic manipulation with Ti-plasmids would have only limited use if the transformed plant cells would be tumor cells that would not regenerate into normal plants. Therefore, it seems obligatory to free the Ti-plasmid from its oncogenes or to inactivate these genes, which inhibit the normal differentiation of plant cells, and then to use such disarmed Ri- or Ti-plasmids for genetic manipulation experiments. T-DNA and other foreign DNA (if freed from *onc*-genes) seem rather stable during differentiation, as well as in mitoses and meioses (see Section IX). A problem in the use of disarmed Ti-plasmid vectors is that transformed plant cells cannot be directly selected because, in contrast to tumor cells, they will not be phytohormone independent. However, they might be selected by the presence of the desired gene, if this is a selectable trait, or by the presence of a selective marker introduced into the Ti-plasmid-based vector. In this respect it is worth noting that the LpDH gene, which is naturally present in octopine Ti-plasmids, can be used for the direct selection of transformed plant cells because it provides transformed cells with resistance to the toxic amino acid analog homoarginine, which it converts into nontoxic homooctopine (Van Slogteren *et al.,* 1982). Alternatively, resistance to the antibiotic G418, to which plant cells are sensitive (Ursic *et al.,* 1981), is encoded by bacterial Km^r determinants that may be developed into a selective marker for plant cells (Colbère-Garapin *et al.,* 1981).

The host range of *A. tumefaciens* and *A. rhizogenes* is confined to dicotyledonous plants (Tamm, 1954; De Cleene and De Ley, 1977). No tumors are formed on monocots, probably because the bacteria cannot attach to their cell walls (Rao *et al.,* 1982). Recently, transformation of plant protoplasts with Ti-plasmid DNA has been described (Draper *et al.,* 1982; Krens *et al.,* 1982), and the introduction of Ti-plasmid DNA via liposomes has also been reported (Dellaporta *et al.,* cited in Howell, 1982). DNA transformation systems may also be developed for monocots, and then Ti-plasmid constructs can be tested in such systems.

Selection for phytohormone independence might not be possible, because monocots probably have a different phytohormone response. However, in such monocot systems transformants might be selected on the basis of homoarginine resistance (Van Slogteren *et al.*, 1982).

Transformation of plant cells with Ti-plasmid DNA occurs with low efficiency, possibly because integration of T-DNA into plant DNA occurs with low frequency. It might therefore be advisable to first introduce a DNA segment with a plant origin of replication and a plant centromere into the *T*-region and to perform plant cell transformation with such constructs, which could independently replicate in plant cells. This would possibly not only increase the frequency of transformation, but would also circumvent potentially deleterious effects of random T-DNA integrations into plant DNA. The plant vector of choice should probably be a small plasmid with the signal sequences necessary to be efficiently transferred from *Agrobacterium* host bacteria to plant cells. A selectable marker and suitable cloning sites should also be present in the vector. Such a vector might even be further improved either by the insertion of a plant replication unit or by the insertion of a piece of plant chromosomal DNA (e.g., rDNA) in order to enhance integration into the plant genome via homologous recombination.

It seems reasonable to assume that Ti-plasmids or derivatives will be used in the near future for the engineering of plants. A number of possible applications of the genetic manipulation of plant cells were recently summarized by Chilton (1980). However, genetic engineering of plant cells probably will become generally applicable only after much research in two areas. First, much more knowledge has to be acquired on the structure and expression of plant genes. This knowledge would allow one to modify isolated genes in such a way that after transfer to plant cells they would be expressed in their new environment. Second, applicability of tissue culture techniques should be extended to more plant species, including the crop plants to be genetically manipulated.

ACKNOWLEDGMENTS

The authors wish to thank all their colleagues for help and for providing information prior to publication. We especially wish to thank Drs. P. W. Laird, G. M. S. Van Slogteren, and R. van Veen for critical reading of the manuscript and for useful suggestions. We would like to thank Ms. S. van den Berg and Mrs. J. B. M. Pruijn-Bruijs for their careful typing of the manuscript.

REFERENCES

Ackermann, C. (1977). Pflanzen aus *Agrobacterium rhizogenes*-tumoren an *Nicotiana tabacum. Plant Sci. Lett.* **8,** 23–30.

Akiyoshi, D. E., Morris, R. O., Hinz, R., Mishke, B. S., Kosuge, T., Garfinkel, D. J., Gordon, M. P., and Nester, E. W. (1983). Cytokinin-auxin balance in crown gall tumors is regulated by specific loci in the T-DNA. *Proc. Natl. Acad. Sci. U.S.A.* **80,** 407–411.

Albinger, G., and Beiderbeck, R. (1977). Übertragung der Fähigkeit zur Wurzelinduktion von *Agrobacterium rhizogenes* auf *A. tumefaciens. Phytopathol. Z.* **90,** 306–310.

Alconero, R. (1980). Crown gall of peaches from Maryland, South Carolina and Tennessee and problems with biological control. *Plant Dis.* **64,** 835–838.

Allen, O. N., and Holding, A. (1974). Genus *Agrobacterium. In* "Bergey's Manual of Determinative Bacteriology" (R. E. Buchanan and N. E. Gibbons, eds.), 8th Ed., pp. 264–267. Williams and Wilkins, Baltimore.

Anand, V. K., and Heberlein, G. T. (1977). Crown gall tumorigenesis in potato tuber tissue. *Am. J. Bot.* **64,** 153–158.

Anderson, A. R., and Moore, L. W. (1979). Host specificity in the genus *Agrobacterium. Phytopathology* **69,** 320–323.

Banerjee, D., Basu, M., Choudhury, I., and Chatterjee, G. C. (1981). Cell surface carbohydrates of *Agrobacterium tumefaciens* involved in adherence during crown gall tumor initiation. *Biochem. Biophys. Res. Commun.* **100,** 1384–1388.

Barnes, W. M. (1977). Plasmid detection and sizing in single colony lysates. *Science* **195,** 393–394.

Beardsley, R. E. (1960). Lysogenicity in *A. tumefaciens. J. Bacteriol.* **80,** 180–187.

Beiderbeck, R. (1973a). Wurzelinduktion an Blättern von *Kalanchöe daigremontiana* durch *Agrobacterium rhizogenes* und der Einfluss von Kinetin auf diesen Prozess. *Z. Pflanzenphysiol.* **68,** 460–467.

Beiderbeck, R. (1973b). Unabhängige Induktion von Teratomen und Wurzeln durch Mischungen verschiedener Stämme von *Agrobacterium. Z. Pflanzenphysiol.* **69,** 163–167.

Beringer, J. E., and Beynon, J. L. (1978). Transfer of the drug-resistance transposon Tn5 to *Rhizobium. Nature (London)* **276,** 633–634.

Bernaerts, M. J., and De Ley, J. (1963). A biochemical test for crown gall bacteria. *Nature (London)* **197,** 406–407.

Bezděk, M., Reich, J., and Tkadleček, L. (1976). Properties of the cured oncogenic strain 37400 of *Agrobacterium tumefaciens. Folia Microbiol.* **21,** 371–377.

Biemann, K., Lioret, C., Asselineau, K., Lederer, E., and Polonski, J. (1960). Sur la structure chimique de la lysopine, nouvel acide aminé isolé de tissu de crown-gall. *Bull. Soc. Chim. Biol.* **42,** 979–991.

Binns, A. N., Wood, H. N., and Braun, A. C. (1981). Suppression of the tumorous state in crown gall teratomas of tobacco. A clonal analysis. *Differentiation* **19,** 97–102.

Bomhoff, G., Klapwijk, P. M., Kester, H. C. M., Schilperoort, R. A., Hernalsteens, J. P., and Schell, J. (1976). Octopine and nopaline synthesis and breakdown genetically controlled by a plasmid of *Agrobacterium tumefaciens. Mol. Gen. Genet.* **145,** 177–181.

Bopp, M., and Resende, F. (1966). Crown-gall-tumoren bei Verschiedenen Arten und Bastarden der *Kalanchoideae. Port. Acta Biol.* **9,** 327–366.

Boucher, C., Bergeron, B., Barate de Bertalmio, M., and Dénarié, J. (1977). Introduction of bacteriophage Mu into *Pseudomonas solanacearum* and *Rhizobium meliloti* using R factor RP4. *J. Gen. Microbiol.* **98,** 253–263.

Boyd, R. J., Hildebrandt, A. C., and Allen, O. N. (1970). Specificity patterns of *Agrobacterium tumefaciens* phages. *Arch. Mikrobiol.* **73**, 324–330.

Braun, A. C. (1943). Studies on tumor inception in the crown-gall disease. *Am. J. Bot.* **30**, 674–677.

Braun, A. C. (1947). Thermal studies on the factors responsible for tumor initiation in crown gall. *Am. J. Bot.* **34**, 234–240.

Braun, A. C. (1948). Studies on the origin and development of plant teratomas incited by the crown-gall bacterium. *Am. J. Bot.* **35**, 511–519.

Braun, A. C. (1951). Recovery of crown-gall tumor cells. *Cancer Res.* **11**, 839–844.

Braun, A. C. (1953). Bacterial and host factors concerned in determining tumor morphology in crown gall. *Bot. Gaz.* **114**, 363–371.

Braun, A. C. (1958). A physiological basis for autonomous growth of crown gall tumor cell. *Proc. Natl. Acad. Sci. U.S.A.* **44**, 344–349.

Braun, A. C. (1959). A demonstration of the recovery of the crown gall tumor cell with the use of complex tumors of single-cell origin. *Proc. Natl. Acad. Sci. U.S.A.* **45**, 932–938.

Braun, A. C., and Wood, H. N. (1976). Suppression of the neoplastic state with the acquisition of specialized functions in cells, tissues, and organs of crown gall teratomas of tobacco. *Proc. Natl. Acad. Sci. U.S.A.* **73**, 496–500.

Bruch, C. W., and Allen, O. N. (1957). Host specificities of four Lotus rhizobiophages. *Can. J. Microbiol.* **3**, 181–189.

Bryan, J., Saeed, N., Fox, D., and Sastry, G. R. K. (1982). R68.45 mediated chromosomal gene transfer in *Agrobacterium tumefaciens*. *Arch. Microbiol.* **131**, 271–277.

Casse, F., Boucher, C., Julliot, J. S., Michel, M., and Dénarié, J. (1979). Identification and characterization of large plasmids in *Rhizobium meliloti* using agarose gel electrophoresis. *J. Gen. Microbiol.* **113**, 229–242.

Chern, C.-K., Hayano, K., Kusaka, I., and Fukui, S. (1976). Dependency of glucose transport on d-glucoside 3-dehydrogenase as a conversion system of glucose to 3-ketoglucose in *Agrobacterium tumefaciens*. *Agric. Biol. Chem.* **40**, 2299–2300.

Chilton, M.-D. (1980). Agrobacterium Ti plasmids as a tool for genetic engineering in plants. *In* "Genetic Engineering of Osmoregulation" (D. W. Rains, R. C. Valentine, and A. Hollaender, eds.), pp. 23–31. Plenum, New York.

Chilton, M.-D., Farrand, S. K., Levin, R. L., and Nester, E. W. (1976). RP4 promotion of transfer of a large *Agrobacterium* plasmid which confers virulence. *Genetics* **83**, 609–618.

Chilton, M.-D., Drummond, M. H., Merlo, D. J., Sciaky, D., Montoya, A. L., Gordon, M. P., and Nester, E. W. (1977). Stable incorporation of plasmid DNA into higher plant cells: The molecular basis of crown gall tumorigenesis. *Cell* **11**, 263–271.

Chilton, M.-D., Montoya, A. L., Merlo, D. J., Drummond, M. H., Nutter, R., Gordon, M. P., and Nester, E. W. (1978a). Restriction endonuclease mapping of a plasmid that confers oncogenicity upon *Agrobacterium tumefaciens* strain B6-806. *Plasmid* **1**, 254–269.

Chilton, M.-D., Drummond, M. H., Merlo, D. J., and Sciaky, D. (1978b). Highly conserved DNA of Ti-plasmids overlaps T-DNA maintained in plant tumors. *Nature (London)* **275**, 147–149.

Chilton, M.-D., Saiki, R. K., Yadav, N., Gordon, M. P., and Quétier, F. (1980). T-DNA from *Agrobacterium* Ti plasmid is in the nuclear DNA fraction of crown gall tumor cells. *Proc. Natl. Acad. Sci. U.S.A.* **77**, 4060–4064.

Chilton, M.-D., Tepfer, D. A., Petit, A., David, C. C., Delbert, F., and Tempé, J. (1982). *Agrobacterium rhizogenes* inserts T-DNA into plant roots. *Nature (London)* **295**, 432–434.

Clark, A. G. (1972). A comparison of the properties of *Agrobacterium tumefaciens* with those of other phytopathogens, symbionts and saprophytes. *Phytopathol. Z.* **74**, 74–83.

Clark, A. G. (1974). Indole acetic acid production by *Agrobacterium* and *Rhizobium* species. *Microbios* **11A**, 29–35.

Clewell, D. B., and Helinski, D. R. (1969). Supercoiled circular DNA-protein complex in *E. coli*: Purification and induced conversion to an open circular DNA form. *Proc. Natl. Acad. Sci. U.S.A.* **62**, 1159–1166.

Colak, Ö., and Beiderbeck, R. (1979). Die Übertragung der Fähigkeit zur Tumorinduktion von *Agrobacterium tumefaciens* auf *A. rhizogenes*. *Phytopath. Z.* **96**, 268–272.

Colbère-Garapin, F., Horodniceanu, F., Kourilsky, P., and Garapin, A. (1981). A new dominant hybrid selective marker for higher eukaryotic cells. *J. Mol. Biol.* **150**, 1–14.

Cole, M. A., and Elkan, G. H. (1973). Transmissible resistance to penicillin G, neomycin, and chloramphenicol in *Rhizobium japonicum*. *Antimicrob. Agents Chemother.* **4**, 248–253.

Conn, H. J., Bottcher, E. J., and Randall, C. (1945). The value of bacteriophage in classifying certain soil bacteria. *J. Bacteriol.* **49**, 359–373.

Cooksey, D. A., and Moore, L. W. (1980a). Biological control of crown gall with fungal and bacterial antagonists. *Phytopathology* **70**, 506–509.

Cooksey, D. A., and Moore, L. W. (1980b). An agrocin mutant of *Agrobacterium radiobacter* K84 and biological control of crown gall. *Phytopathology* **71**, 104.

Cooksey, D. A., and Moore, L. W. (1982). High frequency spontaneous mutations to agrocin 84 resistance in *Agrobacterium tumefaciens* and *A. rhizogenes*. *Physiol. Plant Pathol.* **20**, 129–135.

Costantino, P., Hooykaas, P. J. J., Den Dulk-Ras, H., and Schilperoort, R. A. (1980). Tumor formation and rhizogenicity of *Agrobacterium rhizogenes* carrying Ti plasmids. *Gene* **11**, 79–87.

Costantino, P., Mauro, M. L., Micheli, G., Risuleo, G., Hooykaas, P. J. J., and Schilperoort, R. A. (1981). Fingerprinting and sequence homology of plasmids form different virulent strains of *Agrobacterium rhizogenes*. *Plasmid* **5**, 170–182.

Currier, T. C., and Nester, E. W. (1976a). Isolation of covalently closed circular DNA of high molecular weight from bacteria. *Anal. Biochem.* **66**, 431–441.

Currier, T. C., and Nester, E. W. (1976b). Evidence for diverse types of large plasmids in tumor-inducing strains of *Agrobacterium*. *J. Bacteriol.* **126**, 157–165.

Datta, N., and Hedges, R. W. (1972). Host ranges of R factors. *J. Gen. Microbiol.* **70**, 453–460.

De Beuckeleer, M., Lemmers, M., De Vos, G., Willmitzer, L., Van Montagu, M., and Schell, J. (1981). Further insight on the transferred-DNA of octopine crown gall. *Mol. Gen. Genet.* **183**, 283–288.

De Cleene, M., and De Ley, J. (1977). The host range of crown gall. *Bot. Rev.* **42**, 389–466.

De Greve, H., Decraemer, H., Seurinck, J., Van Montagu, M., and Schell, J. (1981). The functional organization of the octopine *Agrobacterium tumefaciens* plasmid pTiB6S3. *Plasmid* **6**, 235–248.

De Picker, A., Van Montagu, M., and Schell, J. (1978). Homologous DNA sequences in different Ti plasmids are essential for oncogenicity. *Nature (London)* **275**, 150–153.

De Picker, A., De Block, M., Inzé, D., Van Montagu, M., and Schell, J. (1980a). IS-like element IS8 in RP4 plasmid and its involvement in cointegration. *Gene* **10**, 329–338.

De Picker, A., De Wilde, M., De Vos, G., De Vos, R., Van Montagu, M., and Schell, J. (1980b). Molecular cloning of overlapping segments of the nopaline Ti-plasmid pTiC58 as a means to restriction endonuclease mapping. *Plasmid* **3**, 193–211.

De Vos, G., De Beuckeleer, M., Van Montagu, M., and Schell, J. (1981). Restriction endonuclease mapping of the octopine tumor-inducing plasmid pTiAch5 of *Agrobacterium tumefaciens*. *Plasmid* **6**, 249–253.

Dhanvantari, B. N. (1978). Characterization of *Agrobacterium* isolates from stone fruits in Ontario. *Can. J. Bot.* **56**, 2309–2311.

Douglas, C., Tanimoto, E., and Halperin, W. (1979). Early detection of octopine in crown-gall tumors of Jerusalem artichoke. *Plant Sci. Lett.* **15**, 89–99.

Draper, J., Davey, M. R., Freeman, J. P., Cocking, E. C., and Cox, B. J. (1982). Ti plasmid homologous sequences present in tissues from *Agrobacterium* plasmid-transformed *Petunia* protoplasts. *Plant Cell Physiol.* **23**, 451–458.

Drummond, M. H., and Chilton, M.-D. (1978). Tumor-inducing (Ti) plasmids of *Agrobacterium* share extensive regions of DNA homology. *J. Bacteriol.* **136**, 1178–1183.

Drummond, M. H., Gordon, M. P., Nester, E. W., and Chilton, M.-D. (1977). Foreign DNA of bacterial plasmid origin is transcribed in crown gall tumours. *Nature (London)* **269**, 535–536.

Dulbecco, R. (1979). Contributions of microbiology to eucaryotic cell biology: New directions for microbiology. *Microbiol. Rev.* **43**, 443–452.

Eckhardt, T. (1978). A rapid method for the identification of plasmid deoxyribonucleic acid in bacteria. *Plasmid* **1**, 584–588.

El-Fiki, F., and Giles, K. L. (1981). *Agrobacterium tumefaciens* in agriculture and research. *Int. Rev. Cytol.* **13**, 47–58.

Ellis, J. G., and Murphy, P. J. (1981). Four new opines from crown gall tumours—their detection and properties. *Mol. Gen. Genet.* **181**, 36–43.

Ellis, J. G., Kerr, A., Tempé, J., and Petit, A. (1979). Arginine catabolism: A new function of both octopine and nopaline Ti-plasmids of *Agrobacterium*. *Mol. Gen. Genet.* **173**, 263–269.

Ellis, J. G., Kerr, A., Petit, A., and Tempé, J. (1982a). Conjugal transfer of nopaline and agropine Ti-plasmids. The role of agrocinopines. *Mol. Gen. Genet.* **186**, 269–274.

Ellis, J. G., Murphy, P. J., and Kerr, A. (1982b). Isolation and properties of transfer regulatory mutants of the nopaline Ti-plasmid pTiC58. *Mol. Gen. Genet.* **186**, 275–281.

Engler, G., Holsters, M., Van Montagu, M., Schell, J., Hernalsteens, J. P., and Schilperoort, R. (1975). Agrocin 84 sensitivity: A plasmid determined property in *Agrobacterium tumefaciens*. *Mol. Gen. Genet.* **138**, 345–349.

Engler, G., Depicker, A., Maenhout, R., Villarroel, R., Van Montagu, M., and Schell, J. (1981). Physical mapping of DNA base sequence homologies between an octopine and a nopaline Ti-plasmid of *Agrobacterium tumefaciens*. *J. Mol. Biol.* **152**, 183–208.

Expert, D., and Tourneur, J. (1982). ψ, a temperate phage of *Agrobacterium tumefaciens*, is mutagenic. *J. Virol.* **42**, 283–291.

Expert, D., Riviere, F., and Tourneur, J. (1982). The state of phage ψ DNA in lysogenic cells of *Agrobacterium tumefaciens*. *Virology* **121**, 82–94.

Farrand, S. K., Kado, C. I., and Ireland, C. R. (1981). Suppression of tumorigenicity by the inc W R-plasmid pSa in *Agrobacterium tumefaciens*. *Mol. Gen. Genet.* **181**, 44–51.

Firmin, J. L., and Fenwick, R. G. (1977). N^2-(1,3 dicarboxypropyl) ornithine in crown gall tumours. *Phytochemistry* **16**, 761–762.

Firmin, J. L., and Fenwick, G. R. (1978). Agropine—a major new plasmid-determined metabolite in crown gall tumours. *Nature (London)* **276**, 842–844.

Galsky, A. G., Wilsey, J. P., and Powell, R. G. (1980). Crown gall tumor disc bioassay. A possible aid in the detection of compounds with antitumor activity. *Plant Physiol.* **65**, 184–185.

Galsky, A. G., Scheppler, J. A., and Cranford, M. S. (1981). Crown gall tumors possess tumor-specific antigenic sites on their cell walls. *Plant Physiol.* **67**, 1195–1197.

Garfinkel, D. J., and Nester, E. W. (1980). *Agrobacterium tumefaciens* mutants affected in crown gall tumorigenesis and octopine catabolism. *J. Bacteriol.* **144**, 732–743.

Garfinkel, D. J., Simpson, R. B., Ream, L. W., White, F. F., Gordon, M. P., and Nester, E. W. (1981). Genetic analysis of crown gall: Fine structure map of the T-DNA by site-directed mutagenesis. *Cell* **27**, 143–153.

Gelvin, S. B., Gordon, M. P., Nester, E. W., and Aronson, A. I. (1981). Transcription of the *Agrobacterium* Ti-plasmid in the bacterium and in crown gall tumors. *Plasmid* **6**, 17–29.

Gelvin, S. B., Thomashow, M. F., McPherson, J. C., Gordon, M. P., and Nester, E. W. (1982). Sizes and map positions of several plasmid DNA-encoded transcripts in octopine-type crown gall tumors. *Proc. Natl. Acad. Sci. U.S.A.* **79**, 76–80.

Genetello, C., Van Larebeke, N., Holsters, M., De Picker, A., Van Montagu, M., and Schell, J. (1977). The Ti-plasmids of *Agrobacterium* as conjugative plasmids. *Nature (London)* **265**, 561–563.

Glogowski, W., and Galsky, A. G. (1978). *Agrobacterium tumefaciens* site attachment as a necessary prerequisite for crown gall tumor formation on potato disks. *Plant Physiol.* **61**, 1031–1033.

Goldmann, A. (1977). Octopine and nopaline dehydrogenases in crown-gall tumors. *Plant Sci. Lett.* **10**, 49–58.

Goldmann, A., Thomas, D. W., and Morel, G. (1969). Sur la structure de la nopaline, métabolite anormale de certaines tumeurs de crown gall. *C. R. Acad. Sci. Paris* **268**, 852–854.

Graham, P. H. (1964). The application of computer techniques to the taxonomy of root-nodule bacteria of legumes. *J. Gen. Microbiol.* **35**, 511–517.

Gresshoff, P. M., Skotnicki, M. L., and Rolfe, B. G. (1979). Crown gall teratoma formation is plasmid and plant controlled. *J. Bacteriol.* **137**, 1020–1021.

Grindley, N. D. F., Humphreys, G. O., and Anderson, E. S. (1973). Molecular studies of R factor compatibility groups. *J. Bacteriol.* **115**, 387–398.

Gurley, W. B., Kemp, J. D., Albert, M. J., Sutton, D. W., and Callis, J. (1979). Transcription of Ti-plasmid-derived sequences in the octopine-type crown gall tumor lines. *Proc. Natl. Acad. Sci. U.S.A.* **76**, 2828–2832.

Guyon, P., Chilton, M.-D., Petit, A., and Tempé, J. (1980). Agropine in "null-type" crown gall tumors: Evidence for generality of the opine concept. *Proc. Natl. Acad. Sci. U.S.A.* **77**, 2693–2697.

Hack, E., and Kemp, J. D. (1977). Comparison of octopine, histopine, lysopine and octopinic acid synthesizing activities in sunflower crown gall tissues. *Biochem. Biophys. Res. Commun.* **78**, 785–791.

Hamada, S. E., Luckey, J. P., and Farrand, S. K. (1979). R-plasmid-mediated chromosomal gene transfer in *Agrobacterium tumefaciens*. *J. Bacteriol.* **139**, 280–286.

Hamilton, R. H., and Chopan, M. N. (1975). Transfer of the tumor inducing factor in *Agrobacterium tumefaciens*. *Biochem. Biophys. Res. Commun.* **63**, 349–354.

Hamilton, R. H., and Fall, M. Z. (1971). The loss of tumor-initiating ability in *Agrobacterium tumefaciens* by incubation at high temperature. *Experientia* **27**, 229–230.

Hansen, J. B., and Olsen, R. H. (1978). Isolation of large bacterial plasmids and characterization of the P2 incompatibility group plasmids pMG1 and pMG5. *J. Bacteriol.* **135,** 227–238.

Harth, G., Koch, H.-G., Darai, G., Schweitzer, S., and Geider, K. (1982). Amplification of a specific Ti-plasmid sequence in mammalian cells as circular DNA after transfection with Ti plasmid of *Agrobacterium tumefaciens. Zentralbl. Bakteriol.* **253,** 18.

Hepburn, A. G., and Hindley, J. (1979). Regions of DNA sequence homology between an octopine and a nopaline Ti-plasmid of *Agrobacterium tumefaciens. Mol. Gen. Genet.* **169,** 163–172.

Hernalsteens, J. P., Villarroel-Mandiola, R., Van Montagu, M., and Schell, J. (1977). Transposition of Tn1 to a broad-host-range drug resistance plasmid. *In* "DNA Insertion Elements, Plasmids and Episomes" (A. I. Bukhari, J. A. Shapiro, and S. L. Adhya, eds.), pp. 179–183. Cold Spring Harbor Laboratory, Cold Spring Harbor, New York.

Hernalsteens, J. P., De Greve, H., Van Montagu, M., and Schell, J. (1978). Mutagenesis by insertion of the drug resistance transposon Tn7 applied to the Ti-plasmid of *Agrobacterium tumefaciens. Plasmid* **1,** 218–255.

Hernalsteens, J. P., Van Vliet, F., De Beuckeleer, M., Depicker, A., Engler, G., Lemmers, M., Holsters, M., Van Montagu, M., and Schell, J. (1980). The *Agrobacterium tumefaciens* Ti-plasmid as a host vector system for introducing foreign DNA in plant cells. *Nature (London)* **287,** 654–656.

Hille, J., and Schilperoort, R. (1981a). Behavior of *inc*Q plasmids in *Agrobacterium tumefaciens. Plasmid* **6,** 360–362.

Hille, J., and Schilperoort, R. (1981b). The use of transposons to introduce well-defined deletions in plasmids: Possibilities for *in vivo* cloning. *Plasmid* **6,** 151–154.

Hille, J., Klasen, I., and Schilperoort, R. (1982). Construction and application of R prime plasmids, carrying different segments of an octopine Ti-plasmid from *Agrobacterium tumefaciens* for complementation of *vir* genes. *Plasmid* **7,** 107–118.

Hille, J., Van Kan, J., Klasen, I., and Schilperoort, R. (1983). Site-directed mutagenesis in *Escherichia coli* of a stable R772:Ti cointegrate from *Agrobacterium tumefaciens. J. Bacteriol.* **154,** 693–701.

Holsters, M., Silva, B., Van Vliet, F., Hernalsteens, J. P., Genetello, C., Van Montagu, M., and Schell, J. (1978a). In vivo transfer of the Ti-plasmid of *Agrobacterium tumefaciens* to *Escherichia coli. Mol. Gen. Genet.* **163,** 335–338.

Holsters, M., Silva, A., Genetello, C., Engler, G., Van Vliet, F., De Block, M., Villarroel, R., Van Montagu, M., and Schell, J. (1978b). Spontaneous formation of cointegrates of the oncogenic Ti-plasmid and the wide-host-range P-plasmid RP4. *Plasmid* **1,** 456–467.

Holsters, M., De Waele, D., Depicker, A., Messens, E., Van Montagu, M., and Schell, J. (1978c). Transfection and transformation of *Agrobacterium tumefaciens. Mol. Gen. Genet.* **163,** 181–187.

Holsters, M., Silva, B., Van Vliet, F., Genetello, C., De Block, M., Dhaese, P., Depicker, A., Inzé, D., Engler, G., Villarroel, R., Van Montagu, M., and Schell, J. (1980). The functional organization of the nopaline *A. tumefaciens* plasmid pTiC58. *Plasmid* **3,** 212–230.

Holsters, M., Villarroel, R., Van Montagu, M., and Schell, J. (1982). The use of selectable markers for the isolation of plant DNA/T-DNA junction fragments in a cosmid vector. *Mol. Gen. Genet.* **185,** 283–289.

Hooykaas, P. J. J. (1979). "The role of plasmid determined functions in the interactions of Rhizobiaceae with plant cells. A genetic approach." Ph.D. thesis, Leiden University.

Hooykaas, P. J. J., Klapwijk, P. M., Nuti, M. P., Schilperoort, R. A., and Rörsch, A. (1977). Transfer of the *Agrobacterium tumefaciens* Ti-plasmid to avirulent agrobacteria and to *Rhizobium ex planta*. *J. Gen. Microbiol.* **98**, 477–484.

Hooykaas, P. J. J., Schilperoort, R. A., and Rörsch, A. (1979a). Agrobacterium tumor inducing plasmids: Potential vectors for the genetic engineering of plants. *In* "Genetic Engineering" (J. Setlow and A. Hollaender, eds.), Vol. 1, pp. 151–179. Plenum, New York.

Hooykaas, P. J. J., Roobol, C., and Schilperoort, R. A. (1979b). Regulation of the transfer of Ti-plasmids of *Agrobacterium tumefaciens*. *J. Gen. Microbiol.* **110**, 99–109.

Hooykaas, P. J. J., Den Dulk-Ras, H., and Schilperoort, R. A. (1980a). Molecular mechanism of Ti-plasmid mobilization by R-plasmids. Isolation of Ti-plasmids with transposon-insertions in *Agrobacterium tumefaciens*. *Plasmid* **4**, 65–74.

Hooykaas, P. J. J., Den Dulk-Ras, H., Ooms, G., and Schilperoort, R. A. (1980b). Interactions between octopine and nopaline plasmids in *Agrobacterium tumefaciens*. *J. Bacteriol.* **143**, 1295–1306.

Hooykaas, P. J. J., Van Brussel, A. A. N., Den Dulk-Ras, H., Van Slogteren, G. M. S., and Schilperoort, R. A. (1981). Sym plasmid of *Rhizobium trifolii* expressed in different rhizobial species and *Agrobacterium tumefaciens*. *Nature (London)* **291**, 351–353.

Hooykaas, P. J. J., Ooms, G., and Schilperoort, R. A. (1982a). Tumors induced by different strains of *Agrobacterium tumefaciens*. *In* "Molecular Biology of Plant Tumors" (G. Kahl and J. Schell, eds.), pp. 373–390. Academic Press, New York.

Hooykaas, P. J. J., Den Dulk-Ras, H., and Schilperoort, R. A. (1982b). Method for the transfer of large cryptic, non-self-transmissible plasmids: *ex planta* transfer of the virulence plasmid of *Agrobacterium rhizogenes*. *Plasmid* **8**, 94–96.

Hooykaas, P. J. J., Peerbolte, R., Regensburg-Tuïnk, A. J. G., De Vries, P., and Schilperoort, R. A. (1982c). A chromosomal linkage map of *Agrobacterium tumefaciens* and a comparison with the maps of *Rhizobium* spp. *Mol. Gen. Genet.* **188**, 12–17.

Hooykaas, P. J. J., Snijdewint, F. G. M., and Schilperoort, R. A. (1982d). Identification of the Sym plasmid of *Rhizobium leguminosarum* strain 1001 and its transfer to and expression in other rhizobia and *Agrobacterium tumefaciens*. *Plasmid* **8**, 73–82.

Hooykaas, P. J. J., Den Dulk-Ras, H., and Schilperoort, R. A. (1982e). Phenotypic expression of mutations in a wide-host range R plasmid in *Escherichia coli* and *Rhizobium meliloti*. *J. Bacteriol.* **150**, 395–397.

Howell, S. H. (1982). Plant molecular vehicles: Potential vectors for introducing foreign DNA into plants. *Annu. Rev. Plant Physiol.* **33**, 609–650.

Huff, M., and Turgeon, R. (1981). Neoplastic potential of trichomes isolated from tobacco crown gall teratomas. *Differentiation* **19**, 93–96.

Jaynes, J. M., and Strobel, G. A. (1981). The position of *Agrobacterium rhizogenes*. *Int. Rev. Cytol.* **13**, 105–125.

Kado, C. I., and Lin, S.-T. (1981). Rapid procedure for detection and isolation of large and small plasmids. *J. Bacteriol.* **145**, 1365–1373.

Kaiss-Chapman, R. W., and Morris, R. O. (1977). *Trans*-zeatin in culture filtrates of *Agrobacterium tumefaciens*. *Biochem. Biophys. Res. Commun.* **76**, 453–459.

Kartašova, T., Huisman, H., and Schilperoort, R. (1981). A coupled transcription-translation system from *Agrobacterium tumefaciens* and its application to study Ti-plasmid expression in vitro. *Nucleic Acids Res.* **9**, 6763–6772.

Keane, P. J., Kerr, A., and New, P. B. (1970). Crown gall of stone fruit. II. Identification and nomenclature of *Agrobacterium* isolates. *Aust. J. Biol. Sci.* **23**, 585–595.

Kemp, J. D. (1977). A new amino acid derivative present in crown gall tumor tissue. *Biochem. Biophys. Res. Commun.* **74**, 862–868.

Kennedy, B. W., and Alcorn, S. M. (1980). Estimates of USA crop losses to prokaryotic plant pathogens. *Plant Dis.* **64**, 674–676.

Kerr, A. (1969). Transfer of virulence between isolates of *Agrobacterium*. *Nature (London)* **223**, 1175–1176.

Kerr, A. (1971). Acquisition of virulence by non-pathogenic isolates of *Agrobacterium radiobacter*. *Physiol. Plant Pathol.* **1**, 241–246.

Kerr, A. (1972). Biological control of crown gall: Seed inoculation. *J. Appl. Bacteriol.* **35**, 493–497.

Kerr, A. (1980). Biological control of crown gall through production of agrocin 84. *Plant Dis.* **64**, 25–30.

Kerr, A., and Htay, K. (1974). Biological control of crown gall through bacteriocin production. *Physiol. Plant Pathol.* **4**, 37–44.

Kerr, A., and Panagopoulos, C. G. (1977). Biotypes of *Agrobacterium radiobacter* var. *tumefaciens* and their biological control. *Phytopathol. Z.* **90**, 172–179.

Kerr, A., and Roberts, W. P. (1976). *Agrobacterium*: Correlations between and transfer of pathogenicity, octopine and nopaline metabolism and bacteriocin 84 sensitivity. *Physiol. Plant Pathol.* **9**, 205–221.

Kerr, A., Manigault, P., and Tempé, J. (1977). Transfer of virulence *in vivo* and *in vitro* in *Agrobacterium*. *Nature (London)* **265**, 560–561.

Kersters, K., De Ley, J., Sneath, P. H. A., and Sackin, M. (1973). Numerical taxonomic analysis of *Agrobacterium*. *J. Gen. Microbiol.* **78**, 227–239.

Klapwijk, P. M. (1979). "The role of octopine in conjugation and plant tumor formation by *Agrobacterium tumefaciens*. Studies on tumour-inducing plasmids." Ph.D. thesis, Leiden University.

Klapwijk, P. M., and Schilperoort, R. A. (1979). Negative control of octopine degradation and transfer genes of octopine Ti-plasmids in *Agrobacterium tumefaciens*. *J. Bacteriol.* **139**, 424–431.

Klapwijk, P. M., and Schilperoort, R. A. (1982). Genetic determination of octopine degradation. *In* "Molecular Biology of Plant Tumors" (G. Kahl and J. Schell, eds.), pp. 475–495. Academic Press, New York.

Klapwijk, P. M., De Jonge, A. J. R., Schilperoort, R. A., and Rörsch, A. (1975). An enrichment technique for auxotrophs of *Agrobacterium tumefaciens* using a combination of carbenicillin and lysozyme. *J. Gen. Microbiol.* **91**, 177–182.

Klapwijk, P. M., Hooykaas, P. J. J., Kester, H. C. M., Schilperoort, R. A., and Rörsch, A. (1976). Isolation and characterization of *Agrobacterium tumefaciens* mutants affected in the utilization of octopine, octopinic acid and lysopine. *J. Gen. Microbiol.* **96**, 155–163.

Klapwijk, P. M., Oudshoorn, M., and Schilperoort, R. A. (1977). Inducible permease involved in the uptake of octopine, lysopine, and octopinic acid by *Agrobacterium tumefaciens* strains carrying virulence-associated plasmids. *J. Gen. Microbiol.* **107**, 1–11.

Klapwijk, P. M., Scheulderman, T., and Schilperoort, R. A. (1978). Coordinated regulation of octopine degradation and conjugative transfer of Ti-plasmids in *Agrobacterium tumefaciens*: Evidence for a common regulatory gene and separate operons. *J. Bacteriol.* **136**, 775–785.

Klapwijk, P. M., Van Beelen, P., and Schilperoort, R. A. (1979). Isolation of a recombination deficient *Agrobacterium tumefaciens* mutant. *Mol. Gen. Genet.* **173,** 171–175.

Klapwijk, P. M., Van Breukelen, J., Koorevaar, K., Ooms, G., and Schilperoort, R. A. (1980). Transposition of Tn904 encoding streptomycin resistance into the octopine Ti-plasmid of *Agrobacterium tumefaciens. J. Bacteriol.* **141,** 129–136.

Klee, H. J., Gordon, M. P., and Nester, E. W. (1982). Complementation analysis of *Agrobacterium tumefaciens* Ti-plasmid mutations affecting oncogenicity. *J. Bacteriol.* **150,** 327–331.

Klein, R. M., and Tenenbaum, I. L. (1955). A quantitative bioassay for crown-gall tumor formation. *Am. J. Bot.* **42,** 709–712.

Knauf, V. C., and Nester, E. W. (1982). Wide host range cloning vectors: A cosmid clone bank of an *Agrobacterium* Ti-plasmid. *Plasmid* **8,** 45–54.

Koekman, B. P., Ooms, G., Klapwijk, P. M., and Schilperoort, R. A. (1979). Genetic map of an octopine Ti-plasmid. *Plasmid* **2,** 347–357.

Koekman, B. P., Hooykaas, P. J. J., and Schilperoort, R. A. (1980). Localization of the replication control region on the physical map of the octopine Ti-plasmid. *Plasmid* **4,** 184–195.

Koekman, B. P., Hooykaas, P. J. J., and Schilperoort, R. A. (1982). A functional map of the replicator region of the octopine Ti-plasmid. *Plasmid* **7,** 119–132.

Krens, F. A., Molendijk, L., Wullems, G. J., and Schilperoort, R. A. (1982). *In vitro* transformation of plant protoplasts with Ti-plasmid DNA. *Nature (London)* **296,** 72–74.

Kurkdjian, A., Manigault, P., and Beardsley, R. (1974). The pea seedling as a model of normal and abnormal morphogenesis. *Am. Biol. Teach.* **36,** 13–20.

Langley, R. A., and Kado, C. I. (1972). Studies on *Agrobacterium tumefaciens*. Conditions for mutagenesis by N-methyl-N'-nitro-N-nitrosoguanidine and relationships of *A. tumefaciens* mutants to crown gall tumor induction. *Mutat. Res.* **14,** 277–286.

Ledeboer, A. M. (1978). "Large plasmids in Rhizobiaceae. Studies on the transcription of the tumour inducing plasmid from *Agrobacterium tumefaciens* in sterile crown gall tumour cells." Ph.D. thesis, University of Leiden.

Ledeboer, A. M., Krol, A. J. M., Dons, J. J. M., Spier, F., Schilperoort, R. A., Zaenen, I., Van Larebeke, N., and Schell, J. (1976). On the isolation of Ti plasmid from *Agrobacterium tumefaciens. Nucleic Acids Res.* **3,** 449–463.

Leemans, J., Shaw, C., Deblaere, R., De Greve, H., Hernalsteens, J. P., Maes, M., Van Montagu, M., and Schell, J. (1981). Site-specific mutagenesis of *Agrobacterium* Ti plasmids and transfer of genes to plant cells. *J. Mol. Appl. Genet.* **1,** 149–164.

Leemans, J., Deblaere, R., Willmitzer, L., De Greve, H., Hernalsteens, J. P., Van Montagu, M., and Schell, J. (1982). Genetic identification of functions of t_1-DNA transcripts in octopine crown galls. *EMBO J.* **1,** 147–152.

Lemmers, M., De Beuckeleer, M., Holsters, M., Zambryski, P., Depicker, A., Hernalsteens, J. P., Van Montagu, M., and Schell, J. (1980). Internal organization, boundaries and integration of Ti-plasmid DNA in nopaline crown gall tumours. *J. Mol. Biol.* **144,** 353–376.

Lin, B.-C., and Kado, C. I. (1977). Studies on *Agrobacterium tumefaciens* VII. Avirulence induced by temperature and ethidium bromide. *Can. J. Microbiol.* **23,** 1554–1561.

Lipetz, J. (1966). Crown gall tumorigenesis II. Relations between wound healing and the tumorigenic response. *Cancer Res.* **26,** 1597–1604.

Lippincott, J. A., and Heberlein, G. T. (1965). The induction of leaf tumors by *Agrobacterium tumefaciens. Am. J. Bot.* **52,** 396–403.

Lippincott, B. B., and Lippincott, J. A. (1969). Bacterial attachment to a specific wound site as an essential stage in tumor initiation by *Agrobacterium tumefaciens*. *J. Bacteriol.* **97**, 620–628.

Lippincott, J. A., and Lippincott, B. B. (1970). Enhanced tumor initiation by mixtures of tumorigenic and nontumorigenic strains of *Agrobacterium*. *Infect. Immun.* **2**, 623–630.

Lippincott, J. A., and Lippincott, B. B. (1978a). Cell walls of crown-gall tumors and embryonic plant tissues lack *Agrobacterium* adherence sites. *Science* **199**, 1075–1078.

Lippincott, J. A., and Lippincott, B. B. (1978b). Tumor initiation complementation on bean leaves by mixtures of tumorigenic and nontumorigenic *Agrobacterium rhizogenes*. *Phytopathology* **68**, 365–370.

Lippincott, J. A., and Lippincott, B. B. (1980). Microbial adherence in plants. *Recept. Recognition Ser. B* **6**, 377–398.

Lippincott, B. B., Margot, J. B., and Lippincott, J. A. (1977). Plasmid content and tumor initiation complementation by *Agrobacterium tumefaciens* IIBNV6. *J. Bacteriol.* **132**, 824–831.

Liu, S. T., and Kado, C. I. (1979). Indoleacetic acid production: A plasmid function of *Agrobacterium tumefaciens* C58. *Biochem. Biophys. Res. Commun.* **90**, 171–178.

Liu, S. T., Perry, K. L., Schardl, C. L., and Kado, C. I. (1982). *Agrobacterium* Ti-plasmid indoleacetic acid gene is required for crown gall oncogenesis. *Proc. Natl. Acad. Sci. U.S.A.* **79**, 2812–2816.

Loper, J. E., and Kado, C. I. (1979). Host range conferred by the virulence-specifying plasmid of *Agrobacterium tumefaciens*. *J. Bacteriol.* **139**, 591–596.

McPherson, J. C., Nester, E. W., and Gordon, M. P. (1980). Proteins encoded by *Agrobacterium tumefaciens* Ti-plasmid DNA (T-DNA) in crown gall tumors. *Proc. Natl. Acad. Sci. U.S.A.* **77**, 2666–2670.

Marton, L., Wullems, G. J., Molendijk, L., and Schilperoort, R. A. (1979). *In vitro* transformation of cultured cells from *Nicotiana tabacum* by *Agrobacterium tumefaciens*. *Nature (London)* **277**, 129–131.

Matthysse, A. G., Wyman, P. M., and Holmes, K. V. (1978). Plasmid-dependent attachment of *Agrobacterium tumefaciens* to plant tissue culture cells. *Infect. Immun.* **22**, 516–522.

Matthysse, A. G., Holmes, K. V., and Gurlitz, R. H. G. (1981). Elaboration of cellulose fibrils by *Agrobacterium tumefaciens* during attachment to carrot cells. *J. Bacteriol.* **145**, 583–595.

Matzke, A. J. M., and Chilton, M.-D. (1981). Site-specific insertion of genes into T-DNA of the *Agrobacterium* tumor-inducing plasmid: An approach to genetic engineering of higher plant cells. *J. Mol. Appl. Genet.* **1**, 39–49.

Memelink, J., Wullems, G. J., and Schilperoort, R. A. (1983). Nopaline T-DNA is maintained during regeneration and generative propagation of transformed tobacco plants. *Mol. Gen. Genet.* **190**, 516–522.

Ménagé, A., and Morel, G. (1964). Sur la présence d'octopine dans les tissues de crown-gall. *C. R. Acad. Sci. Paris Ser. D* **259**, 4795–4796.

Ménagé, A., and Morel, G. (1965). Sur la présence d'un acide aminé nouveau dans le tissu de crown-gall. *C. R. Soc. Biol.* **159**, 561–562.

Merlo, D. J., and Nester, E. W. (1977). Plasmids in avirulent strains of *Agrobacterium*. *J. Bacteriol.* **129**, 76–80.

Milani, V. J., and Heberlein, G. T. (1972). Transfection in *A. tumefaciens. J. Virol.* **10**, 17–22.

Miller, C. O. (1974). Ribosyl-*trans*-zeatin, a major cytokinin produced by crown gall tumor tissue. *Proc. Natl. Acad. Sci. U.S.A.* **71**, 334–338.

Mitter, E., and Beiderbeck, R. (1980). Symptomausprägung an *Catharanthus*-sprossen nach Infektion mit *Agrobacterium rhizogenes. Z. Pflanzenphysiol.* **100**, 311–317.

Montoya, A. L., Chilton, M.-D., Gordon, M. P., Sciaky, D., and Nester, E. W. (1977). Octopine and nopaline metabolism in *Agrobacterium tumefaciens* and crown gall tumor cells: Role of plasmid genes. *J. Bacteriol.* **129**, 101–107.

Montoya, A. L., Moore, L. W., Gordon, M. P., and Nester, E. W. (1978). Multiple genes coding for octopine-degrading enzymes in *Agrobacterium. J. Bacteriol.* **136**, 909–915.

Moore, L. W., and Cooksey, D. A. (1981). Biology of *Agrobacterium tumefaciens*: Plant interactions. *Int. Rev. Cytol.* **13**, 15–46.

Moore, L., Warren, G., and Strobel, G. (1979). Involvement of a plasmid in the hairy root-disease of plants caused by *Agrobacterium rhizogenes. Plasmid* **2**, 617–626.

Munnecke, D. E., Chandler, P. A., and Starr, M. P. (1963). Hairy root (*Agrobacterium rhizogenes*) of field roses. *Phytopathology* **53**, 788–799.

Murai, N., and Kemp, J. D. (1982a). T-DNA of pTi15955 from *Agrobacterium tumefaciens* is transcribed into a minimum of seven polyadenylated RNAs in a sunflower crown gall tumor. *Nucleic Acids Res.* **10**, 1679–1689.

Murai, N., and Kemp, J. D. (1982b). Octopine synthase messenger RNA isolated from sunflower crown gall callus is homologous to the Ti-plasmid of *Agrobacterium tumefaciens. Proc. Natl. Acad. Sci. U.S.A.* **79**, 86–90.

Murooka, Y., and Harada, T. (1979). Expansion of the host range of coliphage P1 and gene transfer from enteric bacteria to other gram-negative bacteria. *Appl. Environ. Microbiol.* **38**, 754–757.

Murooka, Y., Takizawa, N., and Harada, T. (1981). Introduction of bacteriophage Mu into bacteria of various genera and intergeneric gene transfer by RP4:Mu. *J. Bacteriol.* **145**, 358–368.

Murphy, P. J., and Roberts, W. P. (1979). A basis for agrocin 84 sensitivity in *Agrobacterium radiobacter. J. Gen. Microbiol.* **114**, 207–213.

Murphy, P. J., Tate, M. E., and Kerr, A. (1981). Substituents at N^6 and C-5′ control selective uptake and toxicity of the adenine-nucleotide bacteriocin, agrocin 84, in agrobacteria. *Eur. J. Biochem.* **115**, 539–543.

New, P. B., and Kerr, A. (1971). A selective medium for *Agrobacterium radiobacter* biotype 2. *J. Appl. Bacteriol.* **34**, 233–236.

New, P. B., and Kerr, A. (1972). Biological control of crown gall: Field observations and glasshouse experiments. *J. Appl. Bacteriol.* **35**, 279–287.

Nuti, M. P., Ledeboer, A. M., Lepidi, A. A., and Schilperoort, R. A. (1977). Large plasmids in different *Rhizobium* species. *J. Gen. Microbiol.* **100**, 241–248.

Nuti, M. P., Ledeboer, A. M., Durante, M., Nuti-Rochi, V., and Schilperoort, R. A. (1980). Detection of Ti-plasmid sequences in infected tissues by *in situ* hybridization. *Plant. Sci. Lett.* **18**, 1–6.

Ooms, G. (1981). "Molecular biologic studies on *Agrobacterium tumefaciens* and crown gall plant tumors." Ph.D. thesis, Leiden University.

Ooms, G., Klapwijk, P. M., Poulis, J. A., and Schilperoort, R. A. (1980). Characterization of Tn904 insertions in octopine Ti-plasmid mutants of *Agrobacterium tumefaciens. J. Bacteriol.* **144**, 82–91.

Ooms, G., Hooykaas, P. J. J., Moolenaar, G., and Schilperoort, R. A. (1981). Crown gall plant tumors of abnormal morphology, induced by *Agrobacterium tumefaciens* carrying mutated octopine Ti-plasmids; analysis of T-DNA functions. *Gene* **14**, 33–50.

Ooms, G., Hooykaas, P. J. J., Van Veen, R. J. M., Van Beelen, P., Regensburg-Tuïnk, A. J. G., and Schilperoort, R. A. (1982a). Octopine Ti-plasmid deletion mutants of *Agrobacterium tumefaciens* with emphasis on the right side of the T-region. *Plasmid* **7**, 15–29.

Ooms, G., Bakker, A., Molendijk, L., Wullems, G. J., Gordon, M. P., Nester, E. W., and Schilperoort, R. A. (1982b). T-DNA organization in homogeneous and heterogeneous octopine-type crown gall tissues of *Nicotiana tabacum*. *Cell* **30**, 589–597.

Ooms, G., Molendijk, L., and Schilperoort, R. A. (1982c). Double infection of tobacco plants by two complementing octopine T-region mutants of *Agrobacterium tumefaciens*. *Plant Mol. Biol.* **1**, 217–226.

Otten, L. A. B. M. (1979). "Lysopine dehydrogenase and nopaline dehydrogenase from crown gall tumor tissues." Ph.D. thesis, Leiden University.

Otten, L. (1982). Lysopine dehydrogenase activity as an early marker in crown gall transformation. *Plant Sci. Lett.* **25**, 15–27.

Otten, L. A. B. M., Vreugdenhil, D., and Schilperoort, R. A. (1977). Properties of D(+) lysopine dehydrogenase from crown gall tumour tissue. *Biochim. Biophys. Acta* **485**, 268–277.

Otten, L., De Greve, H., Hernalsteens, J. P., Van Montagu, M., Schieder, O., Straub, J., and Schell, J. (1981). Mendelian transmission of genes introduced into plants by the Ti-plasmids of *Agrobacterium tumefaciens*. *Mol. Gen. Genet.* **183**, 209–213.

Owens, L. D. (1982). Characteristics of teratomas regenerated *in vitro* from octopine-type crown gall. *Plant Physiol.* **69**, 37–40.

Panagopoulos, C. G., and Psallidas, P. G. (1973). Characteristics of Greek isolates of *Agrobacterium tumefaciens* (Smith and Townsend) Conn. *J. Appl. Bacteriol.* **36**, 233–240.

Parker, D. T., and Allen, O. N. (1957). Characteristics of four rhizobiophages active against *Rhizobium meliloti*. *Can. J. Microbiol.* **3**, 651–668.

Perry, K. L., and Kado, C. I. (1982). Characteristics of Ti-plasmids from broad host range and ecologically specific biotype 2 and 3 strains of *Agrobacterium tumefaciens*. *J. Bacteriol.* **151**, 343–350.

Petit, A., and Tempé, J. (1975). Etude du métabolisme des guanidines des tissus de crown-gall par la souche T37 d'*Agrobacterium tumefaciens*. *C. R. Acad. Sci. Paris Ser. D* **281**, 467–469.

Petit, A., and Tempé, J. (1976). Etude du métabolisme de l'homooctopine et de l'un de ses analogues par l'*Agrobacterium tumefaciens*. *C. R. Acad. Sci. Paris Ser. D* **282**, 69–71.

Petit, A., and Tempé, J. (1978). Isolation of *Agrobacterium* Ti-plasmid regulatory mutants. *Mol. Gen. Genet.* **167**, 147–155.

Petit, A., Delhaye, S., Tempé, J., and Morel, G. (1970). Recherches sur les guanidines des tissus de crown-gall. Mise en évidence d'une relation biochimique spécifique entre les souches d'*Agrobacterium tumefaciens* et les tumeurs qu'elles induisent. *Physiol. Veget.* **8**, 205–213.

Petit, A., Tempé, J., Kerr, A., Holsters, M., Van Montagu, M., and Schell, J. (1978). Substrate induction of conjugative activity of *Agrobacterium tumefaciens* Ti-plasmids. *Nature (London)* **271**, 570–572.

Plotkin, G. R. (1980). *Agrobacterium radiobacter* prosthetic valve endocarditis. *Ann. Int. Med.* **93**, 839–840.

Prakash, R. K., and Schilperoort, R. A. (1982). Relationship between nif plasmids of fast-growing *Rhizobium* species and Ti-plasmids of *Agrobacterium tumefaciens. J. Bacteriol.* **149**, 1129–1134.

Prakash, R. K., Schilperoort, R. A., and Nuti, M. P. (1981). Large plasmids of fast-growing rhizobia: Homology studies and location of structural nitrogen fixation (*nif*) genes. *J. Bacteriol.* **145**, 1129–1136.

Rao, S. S., Lippincott, B. B., and Lippincott, J. A. (1982). *Agrobacterium* adherence involves the pectic portion of the host cell wall and is sensitive to the degree of pectin methylation. *Physiol. Plant.* **56**, 374–380.

Regier, D. A., and Morris, R. O. (1982). Secretion of *trans*-zeatin by *Agrobacterium tumefaciens*: A function determined by the nopaline Ti-plasmid. *Biochem. Biophys. Res. Commun.* **104**, 1560–1566.

Riker, A. J. (1923). Some relations of the crown gall organism to its host tissue. *J. Agric. Res.* **25**, 119–132.

Riker, A. J. (1924). Relations of temperature and moisture to the development of crown gall. *Phytopathology* **14**, 30.

Riker, A. J. (1930). Studies on infectious hairy root of nursery apple trees. *J. Agric. Res.* **41**, 507–540.

Riley, P. S., and Weaver, R. E. (1977). Comparison of thirty-seven strains of Vd-3 bacteria with *Agrobacterium radiobacter*: Morphological and physiological observations. *J. Clin. Microbiol.* **5**, 172–177.

Risuleo, G., Battistoni, P., and Costantino, P. (1982). Regions of homology between tumorigenic plasmids from *Agrobacterium rhizogenes* and *Agrobacterium tumefaciens. Plamid* **7**, 45–51.

Rosenberg, C., Casse-Delbart, F., Dusha, I., David, M., and Boucher, C. (1982). Megaplasmids in the plant-associated bacteria *Rhizobium meliloti* and *Pseudomonas solanacearum. J. Bacteriol.* **150**, 402–406.

Roslycky, E. B., Allen, O. N., and McCoy, E. (1962). Phages for *Agrobacterium radiobacter* with reference to host range. *Can. J. Microbiol.* **8**, 71–78.

Sacristàn, M. D., and Melchers, G. (1977). Regeneration of plants from "habituated" and "*Agrobacterium*-transformed" single cell clones of tobacco. *Mol. Gen. Genet.* **152**, 111–117.

Saedi, D., Bruening, G., Kado, C. I., and Dutra, J. C. (1979). Tumor induction by *Agrobacterium tumefaciens* prevented in *Vigna sinensis* seedlings systemically infected by ribonucleic acid viruses. *Infect. Immun.* **23**, 298–304.

Schilperoort, R. A. (1969). "Investigations on plant tumors—Crown Gall. On the biochemistry of tumor induction by *Agrobacterium tumefaciens*." Ph.D. thesis, Leiden University.

Schröder, G., and Schröder, J. (1982). Hybridization selection and translation of T-DNA encoded mRNAs from octopine tumors. *Mol. Gen. Genet.* **185**, 51–55.

Schröder, J., Schröder, G., Huisman, H., Schilperoort, R. A., and Schell, J. (1981a). The mRNA for lysopine dehydrogenase in plant tumor cells is complementary to a Ti-plasmid fragment. *FEBS Lett.* **129**, 166–168.

Schröder, J., Hillebrand, A., Klipp, W., and Pühler, A. (1981b). Expression of plant tumor-specific proteins in minicells of *Escherichia coli*: A fusion protein of lysopine dehydrogenase with chloramphenicol acetyltransferase. *Nucleic Acids Res.* **9**, 5187–5202.

Schroth, M. N., Thompson, J. P., and Hildebrand, D. C. (1965). Isolation of *Agrobacterium tumefaciens-A. radiobacter* group from soil. *Phytopathology* **55**, 645–647.
Schweitzer, S., Blohm, D., and Geider, K. (1980). Expression of Ti-plasmid DNA in *E. coli*: Comparison of homologous fragments cloned from Ti-plasmids of *Agrobacterium* strains C58 and Ach5. *Plasmid* **4**, 196–204.
Schwinghamer, E. A. (1980). A method for improved lysis of some Gram-negative bacteria. *FEMS Microbiol. Lett.* **7**, 157–162.
Sciaky, D., Montoya, A. L., and Chilton, M.-D. (1978). Fingerprints of *Agrobacterium* Ti plasmids. *Plasmid* **1**, 238–253.
Scott, I. M. (1979). Opine content of unorganized and teratomatous tobacco crown gall tissues. *Plant Sci. Lett.* **16**, 239–248.
Semancik, J. S., Grill, L. K., and Civerolo, E. L. (1978). Accumulation of viroid RNA in tumor cells after double infection by *Agrobacterium tumefaciens* and citrus exocortis viroid. *Phytopathology* **68**, 1288–1292.
Sheikholeslam, S., Lin, B. C., and Kado, C. I. (1979). Multiple-size plasmids in *Agrobacterium radiobacter* and *A. tumefaciens*. *Phytopathology* **69**, 54–58.
Simpson, R. B., O'Hara, P. J., Kwok, W., Montoya, A. M., Lichtenstein, C., Gordon, M. P., and Nester, E. W. (1982). DNA from the A6S/2 crown gall tumor contains scrambled Ti-plasmid sequences near its junctions with plant DNA. *Cell* **29**, 1005–1014.
Skinner, F. A. (1977). An evaluation of the nile blue test for differentiating Rhizobia from Agrobacteria. *J. Appl. Bacteriol.* **43**, 91–98.
Skoog, F., and Miller, C. O. (1957). Chemical regulation of growth and organ formation in plant tissues cultured in vitro. *Symp. Soc. Exp. Biol.* **11**, 118–131.
Slota, J. E., and Farrand, S. K. (1982). Genetic isolation and physical characterization of pAgK84, the plasmid responsible for agrocin 84 production. *Plasmid* **8**, 175–186.
Smith, E. F., and Townsend, C. O. (1907). A plant tumor of bacterial origin. *Science* **25**, 671–673.
Spano, L., and Costantino, P. (1982). Regeneration of plants from callus cultures of roots induced by *Agrobacterium rhizogenes* on tobacco. *Z. Pflanzenphysiol.* **106**, 87–92.
Spano, L., Wullems, G. J., Schilperoort, R. A., and Costantino, P. (1981). Hairy root: In vitro growth properties of tissues induced by *Agrobacterium rhizogenes* on tobacco. *Plant Sci. Lett.* **23**, 299–305.
Spano, L., Pomponi, M., Costantino, P., Van Slogteren, G. M. S., and Tempé, J. (1982). Identification of T-DNA in the root-inducing plasmid of the agropine type *Agrobacterium rhizogenes* 1855. *Plant Mol. Biol.* **1**, 291–300.
Stapp, C. (1927). Der bakterielle Pflanzenkrebs und seine Beziehungen zum tierischen und menschlichen Krebs. *Ber. Dtsch. Bot. Ges.* **45**, 480–502.
Stapp, C. (1938). Der Pflanzenkrebs und sein Erreger *Pseudomonas tumefaciens* VI. Mitteilung. *Asparagus sprengeri* Rgl. und *Phaseolus vulgaris* L. als Wirtspflanzen. *Zentralbl. Bakteriol.* **99**, 116–123.
Starr, M. P. (1946). The nutrition of phytopathogenic bacteria II. The genus *Agrobacterium*. *J. Bacteriol.* **52**, 187–194.
Stonier, T. (1960). *Agrobacterium tumefaciens* Conn. II. Production of an antibiotic substance. *J. Bacteriol.* **79**, 889–898.
Süle, S. (1978). Biotypes of *Agrobacterium tumefaciens* in Hungary. *J. Appl. Bacteriol.* **44**, 207–213.
Süle, S., and Kado, C. I. (1980). Agrocin resistance in virulent derivatives of *Agrobacterium tumefaciens* harboring the pTi plasmid. *Physiol. Plant Pathol.* **17**, 347–356.
Sutton, D. W., Kemp, J. D., and Hack, E. (1978). Characterization of the enzyme respon-

sible for nopaline and ornaline synthesis in sunflower crown gall tissues. *Plant Physiol.* **62,** 363–367.

Tait, R. C., Lundquist, R. C., and Kado, C. I. (1982). Genetic map of the crown gall suppressive *inc*W plasmid pSa. *Mol. Gen. Genet.* **186,** 10–15.

Tamm, B. (1954). Experimentelle Untersuchungen über die Verbreitung des bakteriellen Pflanzenkrebses und das Auftreten von Sekundärtumoren. *Arch. Mikrobiol.* **20,** 273–292.

Tate, M. E., Murphy, P. J., Roberts, W. P., and Kerr, A. (1979). Adenine N[6]-substituent of agrocin 84 determines its bacteriocin-like specifity. *Nature (London)* **280,** 697–699.

Telford, J., Boseley, P., Schaffner, W., and Birnstiel, M. (1977). Novel screening procedure for recombinant plasmids. *Science* **195,** 391–393.

Tempé, J., and Goldmann, A. (1982). Occurrence and biosynthesis of opines. In "Molecular Biology of Plant Tumors" (G. Kahl and J. Schell, eds.), pp. 427–449. Academic Press, New York.

Tempé, J., and Petit, A. (1982). Opine utilization by *Agrobacterium*. In "Molecular Biology of Plant Tumors" (G. Kahl and J. Schell, eds.), pp. 451–459. Academic Press, New York.

Tempé, J., Petit, A., Holsters, M., Van Montagu, M., and Schell, J. (1977). Thermosensitive step associated with transfer of the Ti plasmid during conjugation: Possible relation to transformation in crown gall. *Proc. Natl. Acad. Sci. U.S.A.* **74,** 2848–2849.

Tempé, J., Guyon, P., Petit, A., Ellis, J. G., Tate, M. E., and Kerr, A. (1980). Preparation and properties of new catabolic substrates for two types of oncogenic plasmids in *Agrobacterium tumefaciens*. *C. R. Acad. Sci. Paris Ser. D* **290,** 1173–1176.

Tepfer, D. (1983). The potential uses of *Agrobacterium rhizogenes* in the genetic engineering of higher plants: Nature got there first. In "Genetic Engineering in Eukaryotes" (P. F. Lurguin and A. Kleinhofs, eds.), vol. 61, pp. 153–164. Plenum, New York.

Tepfer, D. A., and Tempé, J. (1981). Production d'agropine par des racines formées sous l'action d'*Agrobacterium rhizogenes,* souche A4. *C. R. Acad. Sci.* **292,** 153–156.

Thomas, C. M. (1981). Molecular genetics of broad host range plasmid RK2. *Plasmid* **5,** 10–19.

Thomas, C. M., Hussain, A. A. K., and Smith, C. A. (1982). Maintenance of broad host range plasmid RK2 replicons in *Pseudomonas aeruginosa*. *Nature (London)* **298,** 674–676.

Thomashow, M. F., Nutter, R., Montoya, A. L., Gordon, M. P., and Nester, E. W. (1980a). Integration and organization of Ti plasmid sequences in crown gall tumors. *Cell* **19,** 729–739.

Thomashow, M. F., Nutter, R., Postle, K., Chilton, M.-D., Blattner, F. R., Powell, A., Gordon, M. P., and Nester, E. W. (1980b). Recombination between higher plant DNA and the Ti plasmid of *Agrobacterium tumefaciens*. *Proc. Natl. Acad. Sci. U.S.A.* **77,** 6448–6452.

Thomashow, M. F., Panagopoulos, C. G., Gordon, M. P., and Nester, E. W. (1980c). Host range of *Agrobacterium tumefaciens* is determined by the Ti plasmid. *Nature (London)* **283,** 794–796.

Thomashow, M. F., Knauf, V. C., and Nester, E. W. (1981). The relationship between the limited and wide host range octopine-type Ti plasmids of *Agrobacterium tumefaciens*. *J. Bacteriol.* **146,** 484–493.

Turgeon, R. (1981). Structure of grafted crown-gall teratoma shoots of tobacco: Regulation of transformed cells. *Planta* **153**, 42–48.

Turgeon, R., Wood, H. N., and Braun, A. C. (1976). Studies on the recovery of crown gall tumor cells. *Proc. Natl. Acad. Sci. U.S.A.* **73**, 3562–3564.

Ursic, D., Kemp, J. D., and Helgeson, J. P. (1981). A new antibiotic with known resistance factors, G418, inhibits plant cells. *Biochem. Biophys. Res. Commun.* **101**, 1031–1037.

Van Brussel, A. A. N., Tak, T., Wetselaar, A., Pees, E., and Wijffelman, C. A. (1982). Small Leguminosae as test plants for nodulation of *Rhizobium leguminosarum* and other rhizobia and agrobacteria harbouring a leguminosarum Sym-plasmid. *Plant Sci. Lett.* **27**, 317–325.

Van Larebeke, N., Engler, G., Holsters, M., Van den Elsacker, S., Zaenen, I., Schilperoort, R. A., and Schell, J. (1974). Large plasmid in *Agrobacterium tumefaciens* essential for crown-gall inducing ability. *Nature (London)* **252**, 169–170.

Van Larebeke, N., Genetello, C., Schell, J., Schilperoort, R. A., Hermans, A. K., Hernalsteens, J. P., and Van Montagu, M. (1975). Acquisition of tumour-inducing ability by non-oncogenic agrobacteria as a result of plasmid transfer. *Nature (London)* **255**, 742–743.

Van Larebeke, N., Genetello, C., Hernalsteens, J. P., De Picker, A., Zaenen, I., Messens, E., Van Montagu, M., and Schell, J. (1977). Transfer of Ti plasmids between *Agrobacterium* strains by mobilisation with the conjugative plasmid RP4. *Mol. Gen. Genet.* **152**, 119–124.

Van Slogteren, G. M. S. (1983). "The characterization of octopine crown gall tumor cells. A study on the expression of *Agrobacterium tumefaciens*-derived T-DNA genes." Ph.D. thesis, Leiden University.

Van Slogteren, G. M. S., Hooykaas, P. J. J., Planqué, K., and De Groot, B. (1982). The lysopinedehydrogenase gene used as a marker for the selection of octopine crown gall cells. *Plant Mol. Biol.* **1**, 133–142.

Van Vliet, F., Silva, B., Van Montagu, M., and Schell, J. (1978). Transfer of RP4::Mu plasmids to *Agrobacterium tumefaciens*. *Plasmid* **1**, 446–455.

Vervliet, G., Holsters, M., Teuchy, H., Van Montagu, M., and Schell, J. (1975). Characterization of different plaque-forming and defective temperate phages in *Agrobacterium* strains. *J. Gen. Virol.* **26**, 33–48.

Watson, B., Currier, T. C., Gordon, M. P., Chilton, M.-D., and Nester, E. W. (1975). Plasmid required for virulence of *Agrobacterium tumefaciens*. *J. Bacteriol.* **123**, 255–264.

Whatley, M. H., Bodwin, J. S., Lippincott, B. B., and Lippincott, J. A. (1976). Role for *Agrobacterium* cell envelope lipopolysaccharide in infection site attachment. *Infect. Immun.* **13**, 1080–1083.

Whatley, M. H., Margot, J. B., Schell, J., Lippincott, B. B., and Lippincott, J. A. (1978). Plasmid and chromosomal determination of *Agrobacterium* adherence specificity. *J. Gen. Microbiol.* **107**, 395–398.

White, F. F., and Nester, E. W. (1980a). Hairy root: Plasmid encodes virulence traits in *Agrobacterium rhizogenes*. *J. Bacteriol.* **141**, 1134–1141.

White, F. F., and Nester, E. W. (1980b). Relationship of plasmids responsible for hairy root and crown gall tumorigenicity. *J. Bacteriol.* **144**, 710–720.

White, F. F., Ghidossi, G., Gordon, M. P., and Nester, E. W. (1982). Tumor induction by *Agrobacterium rhizogenes* involves the transfer of plasmid DNA to the plant genome. *Proc. Natl. Acad. Sci. U.S.A.* **79**, 3193–3197.

White, L. O. (1972). The taxonomy of the crown-gall organism *Agrobacterium tumefaciens* and its relationship to Rhizobia and other Agrobacteria. *J. Gen. Microbiol.* **72**, 565–574.

Willmitzer, L., De Beuckeleer, M., Lemmers, M., Van Montagu, M., and Schell, J. (1980). DNA from Ti plasmid present in nucleus and absent from plastids of crown gall plant cells. *Nature (London)* **287**, 359–361.

Willmitzer, L., Otten, L., Simons, G., Schmalenbach, W., Schröder, J., Schröder, G., Van Montagu, M., De Vos, G., and Schell, J. (1981a). Nuclear and polysomal transcripts of T-DNA in octopine crown gall suspension and callus cultures. *Mol. Gen. Genet.* **182**, 255–262.

Willmitzer, L., Schmalenbach, W., and Schell, J. (1981b). Transcription of T-DNA in octopine and nopaline crown gall tumours is inhibited by low concentrations of α-amanitin. *Nucleic Acids Res.* **9**, 4801–4812.

Willmitzer, L., Sanchez-Serrano, J., Buschfeld, E., and Schell, J. (1982a). DNA from *Agrobacterium rhizogenes* is transferred to and expressed in axenic hairy root plant tissues. *Mol. Gen. Genet.* **186**, 16–22.

Willmitzer, L., Simons, G., and Schell, J. (1982b). The t_1-DNA in octopine crown-gall tumors codes for seven well-defined polyadenylated transcripts. *EMBO J.* **1**, 139–146.

Willmitzer, L., Dhaese, P., Schreier, P. H., Schmalenbach, W., Van Montagu, M., and Schell, J. (1983). Size, location and polarity of T-DNA encoded transcripts in nopaline crown gall tumors; common transcripts in octopine and nopaline tumors. *Cell* **32**, 1045–1056.

Windass, J. D., Worsey, M. J., Pioli, E. M., Pioli, D., Barth, P. T., Atherton, K. T., Dart, E. C., Byrom, D., Powell, K., and Senior, P. J. (1980). Improved conversion of methanol to single-cell protein by *Methylophilus methylotrophus*. *Nature (London)* **287**, 396–401.

Wood, H. N., and Braun, A. C. (1965). Studies on the net uptake of solutes by normal and crown gall tumor cells. *Proc. Natl. Acad. Sci. U.S.A.* **54**, 1532–1538.

Wood, H. N., Binns, A. N., and Braun, A. C. (1978). Differential expression of oncogenicity and nopaline synthesis in intact leaves derived from crown gall teratomas of tobacco. *Differentiation* **11**, 175–180.

Wullems, G. J., Molendijk, L., Ooms, G., and Schilperoort, R. A. (1981a). Differential expression of crown gall tumor markers in transformants obtained after *in vitro Agrobacterium tumefaciens*-induced transformation of cell wall regenerating protoplasts derived from *Nicotiana tabacum*. *Proc. Natl. Acad. Sci. U.S.A.* **78**, 4344–4348.

Wullems, G. J., Molendijk, L., Ooms, G., and Schilperoort, R. A. (1981b). Retention of tumor markers in F1 progeny plants from in vitro induced octopine and nopaline tumor tissues. *Cell* **24**, 719–727.

Yadav, N. S., Postle, K., Saiki, R. K., Thomashow, M. F., and Chilton, M.-D. (1980). T-DNA of a crown gall teratoma is covalently joined to host plant DNA. *Nature (London)* **287**, 458–461.

Yadav, N. S., Vanderleyden, J., Bennett, D. R., Barnes, W. M., and Chilton, M.-D. (1982). Short direct repeats flank the T-DNA on a nopaline Ti plasmid. *Proc. Natl. Acad. Sci. U.S.A.* **79**, 6322–6326.

Yang, F., and Simpson, R. B. (1981). Revertant seedlings from crown gall tumors retain a portion of the bacterial Ti-plasmid DNA sequences. *Proc. Natl. Acad. Sci. U.S.A.* **78**, 4151–4155.

Yang, F., McPherson, J. C., Gordon, M. P., and Nester, E. W. (1980a). Extensive transcription of foreign DNA in a crown gall teratoma. *Biochem. Biophys. Res. Commun.* **92**, 1273–1277.

Yang, F., Montoya, A. L., Merlo, D. J., Drummond, M. H., Chilton, M.-D., Nester, E. W., and Gordon, M. P. (1980b). Foreign DNA sequences in crown gall teratomas and their fate during the loss of the tumorous traits. *Mol. Gen. Genet.* **177**, 707–714.

Young, I. G., and Poulis, M. I. (1978). Conjugal transfer of cloning vectors derived from ColE1. *Gene* **4**, 175–179.

Zaenen, I., Van Larebeke, N., Teuchy, H., Van Montagu, M., and Schell, J. (1974). Supercoiled circular DNA in crown gall-inducing *Agrobacterium* strains. *J. Mol. Biol.* **86**, 109–127.

Zambryski, P., Holsters, M., Kruger, K., Depicker, A., Schell, J., Van Montagu, M., and Goodman, H. M. (1980). Tumor DNA structure in plant cells transformed by *A. tumefaciens*. *Science* **209**, 1385–1391.

Zambryski, P., Depicker, A., Kruger, K., and Goodman, H. M. (1982). Tumor induction by *Agrobacterium tumefaciens*: Analysis of the boundaries of T-DNA. *J. Mol. Appl. Genet.* **1**, 361–370.

Zimmerer, R. P., Hamilton, R. H., and Pootjes, C. (1966). Isolation and morphology of temperate *Agrobacterium tumefaciens* bacteriophage. *J. Bacteriol.* **92**, 746–750.

INDEX

A

Achlya, organization of mtDNA in, 46
Actin, genes coding for, 128
Agrobacterium
  detection and isolation of plasmids in,
    218–219
  relationship between Ti-, Ri- and other
    plasmids
    DNA homology studies, 242–244
    genetic analysis and complementa-
      tion studies of virulence func-
      tions, 244–246
    incompatibility studies, 239–242
Agrobacterium tumefaciens, genetic sys-
    tems for, 225–226
  chromosomal linkage maps of, 230–231
  introduction of plasmids from other
    bacterial species and their ex-
    pression, 227–230
  mutagenesis and isolation of mutants,
    226
  transformation, transduction, and con-
    jugation in, 226–227
Aspergillus, amplified mtDNAs in, 54–56

B

Bacteria, taxonomy, crown galls and,
    210–213
Barley, genes coding for seed storage pro-
    teins of, 125

C

Chlamydomonas, linear mitochondrial
    genome of, 28

Chromatin
  general features in higher plants
    histones, 157–164
    nonhistone chromatin proteins,
      164–170
    nucleosomes and higher order struc-
      ture, 154–157
  structure, gene regulation and,
    145–154
  transcriptionally active
    DNA modification and, 190–192
    fractionation and nuclease sen-
      sitivity, 171–174
    nonhistone chromosomal proteins
      and, 186–190
    role of histones in determining struc-
      ture, 174–186
    visualization by electron microscopy,
      170–171
Crown gall cells, suppression of tumorous
    character of
  fate of T-DNA in plants regenerated
    from tumor cells, 258–263
  T-DNA mutations, 252–257
Crown gall tumorigenesis
  bacterial taxonomy and, 210–213
  T-DNA and
    expression of, 250–252
    structure of, 246–250
  tumor induction, 213–218

D

Deoxyribonucleic acid
  homology studies, between Ti-, Ri- and
    other *Agrobacterium* plasmids,
    242–244